Frontiers in Molecular Biology

SERIES EDITORS

B. D. Hames

*Department of Biochemistry
and Molecular Biology
University of Leeds, Leeds LS2 9JT, UK*

D. M. Glover

*Department of Genetics,
University of Cambridge, UK*

TITLES IN THE SERIES

Patterning in Vertebrate Development

EDITED BY

Cheryll Tickle

Division of Cell and Development Biology
School of Life Sciences
University of Dundee
MSI/WTB Complex
Dow Street
Dundee DD1 5EH
UK

OXFORD

UNIVERSITY PRESS

OXFORD

UNIVERSITY PRESS

Great Clarendon Street, Oxford OX2 6DP

Oxford University Press is a department of the University of Oxford
It furthers the University's objective of excellence in research, scholarship,
and education by publishing worldwide in

Oxford New York

Auckland Bangkok Buenos Aires Cape Town Chennai
Dar es Salaam Delhi Hong Kong Istanbul Karachi Kolkata
Kuala Lumpur Madrid Melbourne Mexico City Mumbai Nairobi
São Paulo Shanghai Taipei Tokyo Toronto

Oxford is a registered trade mark of Oxford University Press
in the UK and in certain other countries

Published in the United States
by Oxford University Press Inc., New York

A catalogue record for this title is available from the British Library

Library of Congress Cataloging in Publication Data

Patterning in vertebrate development / edited by Cheryll Tickle
(Frontiers in molecular biology ; 41)
1. Vertebrates–Development. 2. Pattern formation (Biology) I. Tickle, Cheryll II.
Series.

QL959 .P43 2002   571.8'16–dc21   2002029065

ISBN 0 19 963870 5 (Hbk.)
ISBN 0 19 963869 1 (Pbk.)

10 9 8 7 6 5 4 3 2 1

Typeset by Footnote Graphics, Warminster, Wilts

Printed in Great Britain acid free paper by
The Bath Press, Avon

Contents

8 Patterning of the neural crest 166

C. A. ERICKSON

9 Insights into the molecular basis of vertebrate forelimb and hindlimb identity 198

YASUHIKO KAWAKAMI, TOHRU TSUKUI, JENNIFER K. NG, AND JUAN CARLOS
IZPISUA-BELMONTE

10 Evolutionary aspects of vertebrate patterning   214

SEBASTIAN M. SHIMELD

Contributors

M. DAVEY
Division of Cell and Developmental Biology, School of Life Sciences, University of Dundee, MSI/WTB Complex, Dow Street, Dundee DD1 5EH, UK.

C. A. ERICKSON
Section of Molecular and Cellular Biology, University of California-Davis, CA 95616, USA.

ANTHONY GRAHAM
MRC Centre for Developmental Neurobiology, New Hunt's House, King's College London, Guys Campus, London SE1 1UL, UK.

SARAH GUTHRIE
MRC Centre for Developmental Neurobiology, New Hunt's House, King's College London, Guy's Campus, SE1 1UL, UK.

CLEMENTINE HOFMANN
GSF Research Center, Institute of Mammalian Genetics, 85764 Munich, Germany; Max-Planck-Institute of Psychiatry, 80804 Munich, Germany.

JUAN CARLOS IZPISÚA-BELMONTE
The Salk Institute for Biological Studies, Gene Expression Laboratory, 10010 North Torrey Pines Road, La Jolla, CA 92037–1099, USA.

YASUHIKO KAWAKAMI
The Salk Institute for Biological Studies, Gene Expression Laboratory, 10010 North Torrey Pines Road, La Jolla, CA 92037–1099, USA.

M. KERSZBERG
Neurobiologie Moleculaire, Institut Pasteur, Paris F-75724, Cedex 15, France.

IVOR MASON
MRC Centre for Developmental Neurobiology, New Hunt's House, King's College London, Guys Campus, London SE1 1UL, UK.

JENNIFER K. NG
The Salk Institute for Biological Studies, Gene Expression Laboratory, 10010 North Torrey Pines Road, La Jolla, CA 92037–1099, USA.

SEBASTIAN M. SHIMELD
School of Animal and Microbial Sciences, The University of Reading, PO Box 228, Reading RG6 6AJ, UK.

J.C. SMITH
Wellcome Trust/Cancer Research UK Institute, Tennis Court Road, Cambridge CB2 1QR, UK.

KATE G. STOREY
Division of Cell and Developmental Biology, School of Life Sciences,
University of Dundee MSI/WTB Complex, Dow Street, Dundee DD1 5EH, UK.

C. TICKLE
Division of Cell and Developmental Biology, School of Life Sciences,
University of Dundee, MSI/WTB Complex, Dow Street, Dundee DD1 5EH, UK.

TOHRU TSUKUI
Department of Biochemistry, Saitama Medical Schoo, Moroyama, Iruma-gun,
Saitama 350-0495, Japan

R. WHITE.
Division of Developmental Biology, National Institute for Medical Research,
The Ridgeway, Mill Hill, London NW7 1AA, UK.

L. WOLPERT
Department of Anatomy and Developmental Biology, University College London,
London WC1E 6BT, UK.

Plates

Plate 1 Model of the mechanism of localized Wnt pathway activation during dorso-ventral axis specification in *Xenopus*. Dsh associates with a specific class of vesicles at the vegetal pole and these vesicles are transported dorsally along the subcortical microtubule array during cortical rotation. This translocation contributes to the asymmetrical distribution of Dsh along the dorsoventral axis and the localized activation of a maternal Wnt signalling pathway. Activation of Wnt signalling leads to the downregulation of GSK3 activity, thereby promoting the stabilization of β-catenin. β-Catenin then accumulates in dorsal nuclei where, in combination with XTcf-3, it activates the transcription of dorsal-specific regulatory genes. (Adapted with permission from ref. 51.)

Plate 2 *Brachyury* and its targets. Comparison of expression patterns of *Xbra* (A–B) and *Bix1* (C–D) at early gastrula stage 10. *Xbra* is expressed in only the equatorial zone of the embryos, whereas *Bix1* is expressed in both the marginal zone and the vegetal hemisphere. (Reproduced with permission from ref. 106.)

Plate 3 Schematic of the hindbrain. The rhombomeres are labelled r1 through to r7 from anterior (A) to posterior (P). The motor nuclei of the hindbrain are labelled and coloured: trochlear (IVth) is labelled blue; the trigeminal (Vth) red; the facial (VIIth) green; the contral vestibuloacoustic (CVA) pink; the abducens (VIth) yellow; the glossopharyngeal (IXth). The expression domains of *Krox-20* and *Kreisler* are shown in blue, and the expression patterns of the *Hox* genes in orange, with regions of elevated expression shown in red.

Plate 4 Collagen-gel co-culture, used to demonstrate the influence of branchial arch-derived chemoattractants on hindbrain motor neurons. Cranial motor axons were retrogradely labelled using fluorescent tracers, and hindbrain explants containing these neurons were co-cultured with explants of branchial arch (BA) at a distance. Motor axons normally emerge from the lateral sides of the explant, i.e. from the right and left, since they grow away from the midline tissue of the floor plate (FP) under the influence of chemorepulsion. In the presence of a branchial arch explant, many axons still emerge laterally, but some exit the explant adjacent to the branchial arch explant and turn towards this tissue.

Plate 5 'Stripe assay' used to demonstrate avoidance by temporal axons of posterior tectal membranes. Alternating lanes consisting of membranes from anterior and posterior tectal thirds were prepared on a nucleopore filter. To discriminate between the two membrane types for later analyses, membranes derived from the posterior tectum were marked with fluorescein isothiocyanate (FITC)-labelled fluorescent beads. Retinal stripes were stained with N-4-4-(4-didecylaminostryryl) M-methyl-pyridium iodide (DiAsp) before arranging them perpendicular to the membrane stripes. Temporal retinal axons show a clear striped outgrowth and avoid membrane lanes containing posterior membranes, whereas nasal axons do not discriminate between these lanes. (Figure courtesy of Dr Uwe Drescher.)

Plate 6 Schematic depiction of the migratory pathways of the trunk neural crest. (a) Neural crest cells (dark green) detach from the dorsal surface of the neural tube (light green) and migrate ventrally between the neural tube and somites (pink). (b) When they reach the somites, they invade along the developing myotome. (c) The ventrally migrating neural crest cells differentiate into the neurons and glial cells of the sensory and sympathetic ganglia, and spread out along the ventral root motor fibres, where they differentiate into glial cells. Then, 24 h after migrating ventrally, another population of neural crest cells migrates dorsolaterally beneath the ectoderm and these differentiate into the pigment cells of the skin. (Figure courtesy of Martha Spence. Modification of Fig. 1 in ref. 1.)

Plate 7 *In situ* hybridization in a stage-20 chick embryo revealing the expression of FoxD3. FoxD3 is a transcription factor that is expressed by all neural crest cells except melanoblasts. This embryo demonstrates the contribution of the neural crest to the head cranial ganglia and the segmental migration of the neural crest through the somites in the trunk. (Micrograph courtesy of Robert Kos.)

Plate 8 Pertubation of ephrin function in trunk tissue pieces results in the loss of segmental migration through the somites, and the premature migration into the dorsolateral path. (Control) Pieces of stage-13 trunk were placed in culture; after 24 hours neural crest cells (labelled dark brown) had migrated normally through the anterior half of each somite, but had not yet invaded the dorsolateral path. (Ephrin–Fc) When the explants are treated with a fusion protein comprising the Eph-receptor binding portion, the Eph receptors on the neural crest cells are occupied with soluble ligand and so are insensitive to ephrins in their pathways. Consequently, the segmental migration through somites was pertubed and neural crest cells invaded the dorsolateral path precociously. (Micrograph courtesy of Alicia Santiago.)

Abbreviations

AChE	acetylcholinesterase
aei	*after eight* (gene)
ANR	anterior neural ridge
ANT-C	*Antennapedia complex*
AP	anteroposterior
AS-C	*achaete-scute* complex
AVE	anterior visceral endoderm
BDGF	brain-derived growth factor
BDNF	brain-derived neurotrophic factor
bea	*beamter* (gene)
bHLH	basic helix–loop–helix
BIG	Brachyury-induced gene
Bix	Brachyury-induced homeobox-containing gene
BMP	bone morphogenetic protein
BMPR	bone morphogenetic protein receptor
BX-C	*Bithorax* complex
CAM	cell-adhesion molecule
CPSG	chondroitin-6-sulfate proteoglycan
CT	column of Terni
CVA	contralateral vestibuloacoustic
DCC	deleted in colorectal carcinoma
DE	distal element
des	*deadly seven* (gene)
DiAsp	N-4-4-(4-didecylaminostryryl) M-methyl-pyridium iodide
DiI	1,1′-dioctadecyl-3,3,3′,3′-tetramethyl carbocyanine
Dll	Distalless
Dll1	Delta-like1
Dll3	Delta-like3
dn	dominant-negative
DPC	days' postcoitus
dpp	decapentaplegic (gene)
DRG	dorsal root ganglion
DV	dorsoventral
E(spl)	Enhancer of Split
ECM	extracellular matrix
eFGF	the *Xenopus* homologue of *FGF4*
emc	extramacrochaete

Ena	Enabled
endo N	endoneuraminidase
ephrins	Eph-receptor interacting proteins
ES	embryonic stem (cells)
FGF	fibroblast growth factor
FITC	fluorescein isothiocyanate
flik	*follistatin*-like (gene)
fss	*fused somites* (gene)
GAP	GTPase-activating proteins
GBP	GSK3-binding protein
GDI	guanine nucleotide dissociation inhibitors
GEF	guanine exchange factors
GFP	Green Fluorescent Protein
GPI	glycosyl-phosphatidylinositol
GSK3b	glycogen synthase kinase-3b
HES-1	*Hairy/Espl-1* (gene)
HGF	hepatocyte growth factor
K_d	dissociation constant
l-fng	*lunatic fringe* (gene)
LAR	leucocyte antigen related
lgn	lateral geniculate nucleus
LMC_L	lateral component of the lateral motor column
LMC_M	medial component of the lateral motor column
LPM	lateral plate mesoderm
MBT	mid-blastula transition
MCLP	myosin light-chain phosphatase
MMC_L	lateral component of the medial motor column
MMC_M	medial component of the medial motor column
MMP	matrix metalloprotease
MRF	myogenic [bHLH] regulatory factors
NCAM	neural cell-adhesion molecule
NCAM	neural cell-adhesion molecule
NgCAM	neuron–glia cell adhesion molecule
NGF	nerve growth factor
NrCAM	neuronal cell adhesion molecule
NRSF	neuron-restrictive silencing factor
NT	neurotrophin
opb	open brain
PcG	*Polycomb* group of *genes*
PCR	polymerase chain reaction
PE	proximal element
PKCa	protein kinase C-a
PNA	peanut agglutinin
PNM	perinotochordal mesenchyme

PNS	peripheral nervous system
PSA-NCAM	polysialic acid-bearing form of NCAM
pu	*pudgy* (gene)
r1	rhombomere 1
RARE	retinoic-acid response element
RCAS	replication-competent variant of the avian leukaemia virus
REST	RE1-silencing transcription (factor)
RPTP	receptor protein tyrosine phosphatase
SC	spinal chord
Sema3a	Semaphorin-3A
SHH	sonic hedgehog
SIF	small intensively fluorescent (cells)
Sim	single-minded
sog	short gastrulation gene
SSEA	stage-specific embryonic antigen
Su(H)	Suppressor of Hairless
TGF-b	transforming growth factor-b
TdT	terminal deoxynucleotidyl transferase
trk C	tyrosine kinase-C
trxG	*trithorax* group of *genes*
TUNEL	TdT X-dUTP nick end-labelling
uPA	urokinase-type plasminogen activator
UTR	untranslated region
VASP	vasodilator-stimulated phosphoprotein
Vg	vegetal
WASP	Wiskott–Aldrich syndrome protein
Xbra	*Xenopus* homologue of the mouse *Brachyury*, or *T* (tail) gene
ZLI	zona limitans intrathalamica

1 | Patterning and positional information

L. WOLPERT and M. KERSZBERG

1. Introduction

Pattern formation is a key process in embryonic development (1). When a fertilized egg divides, the resulting cells become different and give rise to a variety of structures such as the main body and its appendages. These structures have a well-defined spatial organization—the vertebrate limb is a good example—which is brought about by the process of pattern formation. There are three other related processes: (1) cell differentiation is the process by which different specialized adult cell types such as muscle, cartilage, and neurons are generated; (2) morphogenesis, or a change in form, such as neural tube formation; and (3) growth—it is a general feature of pattern formation that it is initially only specified for any organ on a small scale—rarely greater than more than 100 cell diameters. These processes are related to one another but, in general, pattern formation is the primary event in that it specifies the other processes. Thus patterning in vertebrate limb development specifies where cartilage will form and how it will grow; it also specifies ectodermal cells such as the apical ridge, which controls the overall form of the limb.

2. Positional information

In vertebrate development, the main mechanism for pattern formation involves cell-to-cell interactions via cell signalling (Fig. 1). One class of patterning mechanism is based on the generation by some cells (which may be zygotic or maternal) of positional information, which other cells first read out to acquire a positional value with respect to boundaries, as in a coordinate system. Positional value is then interpreted by cells through a change in their state (2, 3).

A simple formulation comes from the 'French Flag model' of pattern formation (1), in which a set of cells in a line can be blue, red, or white and the pattern is generated by a gradient in the concentration of a substance—a morphogen—the cells responding to different threshold concentrations (Fig. 2). The state resulting from interpretation is determined by the cells' genetic constitutive and developmental history. The change in cell state can result in cell differentiation, change in form, or

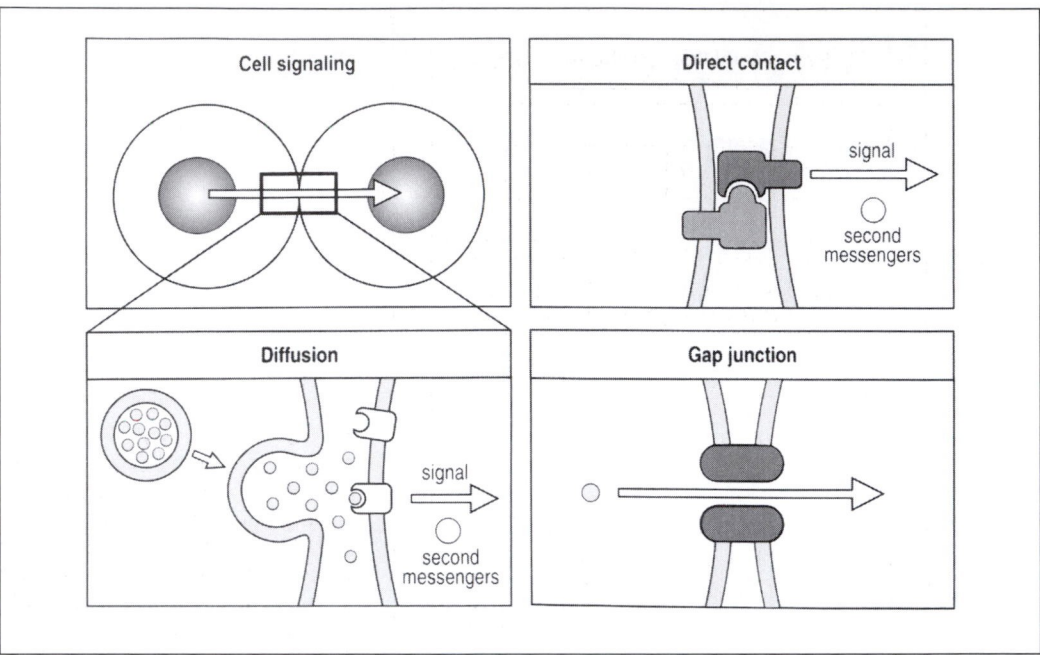

Fig. 1 A signal may be transmitted from one cell to another in three main ways. The signal can be a diffusible molecule, which interacts with a receptor on the target cell surface (top panels), or the signal can be produced by direct contact between two complementary proteins at the cell surfaces (middle panel). Where the signal acts on receptors on the cell membrane, a complex signal transduction may occur in which a cascade of interactions often involving, for example, phosphorylation, leads to specific genes being activated. If the signal involves small molecules, it may pass directly from cell to cell through gap junctions in the plasma membrane (bottom panel). (Taken with permission from ref. 1.)

cell proliferation. A key feature of mechanisms based on positional information is that because the cells in the positional field have different positional values, then, even if they differentiate into the same cell type, say cartilage, these cartilage cells are different from each other and they are non-equivalent as they have different positional values. This non-equivalence can underlie later differences in cell behaviour such as morphogenesis and growth, and will play an important role in systems that can regenerate. There is reason to believe, for example, that regeneration of newt limbs is dependent on different positional values along the limb's proximodistal axis.

In models based on positional information, there is no direct relationship between the positional values and the observed pattern, as the pattern of cell behaviour depends on the interpretation of the positional values. Indeed, the same set of positional values can give rise to an infinite variety of patterns. The positional field of the fore- and hindlimbs is essentially the same, and this is also true for the legs and antennas of *Drosophila*. The differences in their development come from the variations in interpretation due to diverse developmental histories such as their position along the main body axis. How do cells acquire their positional values? Positional fields, i.e. the set of cells that have positional values with respect to the same

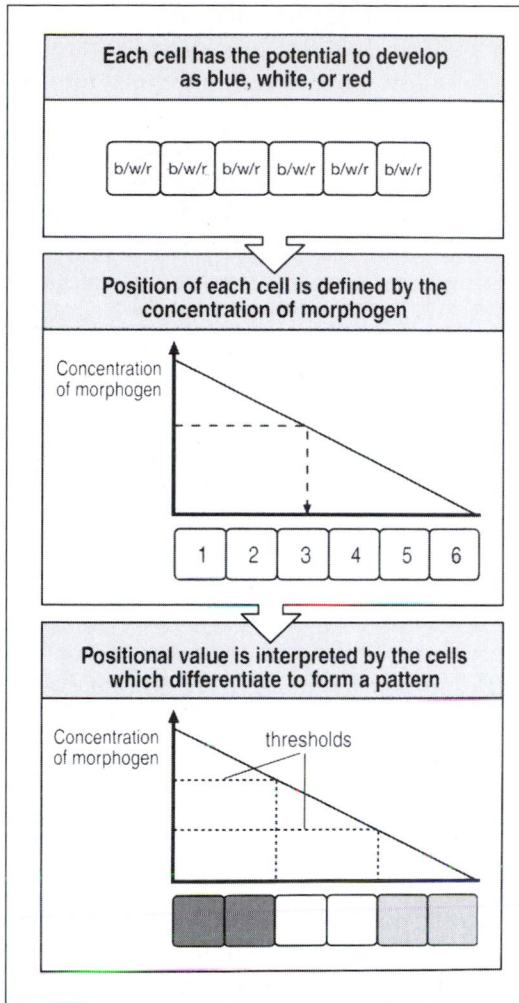

Fig. 2 The 'French Flag model' of pattern formation. Each cell in a line of cells has the potential to develop, from left to right, as blue, white, or red. The line of cells is exposed to a concentration gradient of substance—a morphogen—and each cell acquires a positional value defined by the local concentration of that substance. Each cell then interprets its positional value according to its genetic constitutive and developmental history and differentiates into blue, white, or red, thus forming the French Flag pattern. The basic requirements of this simple model system are that the concentration of substance must be fixed at the two boundaries and must vary linearly between them. Each cell must also contain the necessary information to interpret the positional values. Interpretation of the positional value is based upon different threshold responses to different concentrations of morphogen. Cells that express the same colour may be non-equivalent as they have different positional values. (Taken with permission from ref. 1.)

boundaries, are small, less than about 30 cells along any axis. Thus patterning at any one time occurs with rather few cells, and growth gives rise to the larger structures seen in the adult. A widely proposed model for specifying positional values is based on a graded distribution of a diffusible molecule, a morphogen as in the French Flag model. This morphogen would be released by cells at a boundary and diffuse over the positional field. Positional values would then be specified by the local concentration of the morphogen, whose concentration would decrease with distance from the source. Unfortunately, this simple model of diffusion from a source as the way position is specified neglects the presence of receptors for the morphogens. These affect the diffusion and the effect of the morphogen on the cells. Computer simulations of the process show that receptors close to the source might be expected to bind the morphogen and render it unable to propagate further before they become saturated (4). This gives rise not to a gradient but to a flat distribution of activated receptors near the source, and then a sharp decrease at a distance. However, if the morphogen binds to the receptor with a low kinetic binding constant, only a small fraction of the morphogen will initially be retained by the receptor, so allowing the formation of a smoother gradient of activated receptors. There is also now substantial evidence that various other factors can influence the shape of the gradient by the morphogen up- or downregulating receptor concentration. For example, the gradient in decapentaplegic in the patterning of the *Drosophila* wing is modified by the receptors for the morphogen; in one case the receptors are induced to increase in number by high concentrations of the morphogen and this slows down its lateral diffusion (3, 5).

Other models for setting up a morphogen gradient have thus been proposed. One possibility is that unlike the model just considered, in which there is long-range signalling of the morphogen over the positional field, all the signals may be short-range and the gradient would be set up by a cascade of relays. Thus the signal may only activate an adjacent cell, which will then activate the next cell, and so on. Another model is based on a mechanism in which different receptor subtypes cooperate in passing the morphogen molecules along the cell membrane and then from cell to cell (6). There is also the possibility of the morphogen being passed from cell to cell in vesicles (7). Yet another very different mechanism for setting up a gradient is based on measuring time in a region like a progress zone. For example, if a group of cells are proliferating in a well-defined region such as the progress zone at the tip of the limb or the regressing node region during chick gastrulation, cells are continually leaving the region. Then if the cells can measure the time, T, they have spent in the defined region, and if this becomes frozen as they leave the progress zone, a trail of cells with a gradient of increasing T will ensue. While a variety of candidate morphogens have been proposed for specifying positional information, in most cases the evidence is still not conclusive. The best evidence comes from the development of the early *Drosophila* embryo, but here the gradients are established in effectively a single cell with many nuclei and there are no individual cell membranes (see below) Other good examples are the gradients in decapentaplegic, a member of the transforming growth factor-beta (TGF-β) family, and wingless in the *Drosophila* wing (8). In vertebrates, good candidates are: activin-like molecules in the early *Xenopus* embryo (9, 10); and sonic hedgehog in the developing limb (11) and neural tube (12).

3. Interpretation of positional information

3.1 Introduction

Interpretation of positional information requires the cells to respond in a specific manner to different positional values. If positional value is specified by the concentration of a morphogen then different levels of the morphogen must result in different cell behaviour, including different patterns of gene expression. This implies that cells can respond accurately to threshold levels of morphogens. Just how this is done is not yet known, but there is good evidence that cells can respond to different levels of a morphogen. One of the clearest examples is the differential response of *Xenopus* cells to varying concentrations of a member of the TGF-β family, activin (Fig. 3). Intact blastula tissue from the animal cap of *Xenopus* was exposed to varying concentrations of activin. Low concentrations of which induced the expression of genes like *XBrachyury*, while high concentrations induced genes like *goosecoid*. Only a 2% active type IIa-receptor occupancy by the activin is required to turn on *XBrachyury*, whereas a 6% occupancy is required to induce *goosecoid* expression (13). Once the ligand is bound, signal transduction leads to gene expression. It is also possible that timing plays a role in interpretation (14).

3.2   Prepattern mechanisms

Unlike mechanisms based on positional information, in prepattern models there is a direct relationship between the distribution of the morphogen and the pattern that develops. Thus, a wave-like distribution of a morphogen could give rise to a periodic pattern, with cells differentiating into different cell types depending on whether the concentration of the morphogen corresponded to the peak or base of each wave (15). It is possible that such a prepattern of morphogen could be set up by what is known as a 'reaction–diffusion mechanism'. Although there is no good evidence for such a mechanism in a biological system, it has been shown that a system of interacting and diffusing molecules that activate and inhibit their synthesis can spontaneously set up a wave-like pattern of one of the diffusing molecules. Thus, in principle, it is an attractive mechanism for setting up some periodic patterns.

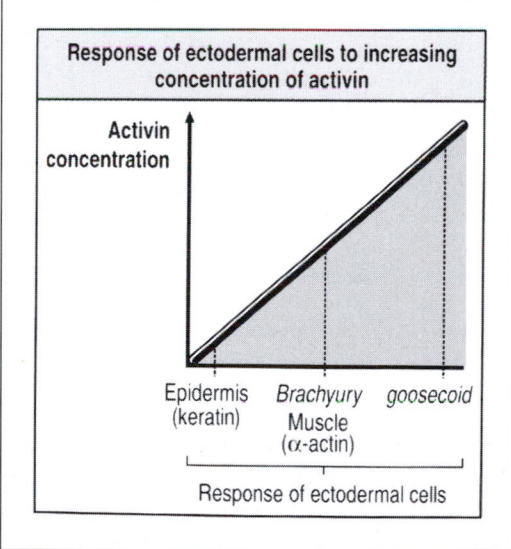

Fig. 3 Graded responses of *Xenopus* tissue to increasing activin concentrations. When cells that would normally form ectoderm are treated with increasing concentrations of activin, particular genes are activated at specific threshold concentrations. (Taken with permission from ref. 1.)

3.3 Insect development

The embryonic development of the fruitfly, *Drosophila*, provides a key model for understanding pattern formation. In the unfertilized egg, *bicoid* mRNA is localized in the anterior end having being laid down there by maternal genes during oogenesis (Fig. 4). After fertilization the mRNA is translated, and the bicoid protein diffuses from the anterior end and forms a concentration gradient along the anteroposterior axis (16). This provides the positional information required for further patterning along this axis. Historically, the bicoid protein gradient provided the first reliable evidence for the existence of the morphogen gradients that had been postulated to control pattern formation.

As the bicoid protein diffuses along the *Drosophila* embryo, it also breaks down—it has a half-life of about 30 minutes—and this breakdown is important in establishing the anteroposterior concentration gradient. The bicoid protein is a transcription factor that acts as a morphogen. It switches on certain zygotic genes at different threshold concentrations, so initiating a new pattern of gene expression along the axis. The gap genes are the first zygotic genes to be expressed along the antero-posterior axis, and they too code for transcription factors. Their expression is

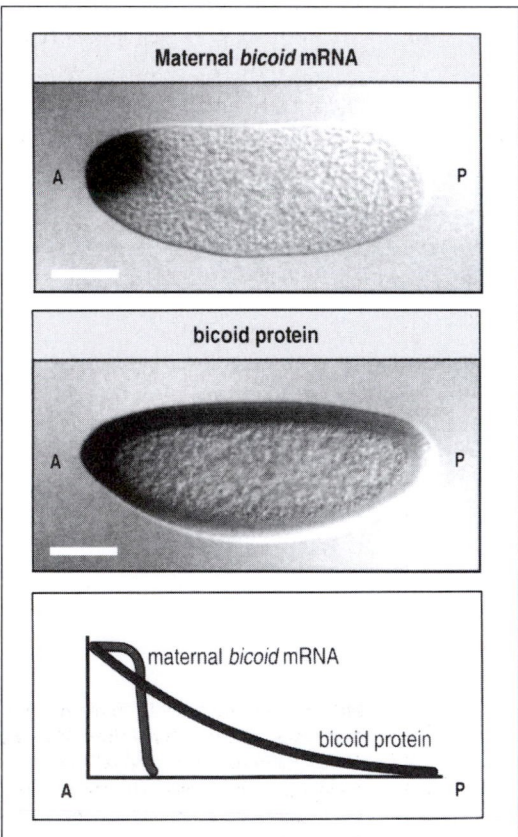

Fig. 4 The distribution of the maternal mRNA for *bicoid* in the egg and the gradient of bicoid protein after fertilization. The *bicoid* mRNA at the anterior end is visualized by *in situ* hybridization (top panel), while the bicoid protein is stained with a labelled antibody (middle panel). Distribution of the mRNA and bicoid protein after translation; diffusion of bicoid protein from its site of synthesis produces an anteroposterior gradient of bicoid protein in the embryo providing positional information (bottom panel). Scale bars = 0.1 mm. (Taken with permission from ref. 1.)

initiated by the anteroposterior gradient of bicoid protein while the embryo is still a syncytial blastoderm—that is when the nuclei have divided but there are no cell membranes to separate the embryo into individual cells. Bicoid protein activates the anterior expression of the gap gene, *hunchback*, above a threshold concentration (Fig. 5). The hunchback protein, which is then translated, is instrumental in switching on the expression of the other gap genes, including *giant*, *Kruppel*, and *knirps*, which are expressed in this order along the anteroposterior axis. As the blastoderm is still acellular at the stage at which the gap genes are expressed, the gap gene proteins can diffuse away from their sites of synthesis. They are short-lived proteins with a half-life of minutes. Their distribution therefore extends only slightly beyond the region in which the gene is expressed, and this typically gives a bell-shaped protein-concentration profile. The hunchback protein is exceptional in this respect: as, although its gene is expressed over a broad anterior region, it has a steep anteroposterior protein gradient because it is only switched on when the bicoid protein is above a threshold concentration. The dorsoventral axis, which is at right angles to the anteroposterior axis, is specified by a set of maternal genes separate from those that specify the anteroposterior axis. But, like the anteroposterior axis, it is initially established in the unfertilized egg in the ovary. The ventral end of the axis is set by the localized deposition of a maternal protein in the extraembryonic vitelline membrane on one side of the egg only; this will become the future ventral region of the embryo. This leads to the localized production of a protein, spaetzle, in the ventral perivitelline space, where it locally activates the Toll receptor protein on the embryo's membrane only in the future ventral region of the embryo. This sends a signal to the adjacent cytoplasm of the embryo that causes a maternal gene product in the adjoining cytoplasm, the dorsal protein, to enter nearby nuclei. This protein,

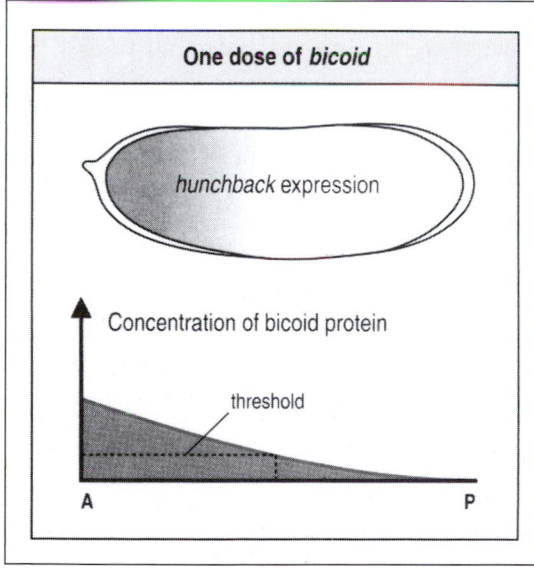

Fig. 5 Maternal bicoid protein controls zygotic *hunchback* gene expression. The expression of *hunchback* is determined by the threshold concentration of bicoid protein. Thus *hunchback* expression is confined to the anterior of the embryo. (Taken with permission from ref. 1.)

encoded by the *dorsal* gene, is a transcription factor with a vital role in organizing the dorsoventral axis. The dorsal protein is initially uniformly distributed along the dorsoventral axis of the embryo. Initially, this protein is restricted to the cytoplasm. However, under the influence of signals from the ventrally activated Toll receptor, it enters nuclei in a graded fashion, with the highest concentration in ventral nuclei and a progressively decreasing concentration in a ventral to dorsal direction, as the Toll signal becomes weaker. Patterning along this axis poses a problem like that of the French Flag and involves interpretation of the dorsal gradient (17). Expression of zygotic genes in localized regions along the dorsoventral axis is thus initially controlled by the graded concentration of the intranuclear dorsal protein, which falls off rapidly in the dorsal half of the embryo; little dorsal protein is found in nuclei above the equator. In the ventral region, dorsal protein has two main functions—it activates certain genes at specific positions in the ventral region; and it represses the activity of other genes, which are therefore only expressed in the dorsal region. In the ventral-most region, where concentrations of intranuclear dorsal protein are highest, the zygotic genes *twist* and *snail* are activated by dorsal protein in a strip of nuclei along the ventral side of the embryo; soon after this, the blastoderm becomes cellular. This ventral strip of cells will form the mesoderm. Another key system in which gradients in positional information play a key role is in the development of the wing imaginal disc.

3.4 Evolution and pattern

Gradients patterning tissues have been found in a variety of tissues. What is striking is the similarity in both the principles involved, as well as in the molecules involved in signalling. The same sets of protein signals are used again and again and include members of the TGF-β family, the hedgehog and wingless families. From the viewpoint of evolution, positional information provides a very suitable system for the generation of novel patterns. Since there is little relation between the positional values of the cells and the expressed pattern of cellular differentiation, the same positional mechanisms can be used to generate an enormous variety of different patterns. The differences in pattern will reflect variations in the interpretation of positional values rather than of signalling between cells. Thus the complexity of patterning comes from the cellular response to common signals.

References

1. Wolpert, L., Beddington, R., Jessell, T., Lawrence, P., Meyerowitz, E., and Smith, J. (2002) *Principles of development* (2nd edition). Oxford University Press, Oxford.
2. Gurdon, J. B. and Bourillot, P-Y. (2001) Morphogen gradient interpretation. *Nature*, **413**, 797.
3. Neumann, C. and Cohen, S. (1997) Morphogens and pattern formation. *Bioessays*, **19**, 721.
4. Kerszberg, M. (1999) Morphogen propagation and action: toward molecular models. *Semin. Cell Dev. Biol.*, **10**, 297.

5. Podos, S. D. and Ferguson, E. L. (1996) Morphogen gradients new insights from DPP. *Trends Genet.*, **15**, 397.
6. Kerszberg, M. and Wolpert, L. (1998) Mechanisms for positional signalling by morphogen transport: a theoretical study. *J. Theoret. Biol.*, **191**, 103.
7. Entchev, V., Schwabedissen, A., and Gonzalez-Gaitan, M. (2000) Gradient formation of the TGF-beta homolog Dpp. *Cell*, **103**, 981.
8. Strigini, M. and Cohen, S. M. (1999) Formation of morphogen gradients in the *Drosophila* wing. *Semin. Cell Dev. Biol.*, **10**, 335.
9. McDowell, N. and Gurdon, J. B. (1999) Activin as a morphogen in *Xenopus* mesoderm induction. *Semin. Cell Dev. Biol.*, **10**, 311.
10. Dale, L. and Wardle, F. (1999) A gradient of BMP activity specifies dorsal-ventral fates in early *Xenopus* embryos. *Semin. Cell Dev. Biol.*, **10**, 319.
11. Tickle, C. (1999) Morphogen gradients in vertebrate limb development. *Semin. Cell Dev. Biol.*, **10**, 345.
12. Briscoe, J. and Ericson, J. (1999) The specification of neural identity by graded sonic hedgehog signalling. *Semin. Cell. Dev. Biol.*, **10**, 353.
13. Gurdon, J. B., Dyson, S., and St Johnson, D. (1988) Cell's perception of position in a concentration gradient. *Cell*, **95**, 159.
14. Pages, F. and Kerridge, S. (2000) Morphogen gradients: a question of time or concentration? *Trends Genet.*, **16**, 40.
15. Murray, J. D. (1988) How the leopard gets its spots. *Sci. Am.*, **258**, 80.
16. Driever, W. and Nusslein-Volhard, C. (1988) The bicoid protein determines position in the *Drosophila* embryo in a concentration-dependent manner. *Cell*, **54**, 95.
17. Rusch, J. and Levine, M. (1996) Threshold responses to the dorsal regulatory gradient and the subdivision of primary tissue territories in the *Drosophila* embryo. *Curr. Opin. Genet. Dev.*, **6**, 416.

2 | Laying down the vertebrate body plan

C. TICKLE, with illustrations by M. DAVEY

1. Introduction

How do different parts of the body arise in their proper positions in vertebrate embryos? In the first chapter, the principles of pattern formation were reviewed. This chapter outlines how the body plan is laid down, and provides the embryological background for the following chapters that deal in detail with patterning of mesoderm (Chapters 3 and 4), nervous system (Chapters 5, 6, 7, and 8), and limbs (Chapter 9) and the molecules involved. Laying down the body plan is essentially a matter of defining the main body axes: the anteroposterior axis (head to tail axis, sometimes known as rostral–caudal axis) and dorsoventral axis (back to front). Once these two axes have been set up, this also defines right and left (Fig. 1). In addition, the three main body layers—ectoderm, mesoderm, and endoderm—are formed and arranged so that endoderm comes to lie on the inside, ectoderm on the outside, and mesoderm in between. Establishment of these layers in their proper positions requires considerable cell rearrangements, which must be precisely choreographed. Ectoderm gives rise to skin epidermis and dorsally to neural plate, from which the nervous system develops; mesoderm gives rise to notochord in the dorsal midline

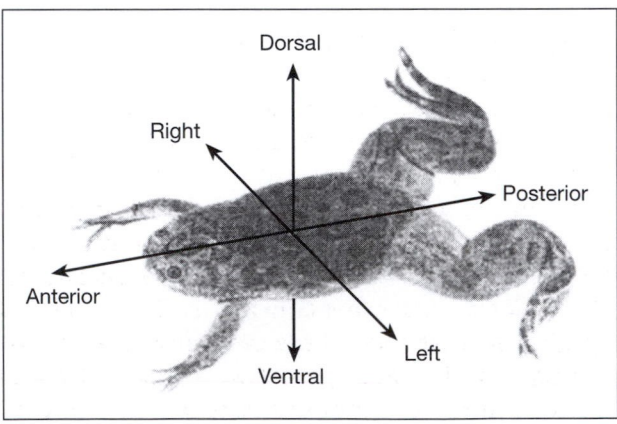

Fig. 1 Main body axes illustrated with respect to an adult *Xenopus* frog.

flanked on either side by somites, then intermediate mesoderm (kidney), lateral mesoderm, and blood; endoderm forms the gut lining.

The problem of laying down the body plan can be viewed in the context of positional information, as outlined in Chapter 1. Thus cells must be informed of their position with respect to these body axes and then have to interpret this information so that they participate in forming the part of the body appropriate to that position.

2. *Xenopus* early development illustrates the general principles

A number of different embryos are important models for understanding vertebrate development. Amphibian and chicken embryos are outstanding models for classical cut-and-paste experimental embryology, while mouse and zebrafish are powerful genetic systems. Therefore we will consider how the body plan of all these embryos is laid down. First of all, however, we will focus on amphibian embryos, in particular on the frog, *Xenopus*, because these embryos are the most convenient for studying the earliest stages of vertebrate development (Chapter 3; see also refs 1–4 for background information).

A newly laid frog egg has a recognizable polarity even prior to fertilization. The upper part of the egg (animal hemisphere) is pigmented, while the lower part (vegetal hemisphere) is more yolky. The egg is arrested at the metaphase stage of the second meiotic division and the first polar body (product of the first meiotic division, containing the same DNA content as the egg but little cytoplasm) remains attached near the animal pole. A number of events are triggered when the egg is fertilized, including completion of the second meiotic division, with accompanying extrusion of the second polar body. The maternal set of chromosomes that results, together with the paternal set from the sperm, establishes the genetic composition of the new individual. Fertilization also results in general activation of the egg, which is followed by a sequence of rapid divisions (cleavage divisions) to give two cells, then four, and so on, thus generating an increasing number of smaller and smaller cells (blastomeres) (Fig. 2).

Prior to fertilization the egg is radially symmetrical, but sperm entry creates a unique point on the egg surface, and this determines, through a series of events, the future dorsoventral body axis (Chapter 3). When the sperm fertilizes the egg, the outer cortical layers of the egg rotate (Fig. 3), which displaces the pigmented animal region of the egg. In some amphibians, a grey crescent can be seen on the side of the egg opposite the site of sperm entry, and this marks the future dorsal side of the embryo. The fertilized egg then undergoes a stereotypical pattern of cleavage divisions related to its polarity. The first two cleavage divisions bisect the animal and vegetal poles and are at right angles to each other (see Fig. 2). The first cleavage goes through the 'grey crescent region' so that each of the two blastomeres contain 'grey crescent' material, while the second cleavage gives two 'dorsal' blastomeres containing 'grey crescent' and two 'ventral' blastomeres outside. The third cleavage is

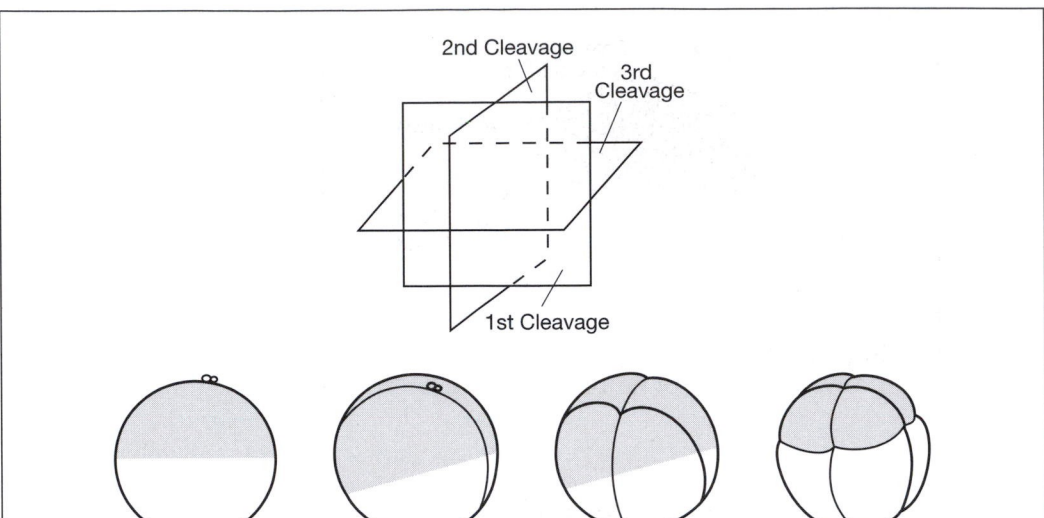

Fig. 2 Diagrams showing stereotypical pattern of cleavage of a fertilized *Xenopus* egg. The upper panel shows how the first three cleavage planes are related to each other; lower panels show the fertilized egg, 2-cell-stage, 4-cell-stage, 8-cell-stage embryos. The animal hemisphere is pigmented; polar bodies shown in the fertilized egg and 2-cell-stage embryo are attached to animal pole.

equatorial, and generates an upper tier of animal hemisphere blastomeres lying on top of a lower tier of vegetal hemisphere blastomeres (see Fig. 2).

These precise patterns of cleavage result in the partitioning of different regions of the fertilized egg into specific blastomeres, and this determines their developmental potential. Thus, although each of the two blastomeres from the 2-cell stage can give rise to a normal embryo, the two 'dorsal' blastomeres of a 4-cell-stage embryo develop into embryos with well-developed dorsal and anterior structures, while the two 'ventral' blastomeres form radially symmetrical embryos lacking dorsal structures (reviewed by Slack (1)). At the 8-cell stage, ablation of either of the two dorsovegetal blastomeres also leads to radially symmetrical embryos. These and other experiments show that dorsoventral determinants localized in the fertilized egg become unevenly distributed during early cleavage stages.

Continued cleavage divisions generate 16, 32, and 64 cells, and so on. When a few hundred cells have been produced, the outer cells start to pump fluid into the inside of the embryo and a cavity (blastocoel) is formed (see Fig. 3). In the animal hemisphere, the blastocoel roof is two cells thick, and fate maps show that the outer layer of cells in this region will give rise to ectodermal tissues—epidermis and neural tissue—while those in the vegetal region will form endoderm. Cells in the inner layer of the equatorial (marginal) region will form mesoderm (Fig. 4; and see refs 5, 6).

Mesoderm formation depends on an interaction between the animal and vegetal regions (see Chapter 3). When the upper portion of the animal hemisphere (animal cap) is removed and cultured on its own, only epidermis forms; while the lower

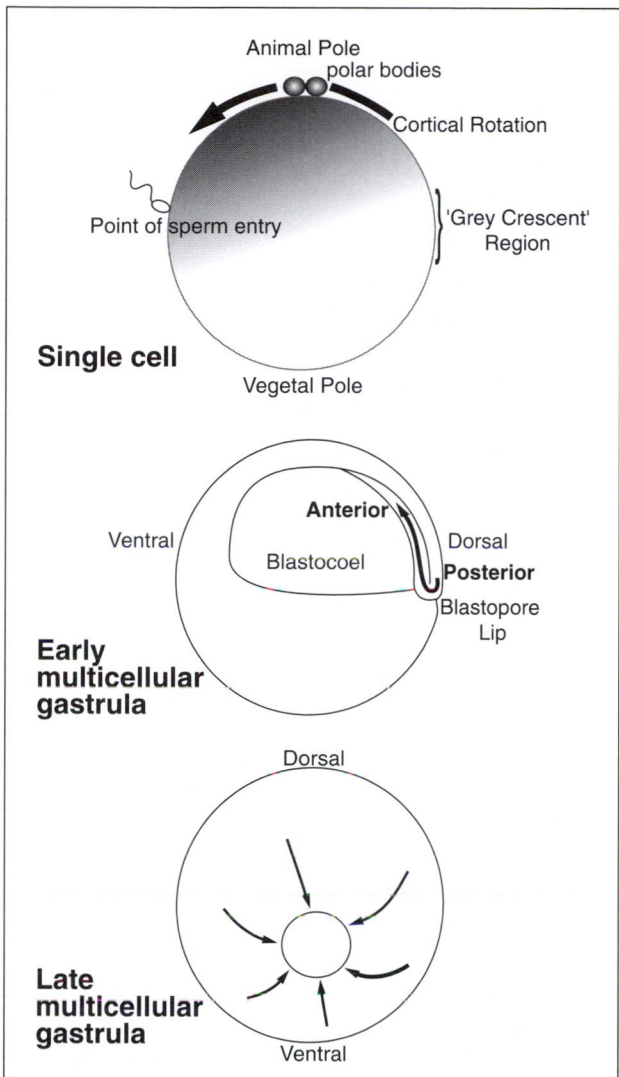

Fig. 3 Relationship between the polarity of a fertilized *Xenopus* egg and the polarity of a multicellular embryo. Both egg and early gastrula are viewed from the side. The fertilized egg is polarized with respect to pigment distribution and polar bodies (see top diagram). Sperm entry causes cortical rotation (bold arrow) and generates a 'grey crescent region' that predicts the dorsal side of the embryo. In early gastrula, an organizer is formed dorsally (dorsal blastopore lip). During gastrulation, mesoderm cells migrate along the roof of the blastocoel (arrow indicates the direction of movement) to establish the anteroposterior axis of the embryo. Late gastrula is viewed from the vegetal pole. The circular blastopore is now formed, and arrows indicate the spreading of the outer layer that will form the ectoderm.

portion of the vegetal hemisphere cultured in isolation forms only endoderm. However, mesoderm forms when both these portions of the embryo are co-cultured. The dorsovegetal region (Nieuwkoop centre) undergoes another interaction with the animal region to induce the dorsal mesoderm that will form the dorsal blastopore lip mesoderm (7). The dorsal blastopore lip is the organizer which produces signals that pattern mesoderm along the dorsoventral axis of the embryo, i.e. to give axial and paraxial mesoderm—notochord and somites—dorsally near the organizer; then intermediate mesoderm, kidney, and muscle; and, finally, ventral mesoderm, including, furthest away from the organizer, blood (see Chapter 3). This signalling activity was first discovered when the organizer from one amphibian embryo was

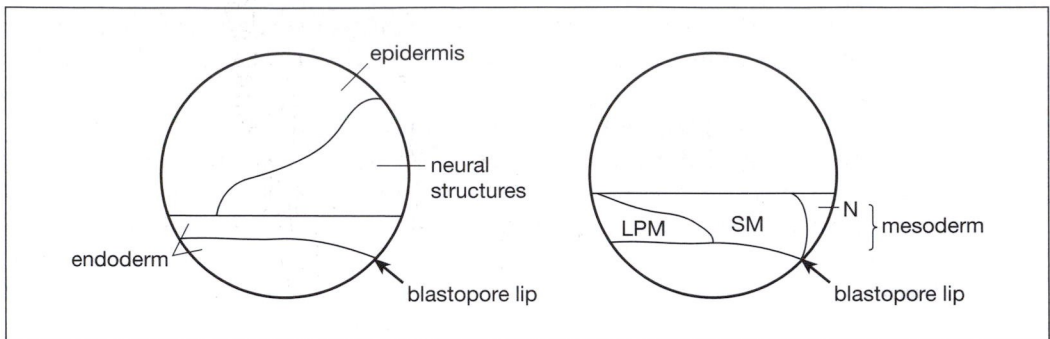

Fig. 4 Fate maps of early *Xenopus* blastula (based on refs 5, 6). The fate map on the left is of the outer cell layer; the fate map on th right is of the inner cell layer. Cells in the animal hemisphere give rise to ectoderm, cells in the equatorial (marginal) region to mesoderm, cells in the vegetal region to endoderm. N, notochord; SM, somitic mesoderm; LPM, lateral plate mesoderm.

transplanted to an ectopic position in another embryo where it dorsalized the host mesoderm. Several models have been suggested to account for the relationship between the type of mesoderm specified along the dorsoventral axis and its distance from the organizer, including those based on gradients and/or timing (see Chapter 1).

The dorsal blastopore lip is the region where gastrulation is initiated. Gastrulation involves considerable cell rearrangements, with mesodermal and endodermal cells moving to inside the embryo (gastrula). As gastrulation proceeds, the blastopore lip extends laterally to create a circular rim where in-tucking of mesoderm and endoderm occurs. At the same time as cells are moving into the inside of the embryo, the outer layer of cells, which will become ectoderm, is expanding (see Fig. 3). The circular blastopore thus becomes smaller and smaller until only a small yolky plug is left that marks the posterior end of the embryo. The complex cell movements and changes in cell shape involved in gastrulation have been documented in some detail (8–10).

The first mesoderm cells to move inside the *Xenopus* embryo form mesoderm at the very anterior of the embryo, while mesoderm cells that move inside later form posterior mesoderm. The anteroposterior specification of the mesoderm is a key event in laying down the body plan. Mesoderm acts as a template for anteroposterior positioning in the whole embryo, and it is through interactions of the mesoderm with both ectoderm and endoderm that these other tissues acquire information about their position along the anteroposterior axis, e.g. different regions of the brain and spinal cord (see Chapters 5, 6, and 7) and different regions of the gut. It was originally suggested that there are separate 'head' and 'tail' organizers—because 'head' structures are induced by a dorsal blastopore lip from an early gastrula, whereas ' tail' structures are induced by a dorsal lip of an older gastrula. Recently, a distinct signalling region in the anterior visceral endoderm has been discovered that specifies the anterior of the mouse embryo (11). Another interpretation of the different outcomes of early versus late blastopore lip grafts is that neural plate is specified first as anterior and then progressively posteriorized. These two models are discussed in Chapter 6.

Cell marking experiments at the blastula stage in *Xenopus* embryos show that cells in the dorsal region of the animal hemisphere will form neural tissue (see Fig. 4). Fate maps, however, only reveal what cells will form if left in place but not whether cells are determined, i.e. committed to a specific developmental pathway. Determination of ectoderm to form neural tissue provides a good example of the difference between fate maps and commitment (1). When cells from this region of an early *Xenopus* gastrula normally fated to form neural structures are transplanted to an ectopic position in another gastrula embryo, they do not develop into neural tissue but instead respond to signals in their new location and participate in forming epidermis (Fig. 5). It is only when these cells are transplanted late in gastrulation, once interactions have occurred between dorsal mesoderm and ectoderm, that they form ectopic neural tissue.

3. Chick, mouse, and zebrafish development

3.1 Early stages in chick, mouse, and zebrafish development

Chick, mouse, and zebrafish eggs vary in their yolkiness, which affects cleavage. Figure 6 compares cleavage in an amphibian embryo with cleavage in these other vertebrates. Both bird and fish eggs have substantial quantities of yolk, and the

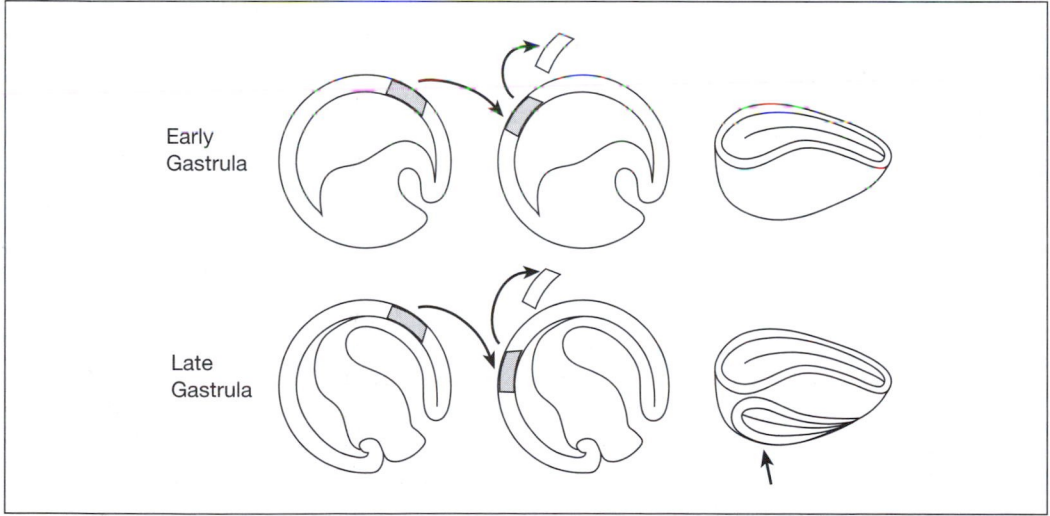

Fig. 5 Outcome of transplantation experiments carried out at early and late gastrula stages. Sections of embryos to show internal tissue arrangements. Cells from the animal hemisphere of an early gastrula fated to form neural structures grafted ventrally in another embryo do not form neural structures. When the same region is transplanted at a late gastrula stage when mesoderm has moved into position underneath, the transplant now forms ectopic nervous system (arrow).

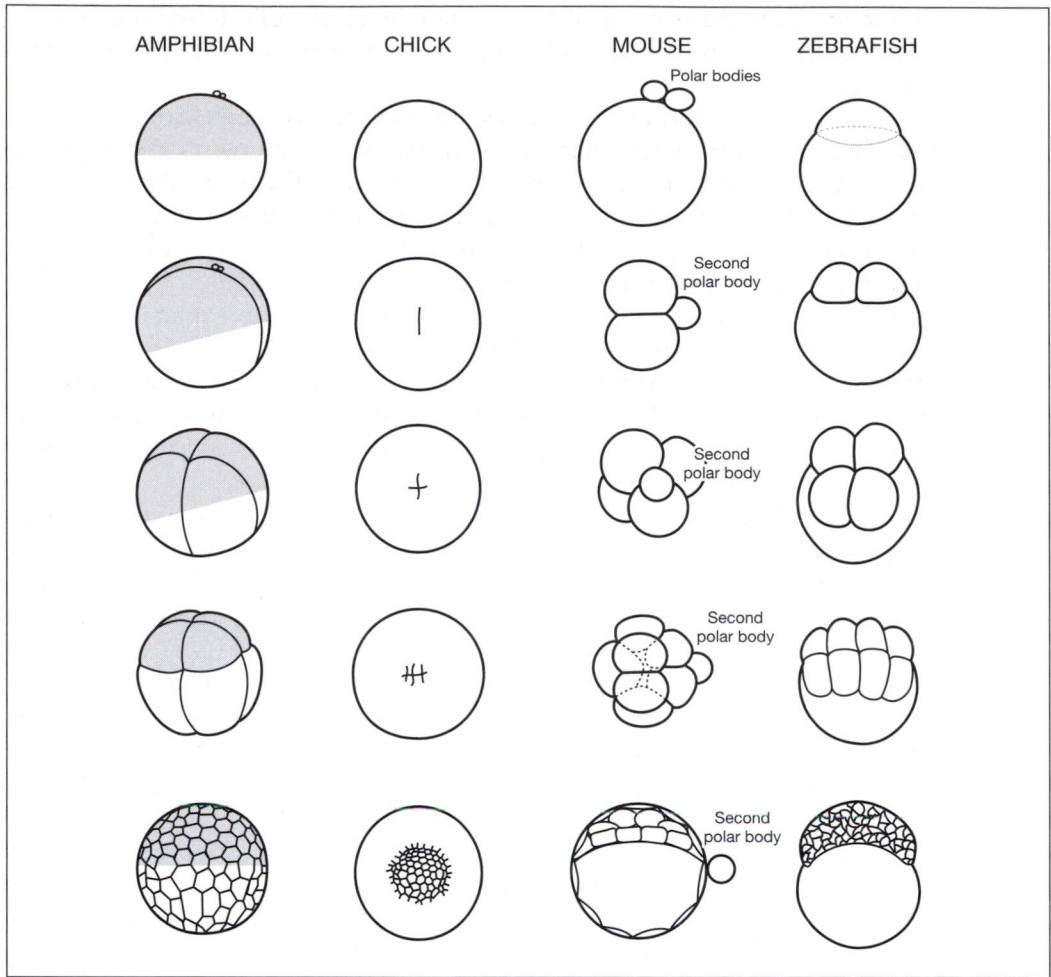

Fig. 6 Comparison of the cleavage stages of fertilized eggs of amphibian, chick, mouse, and zebrafish embryos. Amphibian, mouse, and zebrafish viewed from the side, chicken from above. Note that the different embryos are not drawn at the same scale.

nucleus and cytoplasm are localized at the animal pole. Cleavage divisions then partition this 'cap' of cytoplasm into a number of smaller cells that come to lie on top of the large volume of yolk. In zebrafish, the first five cleavages only partially penetrate into the cap and give rise to 32 blastomeres, which are continuous with the yolk. The sixth cleavage division is horizontal, and generates one outer tier of cells and a syncytial layer below that is still in continuity with the yolk. By the blastoderm stage in chick embryos, a cavity has developed in the central region under the single layer of cells (epiblast) from which the embryo will develop. The region around the epiblast circumference (marginal zone) is still in contact with yolk, and, later on, a layer of hypoblast cells spreads from the marginal zone under the epiblast. The zebrafish blastoderm comprises an outer layer of cells (enveloping layer), a yolk syncytial layer next to the yolk, and, in between, deep cells, that will form the embryo.

In contrast to chick and fish eggs, mouse eggs contain little yolk and therefore cleavage divides the entire fertilized egg into two cells, four cells, eight cells, and so on to give a small ball of cells (see Fig. 6). Between the 8-cell stage and 16-cell stage, the ball of cells undergoes compaction. Cells on the outside maximize their contacts with adjacent cells, which leads to a smooth outline. These 'outside' cells then begin to pump water into the interior to give a fluid-filled cavity (blastocoel), while the small group of 'inside' cells (inner cell mass) comes to lie against one end of the inside wall. The position of the inner cell mass within the blastocyst defines the embryonic versus abembryonic pole (Fig. 7). Outside cells will give rise to the trophoblast and contribute to the placenta; the layer of inner cell mass next to the blastocoel also forms an extraembryonic layer, while the rest of the inner cell mass gives rise to the epiblast from which the embryo develops.

The first signs of axis formation in chick, mouse, and zebrafish embryos is the accumulation of cells at a particular location in the epiblast or blastoderm (Fig. 8). In chick embryos, cells start to accumulate at a particular position in the marginal zone. This thickening is related to the initiation of primitive-streak formation and the future posterior end of the embryo. The primitive streak marks the future dorsal midline of the embryo and begins to extend towards the centre of the epiblast, with the formation of a node at its anterior end. Similarly, in the zebrafish, a thickening known as the 'embryonic shield' appears on one side of the blastoderm. (By this time the blastoderm has spread down to the equator of the egg.) This embryonic shield marks the future dorsal side of the embryo. In the mouse, the axes become

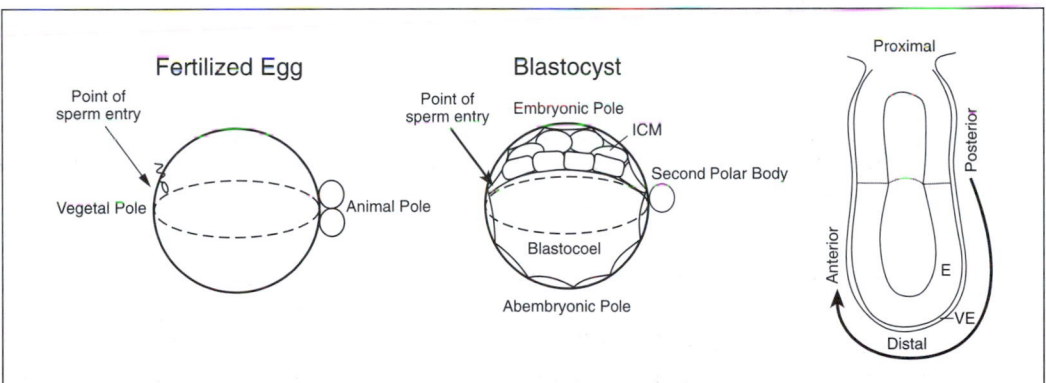

Fig. 7 Relationship between the polarity of a mouse fertilized egg and embryo. In the fertilized egg (left panel), the point of sperm entry determines the plane of the first cleavage division (dashed line). The relationship of this cleavage plane to the blastocyst is shown in the middle panel; polar bodies lie at the border between the inner cell mass and the blastocoel. At the egg cylinder stage (right panel), visceral endoderm that arises next to the polar body moves distally in a posterior to anterior direction (large arrow). Note there is considerable growth following implantation and that the egg cylinder is not drawn to the same scale as the egg and blastocyst. ICM, inner cell mass; E, epiblast; VE, visceral endoderm.

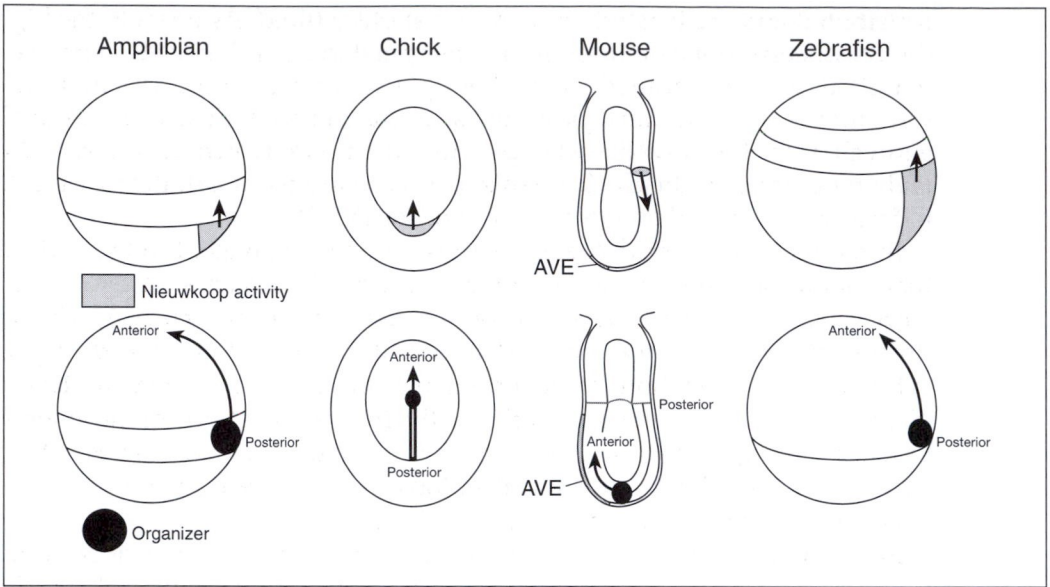

Fig. 8 Comparison between amphibian, chick, mouse, and zebrafish embryos showing induction of the organizer (upper row) and the organizer itself (lower row). In the upper row, Nieuwkoop activity (shaded area) and the arrow indicate where the organizer will form. The lower row shows the organizer (black), with the arrow indicating the movement of dorsal mesoderm cells towards the anterior of the embryo in each case. These dorsal mesoderm cells induce neural plate. In mouse embryos, AVE (anterior visceral endoderm), which produces a 'head' inducer, is indicated. (After Nieto 1999 (20).)

apparent in the postimplantation embryo (egg cylinder stage), by which time it has undergone considerable growth. The epiblast now forms the inner lining of a cup-shaped structure covered with an extraembryonic layer (visceral endoderm; see Fig. 7). Rather like in the chick, the cells of mouse embryos gather at one point on the rim of the cup (future posterior end of the embryo) to form a primitive streak. The streak marks the dorsal midline and then extends towards the bottom of the cup and forms a node. Apart from the geometry, chick and mouse embryos are very similar at these stages.

3.2 Relationship between the polarity of egg and embryo in chicks, mice, and zebrafish

In both chick and fish embryos, relationships between the polarity of the fertilized egg and future embryo are not as well understood as in amphibians. Nevertheless, the localization of determinants within the egg appears to be important in setting up early differences; and, despite differences in embryo morphology, regions that act as organizers can be identified, as can signalling centres that induce them. In chickens, the fertilized egg rotates as it passes down the oviduct before it is laid. This rotation seems to have a similar effect to that of cortical rotation in *Xenopus*. The yolk is redistributed and the blastoderm becomes slightly tilted. As a result, the high point of the blastoderm defines the posterior marginal zone. When cells from the posterior marginal zone are grafted to another spot around the epiblast circumference, the formation of an additional primitive streak is induced. Posterior marginal cells are equivalent to the frog Nieuwkoop centre; the node, which develops at the anterior end of the streak, is the chick organizer and can induce an additional primitive streak to form when grafted to another embryo (see Fig. 8).

In the fertilized zebrafish egg, there is no clear equivalent to the cortical rotation that establishes the dorsal side of the embryo in *Xenopus*. Although there is no correlation between the dorsoventral axis and the second cleavage division plane, dorsoventral asymmetry can be detected at the 32-cell stage and may be due to cytoplasmic localization of determinants in the yolk syncytial layer. Indeed, Nieuwkoop activity can be identified at the future dorsal side of the embryo and this induces the organizer, in this case, the embryonic shield (see Fig. 8). Grafts of embryonic shield lead to the dorsalization of host mesoderm, which is similar to the effect of grafts of dorsal blastopore lip in *Xenopus*.

Early mouse development is highly regulative and therefore it is difficult to see how cytoplasmic determinants could be important. For example, when either animal or vegetal pole cytoplasm is removed from fertilized mouse eggs, they nevertheless develop into normal embryos (12); moreover, aggregation chimera studies have shown that individual blastomeres from early cleavage-stage mouse embryos can contribute to all parts of the embryo and to extraembryonic structures (13). Even at the eight-cell stage when animal versus vegetal components could be important, all blastomeres appear to have the same potential (13). At the 16-cell stage, it has been shown that the position of a cell—either on the inside or on the outside—determines whether it will form an inner cell mass or trophoblast (14).

Recent work has shown, unexpectedly, that the orientation of early cleavage divisions of mouse eggs may play a role in subsequent cell fate, and that the polarity of the fertilized mouse egg may be related to the body axes. In two separate studies, the relationship between the polarity of the egg and the polarity of the blastocyst, and between blastocyst polarity and polarity of the egg cylinder-stage embryo has been established (15, 16).

The point of sperm entry appears to position the plane of the first cleavage division in fertilized mouse eggs (16). The sperm entry point was marked by attaching fluorescent beads to the fertilization cone that forms, transitorily, after fertilization, and these beads provided a way of investigating how the point of sperm entry is related to the axes of the blastocyst. When marked eggs were then observed during cleavage it was found that the beads were very frequently found in the furrow of the first cleavage division, or close by. This suggests that in the mouse, as in *Xenopus*, the point of sperm entry positions the plane of the first cleavage division of the fertilized egg (see Fig. 7). When these embryos were followed through development to the blastocyst, it was found that beads most frequently came to lie near the border between the inner cell mass and the blastocoel cavity. This suggests that the two cells generated at the first cleavage division may normally follow different pathways: one going on to contribute to the inner cell mass, the other to contribute only to the trophoblast.

How can these differences arise, since blastomere isolation experiments and removal of cytoplasm provide no evidence that the localization of cytoplasmic determinants in fertilized mouse eggs is important? Careful observation showed that the two-cell stage cell which inherits the sperm entry site also tends to divide again earlier than the other cell. Thus, it could be this difference in timing that leads to differences between the two cells (16).

The fate of different mouse blastomeres—either near the animal or vegetal pole—can also be traced. The animal pole of fertilized mouse eggs can be recognized because this is where polar bodies are attached. When the fate of blastomeres is traced, either by following animal cells to which the polar body remains attached or by injecting Green Fluorescent Protein (GFP), marked cells are found aligned with the long axis of the inner cell mass (17, 18). Further lineage analysis has shown that extraembryonic cells at the 'animal' end of the blastocyst move distally and extend along the future anteroposterior axis of the embryo (see Fig. 7). A Nieuwkoop centre then arises at the posterior rim of the egg cylinder and this initiates primitive-streak formation (19) (reviewed in ref. 20). As in the chick, the streak marks the dorsal midline of the mouse embryo and the node that develops at the anterior of the streak is the organizer (see Fig. 8). When a labelled mouse node is grafted to another early mouse egg cylinder, this induces the development of a second neural axis and extra somites from the host (21).

3.3 Gastrulation in chick, mouse, and zebrafish

Despite differences in geometry, gastrulation in chick and mouse embryos is very similar. All three germ layers are derived from the single layer of cells in the epiblast. There are fate maps for both chick and mouse embryos at this stage showing what cells in different regions of the epiblast will form (22–24). In the chick, the epiblast is the upper surface of the blastodisc and lies on top of the hypoblast, while, in the mouse, the epiblast lines the egg cylinder and is covered by a layer of visceral endoderm. In both chick and mouse embryos, a primitive streak forms at the future posterior end of the embryo and epiblast cells migrate into the primitive streak. There, cells detach from the epithelium and either migrate as individuals beneath the epiblast to form mesoderm or move deeper to form endoderm (Fig. 9). Endoderm formation in chick and mouse embryos displaces the hypoblast and visceral endoderm, respectively. Cells that remain in the epiblast layer will form ectoderm. At the node, dorsal mesoderm cells migrate anteriorly to form notochord, while the ectoderm overlying this becomes neural plate.

In zebrafish, fate maps at early gastrula stages show that deep cells at the blastoderm margin form endoderm, cells closer to the animal pole form mesoderm, and cells closest to the animal pole, ectoderm. Deep cells give rise to the three body layers by extensive inward movements all around the rim of the extending blastoderm, with cells then converging on the embryonic shield, both dorsally and

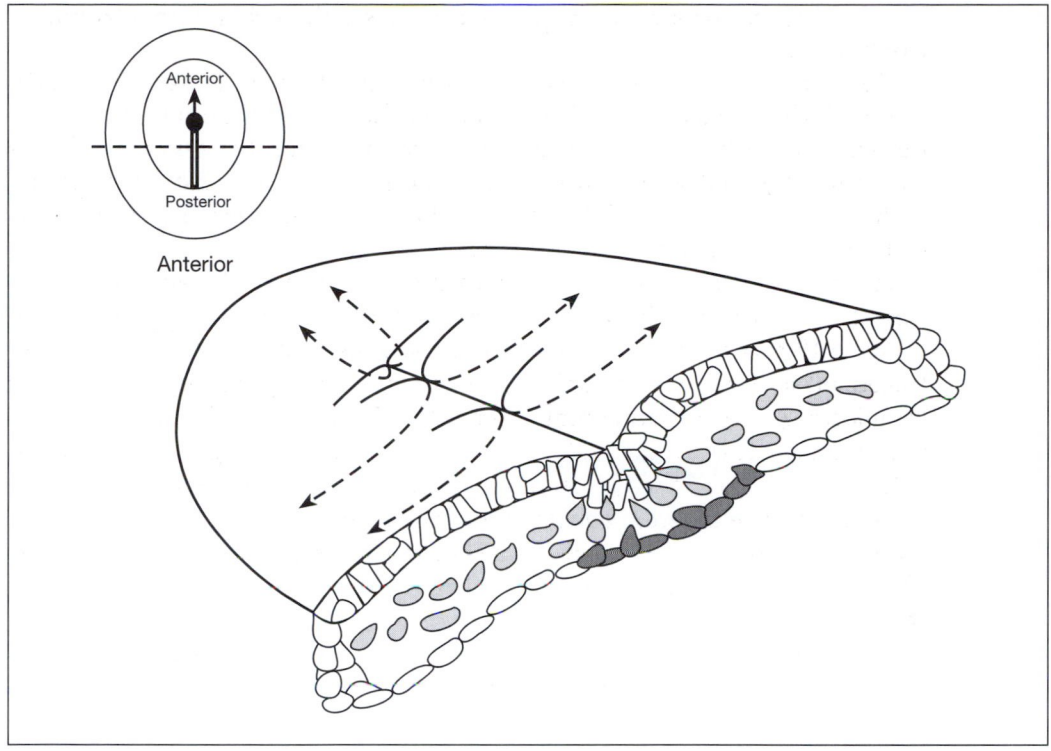

Fig. 9 Gastrulation in a chick embryo. The upper left diagram shows the dorsal view of the primitive streak of a chick embryo (see Fig. 8). The dotted line indicates the level at which the embryo was cut to give view from cut surface shown in the lower diagram. Cells (grey) leave the epiblast and migrate as individuals under the epiblast to form mesoderm; other epiblast cells (black) insert into the hypoblast to form endoderm. Arrows show the direction of cell migration through the streak and node, where mesoderm cells move anteriorly. This process is similar in mice, and the epiblast gives rise to all three body layers.

anteriorly, to establish the anteroposterior axis of the body. There is considerable mixing of cells during this process (25).

4. Conclusions

The way in which the vertebrate body plan is established is worked out best in amphibians. There is a clear relationship between the polarity of the fertilized egg and that of the embryo (see Fig. 3). This is accomplished by the distribution of localized cytoplasmic determinants to specific blastomeres during cleavage, thus establishing early differences between cells. This is followed by a series of inter-actions that eventually results in induction of the organizer in the dorsal region of the embryo. In turn, the organizer then undergoes interactions with neighbouring cells to provide the positional information necessary to establish the body plan. This general picture seems to apply to other vertebrates even though the morphology of the early stages varies considerably. All the vertebrate embryos that we have considered—chick, mouse, zebrafish—show a stereotypical pattern of cleavages that could partition cytoplasmic determinants. There are differences, however, in the time at which differences are established, e.g. early cleavage in amphibians but the blastoderm stage in chick embryos. In mouse embryos there is no evidence for cytoplasmic determinants and, instead, timing seems to be important. There is a series of interactions in all these embryos that lead to formation of the organizer. The identification of the various steps in this process in different vertebrates has been greatly aided by the identification of the molecules involved (see subsequent chapters). In fact, the molecular analysis has revealed similarities in the laying down of the body plan in different vertebrates that would otherwise be difficult to discern. So read on and enjoy!

Acknowledgements

We thank Tara Sheldrake and Juan José Sanz-Ezquerro for their initial help with the figures.

References

1. Slack, J. (1991) *From egg to embryo*. Cambridge University Press, Cambridge.
2. Slack, J. (2001) *Essential developmental biology*. Blackwell Science Ltd, Oxford.
3. Gilbert, S. (2000) *Developmental biology*. Sinauer Associates, Sunderland, Massachusetts.
4. Wolpert, L., Beddington, R., Jessell, T., Lawrence, P., Meyerowitz, E., and Smith, J. (2002) *Principles of development* (2nd edition). Oxford University Press, Oxford.
5. Keller, R. E. (1975) Vital dye mapping of the gastrula and neurula of *Xenopus laevis*. I. Prospective areas and morphogenetic movements of the superficial layer. *Dev. Biol.*, **42**, 222.
6. Keller, R. E. (1976) Vital dye mapping of the gastrula and neurula of *Xenopus laevis*. II. Prospective areas and morphogenetic movements of the deep layer. *Dev. Biol.*, **51**, 118.
7. Nieuwkoop, P. (1977) Origin and establishment of embryonic polar axes in amphibian development. *Curr. Top. Dev. Biol.*, **11**, 115.
8. Keller, R. E. (1980) The cellular basis of epiboly: an SEM study of deep-cell rearrangement during gastrulation in *Xenopus laevis*. *J. Embryol. Exp. Morphol.*, **60**, 201.
9. Keller, R. E., Danilchik, M., Gimlich, R., and Shih, J. (1985) The function and mechanism of convergent extension during gastrulation of *Xenopus laevis*. *J. Embryol. Exp. Morphol.*, **89**(Suppl.), 185.
10. Keller, R. and Danilchik, M. (1988) Regional expression, pattern and timing of convergence and extension during gastrulation of *Xenopus laevis*. *Development*, **103**, 193.
11. Beddington, R. S. and Robertson, E. J. (1999) Axis development and early asymmetry in mammals. *Cell*, **96**, 195.
12. Zernicka-Goetz, M. (1998) Fertile offspring derived from mammalian eggs lacking either animal or vegetal poles. *Development*, **125**, 4803.
13. Kelly, S. J. (1977) Studies of the developmental potential of 4- and 8-cell stage mouse blastomeres. *J. Exp. Zool.*, **200**, 365.
14. Rossant, J. and Vijh, K. M. (1980) Ability of outside cells from preimplantation mouse embryos to form inner cell mass derivatives. *Dev. Biol.*, **76**, 475.

15. Weber, R. J., Pedersen, R. A., Wianny, F., Evans, M., and Zernicka-Goetz, M. (1999) Polarity of the mouse embryo is anticipated before implantation. *Development*, **126**, 5591.

16. Piotrowska, K. and Zernicka-Goetz, M. (2001) Role for sperm in spatial patterning of the early mouse embryo. *Nature*, **409**, 517.

17. Gardner, R. (1997) The early blastocyst is bilaterally symmetrical and its axis of symmetry is aligned with the animal–vegetal axis of the zygote in the mouse. *Development*, **124**, 289.

18. Ciemerych, M. A., Mesnard, D., and Zernicka-Goetz, M. (2000) Animal and vegetal poles of the mouse egg predict the polarity of the embryonic axis, yet are nonessential for development. *Development*, **127**, 3467.

19. Popperl, H., Schmidt, C., Wilson, V., Hume, C. R., Dodd, J., Krumlauf, R., and Beddington, R. S. (1997) Misexpression of Cwnt8C in the mouse induces an ectopic embryonic axis and causes a truncation of the anterior neuroectoderm. *Development*, **124**, 2997.

20. Nieto, M. A. (1999) Reorganizing the organizer 75 years on. *Cell*, **98**, 417.

21. Beddington, R. S. P. (1994) Induction of a second neural axis by the mouse node. *Development*, **120**, 613.

22. Hatada, Y. and Stern, C. D. (1994) A fate map of the epiblast of the early chick embryo. *Development*, **120**, 2879.

23. Smith, J. L. and Schoenwolf, G. C. (1998) Getting organized: new insights into the organizer of higher vertebrates. *Curr. Top. Dev. Biol.*, **40**, 79.

24. Lawson, K. A. and Pedersen, R. A. (1992) Clonal analysis of cell fate during gastrulation and early neurulation in the mouse. *Ciba Found. Symp.*, **165**, 3.

25. Warga, R. M. and Kimmel, C. B. (1990) Cell movements during epiboly and gastrulation in zebrafish. *Development*, **108**, 569.

3 | Patterning the *Xenopus* embryo

J. C. SMITH and R. WHITE

1. Introduction

The problem we wish to address concerns the mechanism by which the vertebrate early gastrula forms from the fertilized egg, with the right genes activated in the right places and with cells poised to undergo the movements of gastrulation. We will concentrate on the amphibian embryo, where the detailed molecular events of early embryogenesis are probably better understood than in any other vertebrate. This understanding is due to what might be regarded as the 'traditional' advantages of the amphibian embryo: its large size, the ease of grafting, the ability to culture cells in a simple salt solution, and the ease with which molecules can be overexpressed by RNA or DNA injection. These advantages have recently been supplemented by the introduction of transgenesis in *Xenopus* (by Amaya and Kroll (1, 2)), allowing, in principle, genes to be expressed in virtually any tissue of interest.

What is lacking is a reliable technique for abolishing gene function in *Xenopus*, and this has been a major handicap in attempting to understand definitively the roles of the various signalling molecules and transcription factors that have been implicated in early development. Some techniques have proved useful, of course, and these include dominant-negative approaches (3, 4), antisense oligonucleotides (5, 6), antisense RNAs (7, 8), antibodies (9), ribozymes (10), secreted forms of cell-surface receptors (11), and, most recently, the use of Morpholino antisense oligodeoxynucleotides (12). Use of these approaches has, on the whole, yielded results that are consistent and which have allowed models to be built describing the events of early embryogenesis. Nagging worries do remain about the specificity of these techniques, but these have been greatly alleviated by results obtained with the zebrafish embryo, which indicate, for example, that the picture obtained from experiments in *Xenopus* concerning dorsoventral patterning of the early embryo is essentially correct.

2. From oocyte to egg: establishing the animal–vegetal axis

The most obvious features of oogenesis in *Xenopus* are growth and the onset of visible polarity. Both features involve yolk. Growth is due to the uptake by the oocyte of blood-borne yolk precursors made originally in the liver. Polarity comes from the way the yolk platelets are distributed, with more yolk, and larger platelets, in the vegetal half of the oocyte. This distribution derives not from a higher uptake of yolk precursors in the vegetal hemisphere, but from the export of yolk platelets out of the animal hemisphere (13). The formation of the yolk gradient is accompanied by the movement of the germinal vesicle to the animal pole of the embryo, while the animal hemisphere also becomes heavily pigmented.

2.1 Localized RNAs

The yolk distribution and pigmentation provide useful markers of polarity for the experimenter. In terms of patterning the embryo, however, the most important process occurring during oogenesis concerns the distribution of RNAs to different regions of the oocyte. The first such RNA to be studied in detail was Vg1, which, as the name implies, is restricted to the vegetal hemisphere of the oocyte and egg (14). Vg1 was isolated through a differential screening approach, and proved to encode a member of the transforming growth factor (TGF)-β family (15). Although discovered some 15 years ago, the role of Vg1 in *Xenopus* development remains obscure. It has been mooted to play a role in mesoderm induction, but this still remains to be proved (see below). However, because injected Vg1 RNA, like the endogenous gene product, becomes localized to the vegetal hemisphere of the *Xenopus* oocyte (16), it has been possible to study the mechanism by which the RNA becomes restricted to the vegetal hemisphere of the embryo. Thus sequences have been identified in the Vg1 3' untranslated region (UTR) that are necessary and sufficient for RNA localization (17, 18), proteins that interact with the Vg1 3' UTR have been isolated (19, 20), and microtubules and the endoplasmic reticulum have been implicated in the translocation of Vg1 RNA to the vegetal cortex (19, 21).

Other vegetally localized RNAs include *Xwnt11* (22), *Bicaudal-C* (23), and *VegT* (also known as *Xombi*, *Antipodean*, and *Brat*) (24–27). The role of maternal *Xwnt11* remains unclear, although in zebrafish embryos zygotic expression of *Wnt11/ Silberblick* is required for normal gastrulation movements (28). It is possible, in *Xenopus*, that a zygotic requirement for *Xwnt11* in gastrulation (29) is supplemented by maternal transcripts, which are preferentially translated in dorsal blastomeres (30).

Of the other vegetally localized transcripts, *Bicaudal-C* is of interest because it can specify endoderm formation (23), while *VegT* is of particular note because ablation of the mRNA by antisense oligonucleotide injections reveals that it is required both for endoderm formation and for the production of mesoderm-inducing signals (6). This is discussed in more detail below.

Finally, this description of differentially distributed maternal RNAs may give the impression that all localized messages are restricted to the vegetal hemisphere of the embryo. This is not true, and indeed the original screen of Rebagliati and colleagues (14) isolated three RNAs restricted to the animal hemisphere and just one—Vg1—enriched in the vegetal hemisphere. It is true, however, that we know very little about what these animal pole RNAs do. One, for example, encodes a subunit of mitochondrial ATPase (31); another encodes a β-subunit of heterotrimeric GTP-binding proteins (32); and yet a third encodes a novel zinc-finger protein (33). The roles of these proteins, and the mechanism by which their RNAs become localized, will no doubt be investigated in the future.

3. Fertilization: establishing the dorsoventral axis

Oogenesis creates a radially symmetrical egg, with RNAs restricted to the animal or the vegetal poles, thus creating an animal–vegetal axis, but no other developmental bias. The second axis, the so-called dorsoventral axis, is defined by the site of sperm entry (see Chapter 2).

It is worth mentioning here that any name for an axis orthogonal to the originally created animal–vegetal axis will be unsatisfactory, because this axis does not translate directly into one of the axes of the swimming tadpole. If pressed, one might say that the new axis corresponds to something between the dorsoventral and anteroposterior axes, but even this would be an oversimplification, because the movements of gastrulation distort the original axes so much. For simplicity, like most authors, we continue to refer to the new axis as the dorsoventral axis, but the reader has been warned.

3.1 Cortical rotation

Xenopus eggs are fertilized by a single sperm, which enters apparently randomly in the animal hemisphere of the egg. The site of sperm entry is usually marked by an area of slightly darker pigmentation, and this side of the embryo always becomes the 'ventral' side of the early gastrula. This specification of the dorsoventral axis occurs through a process known as 'cortical rotation', in which a layer of subcortical cytoplasm rotates with respect to the plasma membrane such that material originally positioned at the vegetal pole becomes translocated to the prospective dorsal side of the embryo (reviewed in ref. 34; Fig. 1). This movement requires the assembly of a parallel array of microtubules (35–37), and is essential for the formation of dorsal structures. If microtubule assembly is disrupted, the resulting embryos lack a dorsal axis; if microtubule stability is increased, embryos become hyperdorsalized (38).

3.2 The β-catenin pathway

Cortical rotation appears to exert its effect by influencing the activity of the Wnt pathway. It has long been known that ventral misexpression of Wnts can dorsalize

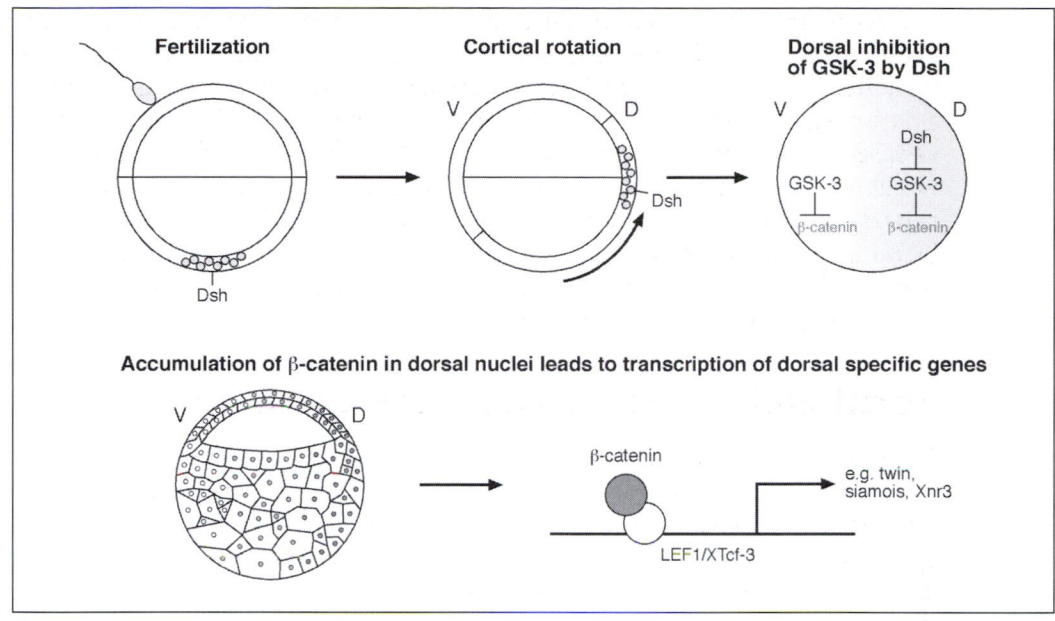

Fig. 1 (See also Plate 1) Model of the mechanism of localized Wnt pathway activation during dorsoventral axis specification in *Xenopus*. Dsh associates with a specific class of vesicles at the vegetal pole and these vesicles are transported dorsally along the subcortical microtubule array during cortical rotation. This translocation contributes to the asymmetrical distribution of Dsh along the dorsoventral axis and the localized activation of a maternal Wnt signalling pathway. Activation of Wnt signalling leads to the downregulation of GSK3 activity, thereby promoting the stabilization of β-catenin. β-Catenin then accumulates in dorsal nuclei where, in combination with XTcf-3, it activates the transcription of dorsal-specific regulatory genes. (Adapted from ref. 51.)

the amphibian embryo (39, 40), as can activation of the canonical Wnt signalling pathway, either by overexpression of β-catenin (41) or by inhibiting the activity of glycogen synthase kinase-3β (GSK3β) (42–44). In addition, depletion of maternal RNA encoding β-catenin results in embryos that lack a dorsal axis (45), as does overexpression of wild-type GSK3β (42–44). The implication from these studies is that in normal embryos the canonical Wnt signalling pathway is active at higher levels in dorsal blastomeres than in ventral ones, a conclusion supported by the observation that β-catenin accumulates in dorsal, but not ventral, nuclei of the cleavage-stage embryo (46). In addition, Dominguez and Green (47) have recently shown that GSK3β protein is depleted on the dorsal side of the embryo (see Fig. 1).

Although there is abundant evidence that the Wnt signalling pathway is necessary for dorsal axis formation, it is not clear how the pathway is activated. The obvious suggestion is that a maternal Wnt family member activates the pathway, the obvious candidate being *Xwnt11*, which is preferentially translated in dorsal blastomeres (30). However, *Xwnt11* is a poor inducer of secondary axes (22) and inhibition of Xwnt11 activity using a dominant-negative construct (29), far from ventralizing embryos, actually promotes the formation of ectopic dorsal structures (48). Furthermore, a dominant-negative version of Dishevelled, a component of the Wnt signalling pathway which acts upstream of GSK3β and β-catenin, does not block axis formation (49). The inability of these interfering constructs to block axis formation does not provide definitive evidence that Wnt and Dishevelled signalling are not involved in axis specification, but they raise the possibility that the Wnt pathway, in the shape of GSK3β and β-catenin, is activated downstream of these components. One route by which this might occur is through the GSK3-binding protein (GBP) (50), ectopic expression of which can bring about the depletion of GSK3β referred to above (47). Another suggestion, however, comes from work of Miller and colleagues (51), who have observed that a GFP-tagged version of Dishevelled associates with small vesicles that are moved to the dorsal side of the embryo during cortical rotation, and that overexpression of Dishevelled can stabilize β-catenin (Fig. 1). These observations suggest that the directed translocation of Dishevelled is responsible for the dorsal accumulation of β-catenin, but they do raise the question of why dominant-negative Dishevelled does not block dorsal axis formation (49). One possibility, suggested by Miller and colleagues (51), is that the dominant-negative construct does not gain access to the correct compartment of the cell to block the function of endogenous Dishevelled, but this requires further investigation and new methods for blocking Dishevelled function in *Xenopus*.

It remains true, however, that β-catenin signalling is involved in dorsal axis determination; we discuss how the nuclear accumulation of β-catenin is converted into specific gene expression in Section 5 below.

4. Mesoderm induction: realizing the animal–vegetal axis

Oogenesis and fertilization create, respectively, the animal–vegetal and dorsoventral axes of the embryo, marked by the vegetal localization of gene products such as VegT and the nuclear localization of β-catenin. The spatial information provided to the embryo in this way is not of sufficient resolution to specify the different cell types and gene-expression domains of the *Xenopus* early gastrula; for this to occur the initial differences are amplified by a series of inductive interactions. These interactions occur during cleavage stages of development, and to describe them it is necessary first to describe these early cell cycles.

4.1 Cleavage divisions in *Xenopus*

The first cleavage in *Xenopus* occurs about 90 minutes after fertilization; microinjection of lineage tracers into one of the two resulting cells indicates that this cleavage divides the embryo essentially into right and left halves. The second cleavage takes place some 30 minutes later and is orthogonal to the first, dividing the embryo into 'dorsal' and 'ventral' halves, while the third, another 30 minutes later, is orthogonal to both and separates four (smaller) animal pole blastomeres from four (larger) vegetal pole cells—cleavage planes tend to be positioned away from the yolkier regions of the cell (see Chapter 2).

A total of 12 of these rapid cleavages occur, in a synchronous fashion and with cleavage planes alternating between equatorial and meridional. During this period, the cell cycles consist just of S and M phases, and RNA synthesis does not occur to any significant level. After the twelfth cleavage, however, in an event referred to as the mid-blastula transition, or MBT (52), synchrony is lost, cell cycles become longer, RNA synthesis begins, and cells acquire some motility. At this point there are 4096 cells and the embryo begins preparations for gastrulation. Fate-mapping experiments show that blastomeres of the vegetal hemisphere will go on to form endoderm, those of the equatorial region will form mesoderm, such as notochord and muscle, and those of the animal pole region will form ectodermal derivatives such as epidermis and neural plate (53, 54).

4.2 Mesoderm induction

The first inductive interaction in *Xenopus* development occurs along the animal–vegetal axis, and involves the induction of mesoderm from equatorial tissue under the influence of a signal (or signals) from vegetal pole blastomeres (55, 56). The interaction was first demonstrated by juxtaposing vegetal pole blastomeres (which, when isolated, give rise to endoderm-like tissue) with cells from the animal pole (which give rise only to ectodermal tissue such as epidermis). Tissue combinations of this sort give rise to large amounts of mesoderm—tissue which neither component would have formed if cultured alone—and lineage-tracing experiments show that the mesoderm derives from the animal pole cells (57). The implication of this experiment is that the mesoderm is normally formed through an interaction of this sort, and, as we shall see below, this is likely to be true.

4.3 Mesoderm-inducing factors

On hearing about mesoderm induction, it is natural to ask about the nature of the signal, but before discussing this it is wise to ask when mesoderm induction occurs. Experiments in which animal pole and vegetal pole tissue were juxtaposed at different developmental stages have recently suggested that the interaction only occurs after the mid-blastula transition (58, 59), although it is possible that a weak signal may commence before this (6, 60, 61). This suggests that the natural inducers are expressed zygotically, and are not represented by maternal RNAs such as Vg1 (62, 63) or maternal proteins such as activin (64), although it is possible that these candidate inducers are present in a latent form before the MBT.

If the suggestion that mesoderm induction occurs after the mid-blastula transition is taken at face value, there are several candidates for the natural mesoderm-inducing signal(s). Most of these are, like activin and Vg1, members of the TGF-β family; it has long been known that activin, for example, is a potent inducer of mesoderm from animal pole tissue (65). The strongest candidates include the nodal-related genes *Xnr1* (66), *Xnr2* (66), and *Xnr4* (67) and *derrière* (68), which is closely related to *Vg1*. All four of these genes are expressed in the vegetal hemisphere and marginal zone of the *Xenopus* blastula, and all are capable of inducing mesodermal cell types from isolated *Xenopus* animal pole tissue (66–68). Furthermore, dominant-negative versions of the proteins, or specific inhibitors of the nodal-related genes such as a carboxyl-terminal fragment of Cerberus, interfere with mesoderm induction by the vegetal hemisphere of the embryo and with mesoderm formation in the intact embryo (68–70).

The candidacy of these factors for the 'natural' mesoderm-inducing signal is strengthened by experiments in which maternal VegT transcripts are ablated from the embryo by antisense oligonucleotide injection. As discussed above, this treatment prevents endoderm formation and the vegetal hemisphere loses the ability to induce mesoderm from juxtaposed animal pole tissue (6). Embryos lacking maternal VegT do not activate the expression of *Xnr1*, *Xnr2*, *Xnr4*, or *derrière*, while ectopic expression of these inducing factors 'rescues' the effect of VegT depletion (71). The *derrière* gene rescues trunk and tail formation, while *Xnr1*, *Xnr2*, and *Xnr4* rescue the head, trunk, and tail (71).

These results suggest that VegT activates the expression of *Xnr1*, *Xnr2*, *Xnr4*, and *derrière*, and indeed VegT can induce the expression of these genes in isolated animal pole regions (72). It is likely that the induction of *Xnr1*, at least, is direct; the *Xnr1* promoter contains at least one T-box binding site (71, 73) and the integrity of these sites is necessary for the maximal activation by VegT of a reporter-construct cloned downstream of the *Xnr1* promoter (73). However, a 3.5-kb reporter-construct bearing these sites does not drive detectable expression in the vegetal hemisphere of transgenic *Xenopus* embryos, and the dorsal expression that is detectable does not require the T-box sites (73). This suggests that other transcription factors and other DNA binding sites are necessary for the correct regulation of *Xnr1*.

Inducing factors such as *Xnr1*, *Xnr2*, *Xnr4*, and *derrière* exert their effects by activating the expression of genes such as *Xbra* (74), *XFKH1 / Pintallavis* (75, 76), and *goosecoid* (77). These members of the TGF-β family act by binding to serine–threonine kinase receptors at the cell surface, which in turn phosphorylate and activate members of the Smad family of signal transduction molecules (78–82). This is not the place to discuss that pathway in detail, but two points are relevant and are discussed below.

4.4   Concentration-dependent effects of inducing factors

First, the activation of a mesoderm-specific gene by members of the TGF-β family is often highly concentration-dependent. The best example of this concerns *Xbra*, the *Xenopus* homologue of the mouse *Brachyury*, or *T*, gene (74). *Xbra* is activated by intermediate doses of the inducing factor activin but not by high concentrations (83–86). This suggests that the correct spatial expression pattern of *Xbra* might be established by diffusion and the establishment of a gradient of activin-like activity, perhaps represented by derrière or a member of the nodal family. Where the concentration of inducing factor is high, near its site of production in the vegetal hemisphere, *Xbra* would not be expressed. Where it is lower, in the marginal zone, the gene would be activated. And where it is lower still, in the animal hemisphere, the gene would again not be expressed, this time because levels are too low. This is an intriguing possibility and more work is required to test the idea, which resembles Wolpert's (87) French Flag model (see Chapter 1). For example, one needs to know how inducing molecules exert long-range effects (see ref. 88) and how the *Xbra* regulatory regions respond to different levels of an inducer so that expression only occurs in a certain window of concentrations (see ref. 86). Work on the latter question is discussed below; for the former it is worth noting that another way of achieving a gradient of an inducing factor, albeit in the dorsoventral and not animal–vegetal axis, is to establish graded levels of expression of the inducing factor. For example, the nodal-related genes are expressed in a gradient in the blastula endoderm, with highest levels dorsally (70).

Attempts to understand the interpretation of a gradient of an inducing factor have also concentrated on *Xbra*, where preliminary experiments demonstrated that 381 base pairs 5' of the transcription start site are sufficient to confer a concentration-dependent response (89). These experiments gave rather variable results, probably due to the fact that reporter DNA was not incorporated into the chromosomes of the injected embryo. The second point referred to above, however, is that it is now possible to make transgenic *Xenopus* embryos in which DNA does become integrated into the host genome (2, 90), and we are now using such embryos to study the activation of *Xbra* (91). In the next few years it should be possible to obtain quite a reasonable understanding of how mesoderm-specific genes such as *Xbra* are activated.

4.5 The functions of mesoderm-specific genes: Brachyury targets

Finally, there is the challenge of investigating the roles and modes of action of the newly induced mesoderm-specific genes. *Xbra* is again of particular interest here, because its function is quite well understood and the challenge is more to understand the molecular basis of its action. Thus, Xbra, like its mouse homologue, is a DNA-binding protein which functions as a transcription activator (4, 92, 93), and, as in the mouse, loss of the transcription activation function of Xbra causes the disruption of posterior mesodermal cell types and an impairment of notochord differentiation (4, 92, 93). There is also a severe disruption of gastrulation movements, and in particular of convergent extension (94). These results indicate that *Xbra* is necessary for normal mesoderm formation, and experiments in *Xenopus* also indicate that the gene is sufficient to allow the formation of mesoderm if it is misexpressed in prospective ectodermal tissue (95–97). With Xbra, therefore, the challenge is to identify the target genes that direct animal pole cells to form mesoderm and permit gastrulation to occur, and we have adopted two approaches to this end.

The first approach involved simple guesswork, from which *eFGF* (the *Xenopus* homologue of *FGF4*) was identified as a candidate Xbra target. The guess was based on the similar expression patterns of the two genes (98), together with the fact that they are involved in an autoregulatory loop in which each maintains expression of the other (99, 100). Use of a dominant-negative construct showed that expression of *eFGF* in the *Xenopus* embryo requires *Xbra* function (101); moreover, sequencing of the *eFGF* 5′ regulatory region identified two elements, one in the 5′ UTR and one positioned approximately 1 kb upstream of the transcription start site, to which Xbra can bind. Both sites were required for full induction of a 2.5-kb *eFGF* promoter-construct by Xbra when assayed in *Xenopus* oocytes (101), but a single site proved to be sufficient to enhance mesoderm-specific expression of a reporter gene following injection into *Xenopus* embryos (101, 102). Together, these experiments suggest that *eFGF* is a bona fide Xbra target, and that one of its roles is to maintain *Xbra* expression.

The second approach has involved a subtractive hybridization screen for genes that are rapidly induced by a hormone-inducible version of Xbra called Xbra-GR (103). One of the genes identified in this screen was *Xwnt11*, which was originally described as a vegetally localized maternal RNA (22), The zygotic expression pattern of *Xwnt11*, however, proved to resemble that of *Xbra*, and Xbra function was required for *Xwnt11* expression during *Xenopus* development (29). Furthermore, although the *Xwnt11* promoter region has not yet been studied, induction of *Xwnt11* by Xbra-GR was found to occur in the absence of protein synthesis, suggesting that activation is indeed direct.

The function of *Xwnt11* during early *Xenopus* development was studied by use of a dominant-negative construct termed dn-Wnt11. Inhibition of Xwnt11 activity by this means proved to disrupt gastrulation movements (29), as was also observed following interference with Xbra function (94). Further experiments revealed that Xwnt11 does not regulate gastrulation through the canonical Wnt signalling pathway involving β-catenin (104), but may instead act in a pathway similar to that involved in establishing planar polarity in *Drosophila* (105).

A second gene isolated in the subtractive hybridization screen was initially named *BIG4* (Brachyury-induced gene 4), but screening of cDNA libraries for full-length versions of the gene resulted in the cloning of four related cDNAs named *Bix1–4* (Brachyury-induced homeobox-containing genes 1–4) (106). As their names imply, the *Bix* genes encode homeodomain-containing proteins whose expression is induced by *Xenopus Brachyury*. The most interesting feature of the *Bix* genes, however, is that, unlike *Xbra*, they are not only expressed in the mesoderm. Rather, they are also activated to high levels in the vegetal region of the embryo (106; see Fig. 2). It transpired that the *Bix* genes are also regulated by the vegetally localized T-box gene *VegT*, and that depletion of maternal VegT RNA prevents the expression of *Bix4* and perhaps the other *Bix* genes as well (102). The *Bix4* 5′ regulatory region contains T-box binding sites (106), two of which sites prove to be necessary and sufficient for the mesodermal and endodermal expression of reporter genes driven by the *Bix4* promoter in transgenic *Xenopus* embryos (102).

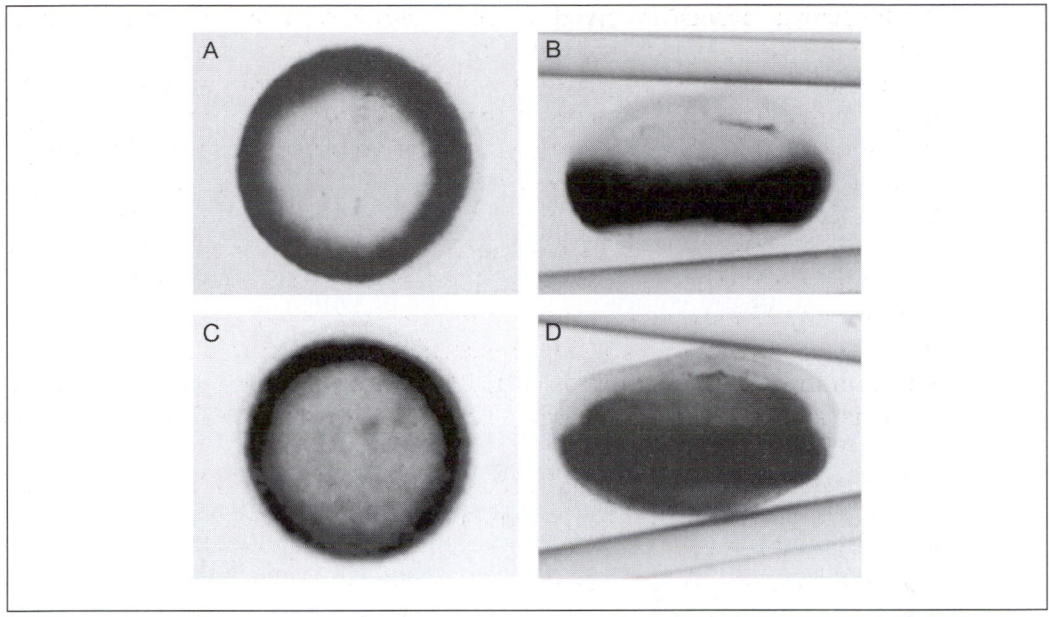

Fig. 2 (See also Plate 2) *Brachyury* and its targets. Comparison of expression patterns of *Xbra* (A–B) and *Bix1* (C–D) at early gastrula stage 10. *Xbra* is expressed in only the equatorial zone of the embryos, whereas *Bix1* is expressed in both the marginal zone and the vegetal hemisphere. (Reproduced from ref. 106.)

Misexpression of *Bix1* in *Xenopus* animal pole regions induces the formation of ventral mesoderm and endoderm, suggesting that the *Bix* family plays a role in the development of both these germ layers. Progress is now being made in elucidating the mechanism by which this occurs; Bix2, for example (also known as Milk (107)) mediates activin-induced transcription by interacting with Smad2, thereby recruiting active Smad2/Smad4 complexes to Bix2 binding sites (108). This provides an interesting example of a transcription factor (VegT) activating the expression of inducing factors (including Xnr1, -2, and -4 and derrière) as well as components of their signal transduction pathway (Bix2). This might have the effect of amplifying the level of signal in the vegetal hemisphere and marginal zone, and further amplification might be achieved through the activation of *Bix* family members by the activin family (106, see also 108).

5. Dorsalization: realizing the dorsoventral axis

As discussed in Section 3 above, the dorsal region of the *Xenopus* embryo is determined by the nuclear accumulation of β-catenin. How is this translated into differential gene expression in this region and thence the formation pattern along the entire dorsoventral axis? Several genes are known to be upregulated in the dorsal region of the embryo through the action of the β-catenin pathway. These include *goosecoid* (77), *siamois* (109), *twin* (110) and *xnr3* (111).

5.1 *Goosecoid, siamois,* and *twin*

The *goosecoid* gene was the first to be discovered whose expression is restricted to the dorsal mesendoderm, or organizer, of the *Xenopus* embryo (77). Expression of *goosecoid* is induced in animal pole tissue by activin and members of the nodal family. However, its maximal expression requires the combined action of Wnt-like and activin-like signals (112) which act, respectively, through a proximal element (PE) and a distal element (DE) in the *goosecoid* promoter (113). The requirement for two signalling systems, one of which (activin-like) has maximal activity in the vegetal hemisphere of the embryo and the other of which (β-catenin) is activated in the dorsal region, may explain why *goosecoid* is activated at highest levels in the dorsovegetal quadrant of the embryo (112).

Although *goosecoid* expression is upregulated by the Wnt pathway, there is no evidence that this induction is direct, and indeed there is no consensus LEF1/TCF3-binding site in the proximal element of the *goosecoid* promoter (113). Rather, it is more likely that the enhancement of *goosecoid* expression is indirect, and mediated through the activation of two related genes called *siamois* and *twin* (109, 110), which are homeobox-containing genes expressed at highest levels in the dorsovegetal region of the early gastrula. *Siamois* and *twin* were both cloned through their abilities to induce secondary axes in *Xenopus* embryos. Neither gene can be induced by mesoderm-inducing factors alone, but they are induced by Wnt signalling, in which case members of the activin family are able to cooperate with the β-catenin pathway to increase levels of *siamois* and *twin* expression (110, 112, 114). As with *goosecoid*, the fact that the maximal expression of *twin* and *siamois* requires both activin-like and β-catenin signalling may explain why the genes are activated at the highest levels in the dorsovegetal quadrant of the embryo (112). There is direct evidence that *twin* is regulated directly by β-catenin signalling; LEF1 protein can interact with consensus LEF1/TCF3 binding sites in the *twin* 5′ regulatory region, and these sites are required for Wnt-mediated induction of a *twin* reporter gene (110).

Evidence that *siamois* and *twin* mediate the enhancement of *goosecoid* expression caused by the Wnt pathway comes from the observation that both genes can activate the expression of *goosecoid* (110, 112) and that *twin* functions by interacting with the proximal element (PE) of the *goosecoid* promoter (110).

What are the functions of *siamois* and *twin*? As stated above, the two genes were cloned through their abilities to induce secondary axes in the *Xenopus* embryo, and the results of experiments interfering with their function in the embryo are consistent with the idea that their normal roles are to determine dorsal tissues. Thus, over-expression of a construct in which the engrailed repressor domain is fused to full-length *siamois* (115) or to the *siamois* homeodomain (116), blocks the formation of dorsal structures and inhibits the expression of dorsal-specific genes such as *goosecoid, noggin, chordin, follistatin, Xlim-1,* and *cerberus* (115, 116).

These results suggest that *siamois* and *twin*, whose expression patterns are established by the combined actions of the activin-like and β-catenin pathways, play key roles in the establishment of dorsal structures in the *Xenopus* embryo. But this cannot be the whole story; the dorsal region of the embryo plays an essential role in patterning the entire dorsoventral axis of the embryo, as was first demonstrated by Spemann and Mangold (117) in their discovery of the 'organizer' (see Chapter 2). These classic experiments showed that transplantation of the region surrounding the dorsal blastopore lip from one embryo to the ventral region of another caused the formation of a secondary axis, with the notochord deriving from the graft but the additional somites deriving from host tissue (see also ref. 118). This suggests that the dorsal region of the embryo produces signals which dorsalize adjacent lateral and ventral cells, and that these signals cannot include siamois and twin because they are transcription factors. Rather, the signals must include additional genes activated by the β-catenin pathway, or they might be downstream of *siamois* and *twin*.

5.2 *Xnr3*

One signalling molecule that is induced by the β-catenin pathway is Xnr3, which is expressed in the superficial layer of Spemann's organizer (111). The *Xnr3* promoter contains a LEF1 binding site, as would be expected for a direct target of β-catenin signalling (119), but this element is not sufficient for the activation of *Xnr3* in response to Wnt signalling. Rather, a second element is also required, which lacks a LEF1 site but does contain the homeodomain recognition sequence 5'-ATTA. It is possible that homeodomain-containing proteins interact with the β-catenin pathway to activate *Xnr3* in the superficial layer of the organizer.

Xnr3 differs from the other *Xenopus* members of the nodal-related family in that it lacks direct mesoderm-inducing activity (111). Nevertheless, it is likely to be involved in patterning the dorsoventral axis of the embryo and also in inducing neural structures. For example, *Xwnt11*, which is coexpressed with *Xnr3* in the organizer epithelium, cooperates with *Xnr3* to dorsalize ventral mesoderm (120). And more recently it has been shown that *Xnr3* can induce isolated animal pole regions to form neural tissue, without the concomitant formation of mesodermal cell types. The two defining activities of the organizer—dorsalization and neural induction (118)—can both, therefore, be mediated by Xnr3.

5.3 Inhibitors of BMP and Wnt signalling

But Xnr3 is not the only organizer-derived molecule that patterns the newly formed mesoderm. The *Xenopus* early gastrula proves to be awash with molecules with powerful ventralizing activity, such as Xwnt-8 (121) and BMP4 (122, 123), and the effect of many dorsalizing molecules expressed in the organizer is to counteract these factors. Prominent among these inhibitors are noggin, chordin, follistatin, cerberus, and Frzb.

The first such inhibitor to be discovered was *noggin*, which was isolated in an expression screen for factors capable of dorsalizing the early embryo (124). *Noggin* is expressed, as might be expected, in the organizer of the *Xenopus* early gastrula, and like *Xnr3* it can mediate both functions of the organizer; treatment of ventral marginal-zone tissue results in dorsalization, marked by the formation of muscle (125), and treatment of prospective ectodermal tissue results in neural differentiation (126).

If searches were ever made for a cell-surface noggin receptor, they would have been in vain, for it turned out that noggin exerts its effects by binding to, and thereby inhibiting the function of, members of the BMP family. Thus, noggin binds strongly to BMP4 (with a K_d of 20 pmol/l) and BMP2 and somewhat less strongly to BMP7 (127). Inhibition of the function of BMP family members thus results in the formation of dorsal tissues.

Might Xnr3 also function by inhibiting the effects of members of the BMP family? At first sight this seems unlikely, because one would imagine that Xnr3 signals, like other members of the TGF-β family, through a dimeric receptor resembling that of activin. It would be unlikely, therefore, to block BMP function by the same mechanism that is used by noggin, chordin, and follistatin. However, Xnr3 can inhibit mesoderm induction by BMP4 and, like the bona fide BMP inhibitors, Xnr3 induces only anterior neural tissue (128). It is possible that Xnr3 inhibits BMP function by binding, but not activating, the BMP receptor. This question requires further investigation.

Chordin was isolated in the course of a screen searching for molecules expressed in the organizer (129). Like *noggin*, *chordin* can function as a neural inducer (130). And like noggin it binds to members of the BMP family, albeit with lower affinity, thereby inhibiting their function (131). The amino acid sequence of chordin reveals similarities with the product of the *Drosophila* gene *short gastrulation* (sog), and there are remarkable parallels in their biological functions. Thus, sog, like chordin, can dorsalize *Xenopus* embryos; and chordin, like sog, can promote ventral development in *Drosophila* (132). The apparently opposite effects of chordin and sog in *Xenopus* and *Drosophila* embryos are consistent with the idea first put forward by Geoffroy St-Hilaire (133)—that arthropods and vertebrates share a common body plan, but that vertebrates turned upside-down during evolution so that ventral structures became dorsal and dorsal structures ventral. The results reflect the degree to which the molecular basis of dorsoventral patterning has been conserved through evolution (134, 135).

If noggin and chordin genuinely mediate the action of Spemann's organizer, they would be expected to be able to exert their inhibitory effects several cell diameters away from their sites of synthesis. Experiments in which the two molecules are misexpressed together with a cell-lineage marker confirm that this is the case; moreover they suggest that a gradient of noggin or chordin might set up a reverse gradient of BMP activity and thereby establish a gradient of positional information which might pattern the dorsoventral axis (136, 137).

Follistatin is also expressed in the organizer of the *Xenopus* embryo and, like noggin and chordin, is able to elicit both dorsalization and neural induction (138). Follistatin has long been known to bind to, and inhibit the action of, activin (139), but only recently has it also been shown to inhibit the effects of members of the BMP family (140, 141). Interestingly, whereas chordin and noggin interfere with the ability of BMP4 to bind the BMP receptor (127, 131), this is not true for follistatin, suggesting that follistatin and chordin employ different mechanisms in inhibiting BMP function (141).

The genes *follistatin*, *chordin*, and *noggin* represent three examples expressed in the organizer region that inhibit the functions of BMP family members and thereby promote dorsalization and neural induction. Another gene expressed in the organizer is *cerberus* (142). Like the other genes, *cerberus* is able to induce ectoderm to form neural tissue (142), but unlike the others, it inhibits the formation of trunk mesoderm (142). It is also capable, unlike the others, of inducing the formation of a complete head (142). These observations suggest that Cerberus might inhibit the action of members of the BMP family, but they also indicate that the molecule has other functions. This conclusion was confirmed by De Robertis and his colleagues (143), who found that Cerberus antagonizes the activity not only of BMP4, but also of the functions of the *nodal* family and of Wnt signals. These activities map to different regions of the Cerberus molecule, and mark it as a highly unusual multifunctional antagonist. Indeed, it has proved possible to express just the C-terminal portion of Cerberus and thereby inhibit the function of just the *nodal*-related genes, thereby obtaining evidence that a gradient of these factors elicits mesoderm induction at blastula stages (70).

It is the ability of Cerberus to inhibit the function of Wnt signals, as well as of BMP family members, that underlies its ability to induce heads; simultaneous inhibition of BMP signalling by means of a truncated receptor, and of Wnt signalling, by means of a dominant-negative Xwnt-8 construct (144), also leads to the formation of additional heads (145). Repression of these two ventralizing signalling pathways therefore establishes the conditions required for head specification.

As we have seen, there are multiple inhibitors of the BMP pathway; is the same true of the Wnt pathway? The identification of the *Frizzled* genes, which encode Wnt receptors, led to the discovery of a secreted protein, Frzb (146, 147), that is related to Frizzled and binds to Xwnt-8, thereby inhibiting its biological function (reviewed in ref. 148). Frzb is expressed in the *Xenopus* organizer, and one can imagine the molecule being secreted and thereby establishing a gradient along the dorsoventral axis of the embryo. This, in turn, might create a reverse gradient of Xwnt-8 activity, in much the same way that chordin and noggin might establish a reverse gradient of BMP activity (136, 137). Another similarity with the BMP system, where noggin and chordin both act as inhibitors, is that Frzb is not the only Wnt inhibitor expressed in the organizer. Another is *Crescent* (149, 150), which belongs to the same family as Frzb but that does, however, have distinct effects following overexpression. Injection of *Crescent* mRNA causes the development of cyclopic embryos, a phenotype presaged by abnormalities in the morphogenetic movements of the anterior midline (149). In this case, it is possible that *Crescent* inhibits the function of *Xwnt11*, which is known to play a role in gastrulation and morphogenesis (28, 29).

6. Conclusions

In this chapter we have reviewed the mechanisms by which the mesoderm of the *Xenopus* embryo is patterned. Inevitably, the story will be more complicated than we recount here, and obviously there are many things we do not understand. The three issues that we should like to address are these. First, do gradients of inducing factors really exist, and if so, how are they converted into differential patterns of gene expression? Second, once the embryo has established different regional patterns of gene expression, how do cells undergo the morphogenetic movements of gastrulation to bring themselves into the right places? And finally, we still do not have a direct link between the patterns of expression of regulatory genes such as *Xbra*, *siamois*, or *goosecoid* and the onset of terminal differentiation, with the expression of genes such as cardiac actin (151–153).

But there is reason for optimism. Although most attempts to deduce gene function in *Xenopus* involve overexpression experiments, and when dominant-negative or antisense approaches are used they can sometimes be open to alternative interpretations, work in zebrafish is confirming the main conclusions that have been drawn. For example, the roles of members of the *nodal* family in mesoderm formation have been confirmed by the discovery that *squint* and *cyclops* encode members of the *nodal* family, and that embryos lacking their gene products fail to form endoderm and most mesoderm (154–156). A similar phenotype is seen in embryos lacking *one-eyed pinhead*, a gene that functions as an essential cell-surface cofactor for nodal signalling (157, 158). In the dorsoventral axis of the embryo, the importance of inhibition of BMP signalling for dorsal development is confirmed by the ventralized phenotype of embryos mutant for *dino*, which encodes chordin (159), and the dorsalized phenotype of embryos mutant for *swirl*, which encodes BMP2 (160, 161), *snailhouse*, which encodes BMP7 (162, 163), and *somitabun*, which encodes Smad5 (164), a member of the Smad family that transduces BMP signalling. And finally, it has recently been found that the zebrafish gene *Silberblick* encodes Wnt11 (28); and, as would be predicted from work in *Xenopus* (29), *Silberblick* mutant embryos are defective in gastrulation movements (28).

Together, these observations bode well for a productive partnership between *Xenopus* and zebrafish in furthering our understanding of the molecular basis of pattern formation in the early embryo.

Acknowledgements

We thank Derek Stemple and Huw Williams for their helpful comments on the manuscript. Work from our laboratory was supported by the Medical Research Council.

References

1. Kroll, K. L. and Amaya, E. (1996) Transgenic Xenopus embryos from sperm nuclear transplantations reveal FGF signaling requirements during gastrulation. *Development*, **122**, 3173.
2. Amaya, E. and Kroll, K. L. (1999) A method for generating transgenic frog embryos. *Methods Mol. Biol.*, **97**, 393.

3. Amaya, E., Musci, T. J., and Kirschner, M. W. (1991) Expression of a dominant negative mutant of the FGF receptor disrupts mesoderm formation in Xenopus embryos. *Cell*, **66**, 257.

4. Conlon, F. L., Sedgwick, S. G., Weston, K. M., and Smith, J. C. (1996) Inhibition of Xbra transcription activation causes defects in mesodermal patterning and reveals auto-regulation of Xbra in dorsal mesoderm. *Development*, **122**, 2427.

5. Heasman, J., Holwill, S., and Wylie, C. C. (1991) Fertilization of cultured Xenopus oocytes and use in studies of maternally inherited molecules. *Methods Cell Biol.*, **36**, 213.

6. Zhang, J., Houston, D. W., King, M. L., Payne, C., Wylie, C., and Heasman, J. (1998) The role of maternal VegT in establishing the primary germ layers in Xenopus embryos. *Cell*, **94**, 515.

7. Steinbeisser, H., Fainsod, A., Niehrs, C., Sasai, Y., and De Robertis, E. M. (1995) The role of gsc and BMP-4 in dorsal–ventral patterning of the marginal zone in Xenopus: a loss-of-function study using antisense RNA. *EMBO J.*, **14**, 5230.

8. Latinkic, B. V. and Smith, J. C. (1999) Goosecoid and Mix.1 repress *Brachyury* expression and are required for head formation in Xenopus. *Development*, **126**, 1769.

9. Glinka, A., Wu, W., Delius, H., Monaghan, A. P., Blumenstock, C., and Niehrs, C. (1998) Dickkopf-1 is a member of a new family of secreted proteins and functions in head induction. *Nature*, **391**, 357.

10. Steinbach, O. C., Wolffe, A. P., and Rupp, R. A. (1997) Somatic linker histones cause loss of mesodermal competence in Xenopus. *Nature*, **389**, 395.

11. Dyson, S. and Gurdon, J. B. (1997) Activin signalling has a necessary function in *Xenopus* early development. *Curr. Biol.*, **7**, 81.

12. Heasman, J., Kofron, M., and Wylie, C. C. (2000) β-Catenin signalling activity dissected in the early *Xenopus* embryo: a novel antisense approach. *Dev. Biol.*, **222**, 124.

13. Danilchik, M. V. and Gerhart, J. C. (1987) Differentiation of the animal–vegetal axis in Xenopus laevis oocytes. I. Polarized intracellular translocation of platelets establishes the yolk gradient. *Dev. Biol.*, **122**, 101.

14. Rebagliati, M. R., Weeks, D. L., Harvey, R. P., and Melton, D. A. (1985) Identification and cloning of localized maternal RNAs from Xenopus eggs. *Cell*, **42**, 769.

15. Weeks, D. L. and Melton, D. A. (1987) A maternal mRNA localized to the vegetal hemisphere in Xenopus eggs codes for a growth factor related to TGF-beta. *Cell*, **51**, 861.

16. Yisraeli, J. K. and Melton, D. A. (1988) The material mRNA Vg1 is correctly localized following injection into Xenopus oocytes. *Nature*, **336**, 592.

17. Mowry, K. L. and Melton, D. A. (1992) Vegetal messenger RNA localization directed by a 340-nt RNA sequence element in Xenopus oocytes. *Science*, **255**, 991.

18. Gautreau, D., Cote, C. A., and Mowry, K. L. (1997) Two copies of a subelement from the Vg1 RNA localization sequence are sufficient to direct vegetal localization in *Xenopus* oocytes. *Development*, **124**, 5013.

19. Deshler, J. O., Highett, M. I., and Schnapp, B. J. (1997) Localization of Xenopus Vg1 mRNA by Vera protein and the endoplasmic reticulum. *Science*, **276**, 1128. [See comments.]

20. Havin, L., Git, A., Elisha, Z., Oberman, F., Yaniv, K., Schwartz, S. P., Standart, N., and Yisraeli, J. K. (1998) RNA-binding protein conserved in both microtubule-and microfilament-based RNA localization. *Genes Dev.*, **12**, 1593.

21. Yisraeli, J. K., Sokol, S., and Melton, D. A. (1990) A two-step model for the localization of maternal mRNA in Xenopus oocytes: involvement of microtubules and microfilaments in the translocation and anchoring of Vg1 mRNA. *Development*, **108**, 289.

22. Ku, M. and Melton, D. A. (1993) *Xwnt-11*, a maternally expressed Xenopus *wnt* gene. *Development*, **119**, 1161.
23. Wessely, O. and De Robertis, E. M. (2000) The Xenopus homologue of Bicaudal-C is a localized maternal mRNA that can induce endoderm formation. *Development*, **127**, 2053.
24. Zhang, J. and King, M. L. (1996) Xenopus VegT RNA is localized to the vegetal cortex during oogenesis and encodes a novel T-box transcription factor involved in mesodermal patterning. *Development*, **122**, 4119.
25. Stennard, F., Carnac, G., and Gurdon, J. B. (1996) The *Xenopus* T-box gene, *Antipodean*, encodes a vegetally localised maternal mRNA and can trigger mesoderm formation. *Development*, **122**, 4179.
26. Lustig, K. D., Kroll, K. L., Sun, E. E., and Kirschner, M. W. (1996) Expression cloning of a Xenopus T-related gene (Xombi) involved in mesodermal patterning and blastopore lip formation. *Development*, **122**, 4001.
27. Horb, M. E. and Thomsen, G. H. (1997) A vegetally localized T-box transcription factor in Xenopus eggs specifies mesoderm and endoderm and is essential for embryonic mesoderm formation. *Development*, **124**, 1689.
28. Heisenberg, C. P., Tada, M., Rauch, G. J., Saude, L., Concha, M. L., Geisler, R., Stemple, D. L., Smith, J. C., and Wilson, S. W. (2000) Silberblick/Wnt11 mediates convergent extension movements during zebrafish gastrulation. *Nature*, **405**, 76.
29. Tada, M. and Smith, J. C. (2000) *Xwnt11* is a target of *Xenopus* Brachyury: regulation of gastrulation movements via Dishevelled, but not through the canonical Wnt pathway. *Development*, **127**, 2227.
30. Schroeder, K. E., Condic, M. L., Eisenberg, L. M., and Yost, H. J. (1999) Spatially regulated translation in embryos: asymmetric expression of maternal Wnt-11 along the dorsal–ventral axis in *Xenopus*. *Dev. Biol.*, **214**, 288.
31. Weeks, D. L. and Melton, D. A. (1987) A maternal mRNA localized to the animal pole of Xenopus eggs encodes a subunit of mitochondrial ATPase. *Proc. Natl Acad. Sci. USA*, **84**, 2798.
32. Devic, E., Paquereau, L., Rizzoti, K., Monier, A., Knibiehler, B., and Audigier, Y. (1996) The mRNA encoding a beta subunit of heterotrimeric GTP-binding proteins is localized to the animal pole of Xenopus laevis oocyte and embryos. *Mech. Dev.*, **59**, 141.
33. De, J., Lai, W. S., Thorn, J., Goldsworthy, S. M., Liu, X., Blackwell, T. K., and Blackshear, P. J. (1999) Identification of four CCCH zinc finger proteins in Xenopus, including a novel vertebrate protein with four zinc fingers and severely restricted expression. *Gene*, **288**, 133.
34. Harland, R. and Gerhart, J. (1997) Formation and function of Spemann's organizer. *Annu. Rev. Cell Dev. Biol.*, **13**, 611.
35. Elinson, R. P. and Rowning, B. (1988) A transient array of parallel microtubules in frog eggs. Potential tracks for a cytoplasmic rotation that specifies the dorso-ventral axis. *Dev. Biol.*, **128**, 185.
36. Rowning, B. A., Wells, J., Wu, M., Gerhart, J. C., Moon, R. T., and Larabell, C. A. (1997) Microtubule-mediated transport of organelles and localization of beta-catenin to the future dorsal side of *Xenopus* eggs. *Proc. Natl Acad. Sci. USA*, **94**, 1224.
37. Houliston, E. and Elinson, R. P. (1991) Patterns of microtubule polymerization relating to cortical rotation in Xenopus laevis eggs. *Development*, **112**, 107.
38. Gerhart, J., Danilchik, M., Doniach, T., Roberts, S., Rowning, B., and Stewart, R. (1989) Cortical rotation of the Xenopus egg: consequences for the anteroposterior pattern of embryonic dorsal development. *Development*, **107**(Suppl.), 37.

39. McMahon, A. P. and Moon, R. T. (1989) Ectopic expression of the proto-oncogene int-1 in Xenopus embryos leads to duplication of the embryonic axis. *Cell*, **58**, 1075.

40. Smith, W. C. and Harland, R. M. (1991) Injected Xwnt-8 RNA acts early in Xenopus embryos to promote formation of a vegetal dorsalizing center. *Cell*, **67**, 753.

41. Guger, K. A. and Gumbiner, B. M. (1995) β-Catenin has Wnt-like activity and mimics the Nieuwkoop signaling center in *Xenopus* dorsal-ventral patterning. *Dev. Biol.*, **172**, 115.

42. He, X., Saint-Jeannet, J.-P., Woodgett, J. R., Varmus, H. E., and Dawid, I. B. (1995) Glycogen synthase kinase-3 and dorsoventral patterning in *Xenopus* embryos. *Nature*, **374**, 617.

43. Pierce, S. B. and Kimelman, D. (1995) Regulation of Spemann organizer formation by the intracellular kinase Xgsk-3. *Development*, **121**, 755.

44. Dominguez, I., Itoh, K., and Sokol, S. Y. (1995) Role of glycogen synthase kinase 3 beta as a negative regulator of dorsoventral axis formation in Xenopus embryos. *Proc. Natl Acad. Sci. USA*, **92**, 8498.

45. Heasman, J., Crawford, A., Goldstone, K., Garner-Hamrick, P., Gumbiner, B., McCrea, P., Kintner, C., Noro, C. Y., and Wylie, C. (1994) Overexpression of cadherins and under-expression of beta-catenin inhibit dorsal mesoderm induction in early Xenopus embryos. *Cell*, **79**, 791.

46. Larabell, C. A., Torres, M., Rowning, B. A., Yost, C., Miller, J. R., Wu, M., Kimelman, D., and Moon, R. T. (1997) Establishment of the dorso-ventral axis in *Xenopus* embryos is presaged by early asymmetries in β-catenin that are modulated by the Wnt signaling pathway. *J. Cell Biol.*, **136**, 1123.

47. Dominguez, I. and Green, J. B. A. (2000) Dorsal downregulation of GSK3b by a non-Wnt-like mechanism is an early molecular consequence of cortical rotation in early *Xenopus* embryos. *Development*, **127**, 861.

48. Kuhl, M., Sheldahl, L. C., Malbon, C. C., and Moon, R. T. (2000) Ca(2+)/Calmodulin-dependent protein kinase II is stimulated by wnt and frizzled homologs and promotes ventral cell fates in Xenopus. *J. Biol. Chem.*, **275**, 12701.

49. Sokol, S. Y. (1996) Analysis of Dishevelled signalling pathways during Xenopus development. *Curr. Biol.*, **6**, 1456.

50. Yost, C., Farr, G. H. 3rd, Pierce, S. B., Ferkey, D. M., Chen, M. M., and Kimelman, D. (1998) GBP, an inhibitor of GSK-3, is implicated in Xenopus development and oncogenesis. *Cell*, **93**, 1031.

51. Miller, J. R., Rowning, B. A., Larabell, C. A., Yang-Snyder, J. A., Bates, R. L., and Moon, R. T. (1999) Establishment of the dorsal-ventral axis in *Xenopus* embryos coincides with the dorsal enrichment of dishevelled that is dependent on cortical rotation. *J. Cell Biol.*, **146**, 427.

52. Newport, J. and Kirschner, M. (1982) A major developmental transition in early Xenopus embryo: I characterization and timing of cellular changes at the midblastula stage. *Cell*, **30**, 675.

53. Dale, L. and Slack, J. M. W. (1987) Fate map for the 32-cell stage of *Xenopus laevis*. *Development*, **99**, 527.

54. Moody, S. A. (1987) Fates of the blastomeres of the 32 cell Xenopus embryo. *Dev. Biol.*, **122**, 300.

55. Nieuwkoop, P. D. (1969) The formation of mesoderm in Urodelean amphibians. I. Induction by the endoderm. *Wilhelm Roux's Arch. EntwMech. Org.*, **162**, 341.

56. Sudarwati, S. and Nieuwkoop, P. D. (1971) Mesoderm formation in the Anuran Xenopus laevis (Daudin). *Wilhelm Roux's Arch. EntwMech. Org.*, **166**, 189.

57. Dale, L., Smith, J. C., and Slack, J. M. W. (1985) Mesoderm induction in *Xenopus laevis*: a quantitative study using a cell lineage label and tissue-specific antibodies. *J. Embryol. Exp. Morphol.*, **89**, 289.

58. Wylie, C., Kofron, M., Payne, C., Anderson, R., Hosobuchi, M., Joseph, E., and Heasman, J. (1996) Maternal beta-catenin establishes a 'dorsal signal' in early Xenopus embryos. *Development*, **122**, 2987.

59. Yasuo, H. and Lemaire, P. (1999) A two-step model for the fate determination of presumptive endodermal blastomeres in Xenopus embryos. *Curr. Biol.*, **9**, 869.

60. Jones, E. A. and Woodland, H. R. (1987) The development of animal caps cells in *Xenopus*: a measure of the start of animal cap competence to form mesoderm. *Development*, **101**, 557.

61. Kimelman, D. and Griffin, K. J. (1998) Mesoderm induction: a postmodern view. *Cell*, **94**, 419. [Comment]

62. Thomsen, G. H. and Melton, D. A. (1993) Processed Vg1 protein is an axial mesoderm inducer in Xenopus. *Cell*, **74**, 433.

63. Weeks, D. L. and Melton, D. A. (1987) A maternal mRNA localized to the vegetal hemisphere in Xenopus eggs codes for a growth factor related to TGF-beta. *Cell*, **51**, 861.

64. Fukui, A., Nakamura, T., Uchiyama, H., Sugino, K., Sugino, H., and Asashima, M. (1994) Identification of activins A, AB and B and follistatin proteins in *Xenopus* embryos. *Dev. Biol.*, **163**, 279.

65. Smith, J. C., Price, B. M. J., Van Nimmen, K., and Huylebroeck, D. (1990) Identification of a potent Xenopus mesoderm-inducing factor as a homologue of activin A. *Nature*, **345**, 732.

66. Jones, C. M., Kuehn, M. R., Hogan, B. L. M., Smith, J. C., and Wright, C. V. E. (1995) Nodal-related signals induce axial mesoderm and dorsalize mesoderm during gastrulation. *Development*, **121**, 3651.

67. Joseph, E. M. and Melton, D. A. (1997) Xnr4: a Xenopus nodal-related gene expressed in the Spemann organizer. *Dev. Biol.*, **184**, 367.

68. Sun, B. I., Bush, S. M., Collins-Racie, L. A., LaVallie, E. R., DiBlasio-Smith, E. A., Wolfman, N. M., McCoy, J. M., and Sive, H. L. (1999) derrière: a TGF-beta family member required for posterior development in Xenopus. *Development*, **126**, 1467.

69. Osada, S. I. and Wright, C. V. (1999) *Xenopus* nodal-related signaling is essential for mesendodermal patterning during early embryogenesis. *Development*, **126**, 3229.

70. Agius, E., Oelgeschlager, M., Wessely, O., Kemp, C., and De Robertis, E. M. (2000) Endodermal Nodal-related signals and mesoderm induction in Xenopus. *Development*, **127**, 1173.

71. Kofron, M., Demel, T., Xanthos, J., Lohr, J., Sun, B., Sive, H., Osada, S., Wright, C., Wylie, C., and Heasman, J. (1999) Mesoderm induction in Xenopus is a zygotic event regulated by maternal VegT via TGFbeta growth factors. *Development*, **126**, 5759.

72. Clements, D., Friday, R. V., and Woodland, H. R. (1999) Mode of action of VegT in mesoderm and endoderm formation. *Development*, **126**, 4903.

73. Hyde, C. E. and Old, R. W. (2000) Regulation of the early expression of the Xenopus nodal-related 1 gene, Xnr1. *Development*, **127**, 1221.

74. Smith, J. C., Price, B. M., Green, J. B. A., Weigel, D., and Herrmann, B. G. (1991) Expression of a Xenopus homolog of *Brachyury* (T) is an immediate-early response to mesoderm induction. *Cell*, **67**, 79.

75. Ruiz i Altaba, A. and Jessell, T. M. (1992) Pintallavis, a gene expressed in the organizer and midline cells of frog embryos: involvement in the development of the neural axis. *Development*, **116**, 81.

76. Kaufmann, E. and Knochel, W. (1996) Five years on the wings of fork head. *Mech. Dev.*, **57**, 3.

77. Cho, K. W. Y., Blumberg, B., Steinbeisser, H., and De Robertis, E. M. (1991) Molecular nature of Spemann's organizer: the role of the Xenopus homeobox gene *goosecoid*. *Cell*, **67**, 1111.

78. Hill, C. S. (1996) Signalling to the nucleus by members of the transforming growth factor-beta (TGF-beta) superfamily. *Cell Signal.*, **8**, 533.

79. Derynck, R. and Feng, X. H. (1997) TGF-beta receptor signaling. *Biochim. Biophys. Acta*, **1333**, F105.

80. Heldin, C. H., Miyazono, K., and ten Dijke, P. (1997) TGF-beta signalling from cell membrane to nucleus through SMAD proteins. *Nature*, **390**, 465.

81. Massagué, J. (1998) TGF-β signal transduction. *Annu. Rev. Biochem.*, **67**, 753.

82. Whitman, M. (1998) Smads and early developmental signaling by the TGFbeta superfamily. *Genes Dev.*, **12**, 2445.

83. Green, J. B. A., New, H. V., and Smith, J. C. (1992) Responses of embryonic Xenopus cells to activin and FGF are separated by multiple dose thresholds and correspond to distinct axes of the mesoderm. *Cell*, **71**, 731.

84. Gurdon, J. B., Harger, P., Mitchell, A., and Lemaire, P. (1994) Activin signalling and response to a morphogen gradient. *Nature*, **371**, 487.

85. Gurdon, J. B., Mitchell, A., and Mahony, D. (1995) Direct and continuous assessment by cells of their position in a morphogen gradient. *Nature*, **376**, 520.

86. Papin, C. and Smith, J. C. (2000) Gradual refinement of activin-induced thresholds requires protein synthesis. *Dev. Biol.*, **217**, 166.

87. Wolpert, L. (1969) Positional information and the spatial pattern of cellular differentiation. *J. Theor. Biol.*, **25**, 1.

88. McDowell, N., Zorn, A. M., Crease, D. J., and Gurdon, J. B. (1997) Activin has direct long-range signalling activity and can form a concentration gradient by diffusion. *Curr. Biol.*, **7**, 671.

89. Latinkic, B. V., Umbhauer, M., Neal, K. A., Lerchner, W., Smith, J. C., and Cunliffe, V. (1997) The *Xenopus Brachyury* promoter is activated by FGF and low concentrations of activin and suppressed by high concentrations of activin and by paired-type homeodomain proteins. *Genes Dev.*, **11**, 3265.

90. Kroll, K. L. and Amaya, E. (1996) Transgenic *Xenopus* embryos from sperm nuclear transplantations reveal FGF signalling requirements during gastrulation. *Development*, **122**, 3173.

91. Lerchner, W., Latinkic, B. V., Remacle, J. E., Huylebroeck, D., and Smith, J. C. (2000) Region-specific activation of the *Xenopus Brachyury* promoter involves active repression in ectoderm and endoderm: a study using transgenic frog embryos. *Development*, **127**, 2729.

92. Kispert, A. and Herrmann, B. G. (1993) The Brachyury gene encodes a novel DNA binding protein. *EMBO J.*, **12**, 3211.

93. Kispert, A., Koschorz, B., and Herrmann, B. G. (1995) The T protein encoded by *Brachyury* is a tissue-specific transcription factor. *EMBO J.*, **14**, 4763.

94. Conlon, F. L. and Smith, J. C. (1999) Interference with Brachyury function inhibits convergent extension, causes apoptosis, and reveals separate requirements in the FGF and activin signalling pathways. *Dev. Biol.*, **213**, 85.

95. Cunliffe, V. and Smith, J. C. (1992) Ectopic mesoderm formation in *Xenopus* embryos caused by widespread expression of a *Brachyury* homologue. *Nature*, **358**, 427.

96. Cunliffe, V. and Smith, J. C. (1994) Specification of mesodermal pattern in *Xenopus laevis* by interactions between Brachyury, noggin and Xwnt-8. *EMBO J.*, **13**, 349.

97. O'Reilly, M. A., Smith, J. C., and Cunliffe, V. (1995) Patterning of the mesoderm in Xenopus: dose-dependent and synergistic effects of Brachyury and Pintallavis. *Development*, **121**, 1351.

98. Isaacs, H. V., Pownall, M. E., and Slack, J. M. W. (1995) eFGF is expressed in the dorsal mid-line of *Xenopus laevis*. *Int. J. Dev. Biol.*, **39**, 575.

99. Isaacs, H. V., Pownall, M. E., and Slack, J. M. W. (1994) eFGF regulates *Xbra* expression during *Xenopus* gastrulation. *EMBO J.*, **13**, 4469.

100. Schulte-Merker, S. and Smith, J. C. (1995) Mesoderm formation in response to *Brachyury* requires FGF signalling. *Curr. Biol.*, **5**, 62.

101. Casey, E. S., O'Reilly, M. A., Conlon, F. L., and Smith, J. C. (1998) The T-box transcription factor Brachyury regulates expression of eFGF through binding to a non-palindromic response element. *Development*, **125**, 3887.

102. Casey, E. S., Tada, M., Fairclough, L., Wylie, C. C., Heasman, J., and Smith, J. C. (1999) *Bix4* is activated directly by VegT and mediates endoderm formation in *Xenopus* development. *Development*, **126**, 4193.

103. Tada, M., O'Reilly, M.-A. J., and Smith, J. C. (1997) Analysis of competence and of *Brachyury* autoinduction by use of hormone-inducible Xbra. *Development*, **124**, 2225.

104. Cadigan, K. M. and Nusse, R. (1997) Wnt signaling: a common theme in animal development. *Genes Dev.*, **11**, 3286.

105. Adler, P. N. (1992) The genetic control of tissue polarity in Drosophila. *BioEssays*, **14**, 735.

106. Tada, M., Casey, E. S., Fairclough, L., and Smith, J. C. (1998) *Bix1*, a direct target of *Xenopus* T-box genes, causes formation of ventral mesoderm and endoderm. *Development*, **125**, 3997.

107. Ecochard, V., Cayrol, C., Rey, S., Foulquier, F., Caillol, D., Lemaire, P., and Duprat, A. M. (1998) A novel Xenopus mix-like gene milk involved in the control of the endomeso-dermal fates. *Development*, **125**, 2577.

108. Germain, S., Howell, M., Esslemont, G. M., and Hill, C. S. (2000) Homeodomain and winged-helix transcription factors recruit activated Smads to distinct promoter elements via a common Smad interaction motif. *Genes Dev.*, **14**, 435.

109. Lemaire, P., Garrett, N., and Gurdon, J. B. (1995) Expression cloning of Siamois, a Xenopus homeobox gene expressed in dorsal-vegetal cells of blastulae and able to induce a complete secondary axis. *Cell*, **81**, 85.

110. Laurent, M. N., Blitz, I. L., Hashimoto, C., Rothbacher, U., and Cho, K. W. (1997) The *Xenopus* homeobox gene twin mediates Wnt induction of goosecoid in establishment of Spemann's organizer. *Development*, **124**, 4905.

111. Smith, W. C., McKendry, R., Ribisi, S. Jr., and Harland, R. M. (1995) A nodal-related gene defines a physical and functional domain within the Spemann organizer. *Cell*, **82**, 37.

112. Crease, D. J., Dyson, S., and Gurdon, J. B. (1998) Cooperation between the activin and Wnt pathways in the spatial control of organizer gene expression. *Proc. Natl Acad. Sci. USA*, **95**, 4398.

113. Watabe, T., Kim, S., Candia, A., Rothbacher, U., Hashimoto, C., Inoue, K., and Cho, K. W. (1995) Molecular mechanisms of Spemann's organizer formation: conserved growth factor synergy between Xenopus and mouse. *Genes Dev.*, **9**, 3038.

114. Carnac, G., Kodjabachian, L., Gurdon, J. B., and Lemaire, P. (1996) The homeobox gene Siamois is a target of the Wnt dorsalisation pathway and triggers organiser activity in the absence of mesoderm. *Development*, **122**, 3055.

115. Kessler, D. S. (1997) Siamois is required for formation of Spemann's organizer. *Proc. Natl Acad. Sci. USA*, **94**, 13017.

116. Fan, M. J. and Sokol, S. Y. (1997) A role for Siamois in Spemann organizer formation. *Development*, **124**, 2581.

117. Spemann, H. and Mangold, H. (1924) Uber inducktion von embryonenanlagen durch implantation artfremder organisatoren. *Wilhelm Roux's Arch. EntwMech. Org.*, **100**, 599.

118. Smith, J. C. and Slack, J. M. W. (1983) Dorsalization and neural induction: properties of the organizer in *Xenopus laevis. J. Embryol. Exp. Morph.*, **78**, 299.

119. McKendry, R., Hsu, S. C., Harland, R. M., and Grosschedl, R. (1997) LEF-1/TCF proteins mediate wnt-inducible transcription from the Xenopus nodal-related 3 promoter. *Dev. Biol.*, **192**, 420.

120. Glinka, A., Delius, H., Blumenstock, C., and Niehrs, C. (1996) Combinatorial signalling by Xwnt-11 and Xnr3 in the organizer epithelium. *Mech. Dev.*, **60**, 221.

121. Christian, J. L., McMahon, J. A., McMahon, A. P., and Moon, R. T. (1991) *Xwnt-8*, a *Xenopus* Wnt-1/int-1-related gene responsive to mesoderm-inducing factors, may play a role in ventral mesodermal patterning during embryogenesis. *Development*, **111**, 1045.

122. Dale, L., Howes, G., Price, B. M., and Smith, J. C. (1992) Bone morphogenetic protein 4: a ventralizing factor in early Xenopus development. *Development*, **115**, 573.

123. Jones, C. M., Lyons, K. M., Lapan, P. M., Wright, C. V. E., and Hogan, B. L. M. (1992) DVR-4 (Bone morphogenetic protein-4) as a posterior-ventralizing factor in *Xenopus* mesoderm induction. *Development*, **115**, 639.

124. Smith, W. C. and Harland, R. M. (1992) Expression cloning of noggin, a new dorsalizing factor localized to the Spemann organizer in Xenopus embryos. *Cell*, **70**, 829.

125. Smith, W. C., Knecht, A. K., Wu, M., and Harland, R. M. (1993) Secreted noggin protein mimics the Spemann organizer in dorsalizing *Xenopus* mesoderm. *Nature*, **361**, 547.

126. Knecht, A. K., Good, P. J., Dawid, I. B., and Harland, R. M. (1995) Dorsal-ventral patterning and differentiation of noggin-induced neural tissue in the absence of mesoderm. *Development*, **121**, 1927.

127. Zimmerman, L. B., De Jesús Escobar, J. M., and Harland, R. M. (1996) The Spemann organizer signal noggin binds and inactivates bone morphogenetic protein 4. *Cell*, **86**, 599.

128. Hansen, C. S., Marion, C. D., Steele, K., George, S., and Smith, W. C. (1997) Direct neural induction and selective inhibition of mesoderm and epidermis inducers by Xnr3. *Development*, **124**, 483.

129. Sasai, Y., Lu, B., Steinbeisser, H., Geissert, D., Gont, L. K., and De Robertis, E. M. (1994) Xenopus chordin: a novel dorsalizing factor activated by organizer-specific homeobox genes. *Cell*, **79**, 779.

130. Sasai, Y., Lu, B., Steinbeisser, H., and De Robertis, E. M. (1995) Regulation of neural induction by the Chd and Bmp-4 antagonistic patterning signals in Xenopus. *Nature*, **376**, 333.

131. Piccolo, S., Sasai, Y., Lu, B., and De Robertis, E. M. (1996) Dorsoventral patterning in Xenopus: inhibition of ventral signals by direct binding of chordin to BMP-4. *Cell*, **86**, 589.

132. Holley, S. A., Jackson, P. D., Sasai, Y., Lu, B., De Robertis, E. M., Hoffmann, F. M., and Ferguson, E. L. (1995) A conserved system for dorsal-ventral patterning in insects and vertebrates involving sog and chordin. *Nature*, **376**, 249.

133. Geoffroy St-Hilaire, E. (1822) Considérations générales sur la vertèbre. *Mém. Mus. Hist. Nat.*, **9**, 89.

134. Jones, C. M. and Smith, J. C. (1995) Revolving vertebrates. *Curr. Biol.*, **5**, 574.
135. De Robertis, E. M. and Sasai, Y. (1996) A common plan for dorsoventral patterning in Bilateria. *Nature*, **380**, 37.
136. Dosch, R., Gawantka, V., Delius, H., Blumenstock, C., and Niehrs, C. (1997) Bmp-4 acts as a morphogen in dorsoventral mesoderm patterning in *Xenopus*. *Development*, **124**, 2325.
137. Jones, C. M. and Smith, J. C. (1998) Establishment of a BMP-4 morphogen gradient by long-range inhibition. *Dev. Biol.* **194**, 12.
138. Hemmati-Brivanlou, A., Kelly, O. G., and Melton, D. A. (1994) Follistatin, an antagonist of activin, is expressed in the Spemann organizer and displays direct neuralizing activity. *Cell*, **77**, 283.
139. Nakamura, T., Takio, K., Eto, Y., Shibai, H., Titani, K., and Sugino, H. (1990) Activin-binding protein from rat ovary is follistatin. *Acta Endocrinol. Copenh.*, **122**, 96.
140. Yamashita, H., ten Dijke, P., Huylebroeck, D., Sampath, T. K., Andries, M., Smith, J. C., Heldin, C.-H., and Miyazono, K. (1995) Osteogenic protein-1 binds to activin type II receptor and induces certain activin-like effects. *J. Cell Biol.*, **130**, 217.
141. Iemura, S.-I., Yamamoto, T. S., Takagi, C., Uchiyama, H., Natsumi, T., Shimisaki, S., Sugino, H., and Ueno, N. (1998) Direct binding of follistatin to a complex of bone-morphogenetic protein and its receptor inhibits ventral and epidermal cell fates in early *Xenopus* embryo. *Proc. Natl Acad. Sci. USA*, **95**, 9337.
142. Bouwmeester, T., Kim, S., Sasai, Y., Lu, B., and De Robertis, E. M. (1996) Cerberus is a head-inducing secreted factor expressed in the anterior endoderm of Spemann's organizer. *Nature*, **382**, 595.
143. Piccolo, S., Agius, E., Leyns, L., Bhattacharyya, S., Grunz, H., Bouwmeester, T., and De Robertis, E. M. (1999) The head inducer Cerberus is a multifunctional antagonist of Nodal, BMP and Wnt signals. *Nature*, **397**, 707.
144. Hoppler, S., Brown, J. D., and Moon, R. T. (1996) Expression of a dominant-negative Wnt blocks induction of MyoD in *Xenopus* embryos. *Genes Dev.*, **10**, 2805.
145. Glinka, A., Wu, W., Onichtchouk, D., Blumenstock, C., and Niehrs, C. (1997) Head induction by simultaneous repression of Bmp and Wnt signalling in *Xenopus*. *Nature*, **389**, 517.
146. Leyns, L., Bouwmeester, T., Kim, S. H., Piccolo, S., and De Robertis, E. M. (1997) Frzb-1 is a secreted antagonist of Wnt signaling expressed in the Spemann organizer. *Cell*, **88**, 747.
147. Wang, S., Krinks, M., Lin, K., Luyten, F. P., and Moos, M. Jr. (1997) Frzb, a secreted protein expressed in the Spemann organizer, binds and inhibits Wnt-8. *Cell*, **88**, 757.
148. Zorn, A. M. (1997) Cell–cell signalling: frog frizbees. *Curr. Biol.*, **7**, R501.
149. Pera, E. M. and De Robertis, E. M. (2000) A direct screen for secreted proteins in *Xenopus* embryos identifies distinct activities for the wnt antagonists crescent and frzb-1. *Mech. Dev.*, **96**, 183.
150. Shibata, M., Ono, H., Hikasa, H., Shinga, J., and Taira, M. (2000) Xenopus crescent encoding a frizzled-like domain is expressed in the Spemann organizer and pronephros. *Mech. Dev.*, **96**, 243.
151. Mohun, T. J., Brennan, S., Dathan, N., Fairman, S., and Gurdon, J. B. (1984) Cell type-specific activation of actin genes in the early amphibian embryo. *Nature*, **311**, 716.
152. Mohun, T. J., Garrett, N., and Gurdon, J. B. (1986) Upstream sequences required for tissue-specific activation of the cardiac actin gene in *Xenopus laevis* embryos. *EMBO J.*, **5**, 3185.

153. Mohun, T. J., Taylor, M. V., Garrett, N., and Gurdon, J. B. (1989) The CArG promoter sequence is necessary for muscle-specific transcription of the cardiac actin gene in Xenopus embryos. *EMBO J.*, **8**, 1153.

154. Feldman, B., Gates, M. A., Egan, E. S., Dougan, S. T., Rennebeck, G., Sirotkin, H. I., Schier, A. F., and Talbot, W. S. (1998) Zebrafish organizer development and germ-layer formation require nodal-related signals. *Nature*, **395**, 181.

155. Feldman, B., Dougan, S. T., Schier, A. F., and Talbot, W. S. (2000) Nodal-related signals establish mesendodermal fate and trunk neural identity in zebrafish. *Curr. Biol.*, **10**, 531.

156. Schier, A. F. and Shen, M. M. (2000) Nodal signalling in vertebrate development. *Nature*, **403**, 385.

157. Schier, A. F., Neuhauss, S. C., Helde, K. A., Talbot, W. S., and Driever, W. (1997) The one-eyed pinhead gene functions in mesoderm and endoderm formation in zebrafish and interacts with no tail. *Development*, **124**, 327.

158. Gritsman, K., Zhang, J., Cheng, S., Heckscher, E., Talbot, W. S., and Schier, A. F. (1999) The EGF-CFC protein one-eyed pinhead is essential for nodal signaling. *Cell*, **97**, 121.

159. Schulte-Merker, S., Lee, K. J., McMahon, A. P., and Hammerschmidt, M. (1997) The zebrafish organizer requires chordino. *Nature*, **387**, 862.

160. Kishimoto, Y., Lee, K. H., Zon, L., Hammerschmidt, M., and Schulte-Merker, S. (1997) The molecular nature of zebrafish swirl: BMP2 function is essential during early dorsoventral patterning. *Development*, **124**, 4457.

161. Nguyen, V. H., Schmid, B., Trout, J., Connors, S. A., Ekker, M., and Mullins, M. C. (1998) Ventral and lateral regions of the zebrafish gastrula, including the neural crest progenitors, are established by a bmp2b/swirl pathway of genes. *Dev. Biol.*, **199**, 93.

162. Dick, A., Hild, M., Bauer, H., Imai, Y., Maifeld, H., Schier, A. F., Talbot, W. S., Bouwmeester, T., and Hammerschmidt, M. (2000) Essential role of Bmp7 (snailhouse) and its prodomain in dorsoventral patterning of the zebrafish embryo. *Development*, **127**, 343.

163. Schmid, B., Furthauer, M., Connors, S. A., Trout, J., Thisse, B., Thisse, C., and Mullins, M. C. (2000) Equivalent genetic roles for bmp7/snailhouse and bmp2b/swirl in dorsoventral pattern formation. *Development*, **127**, 957.

164. Hild, M., Dick, A., Rauch, G. J., Meier, A., Bouwmeester, T., Haffter, P., and Hammerschmidt, M. (1999) The smad5 mutation somitabun blocks Bmp2b signaling during early dorsoventral patterning of the zebrafish embryo. *Development*, **126**, 2149.

4 | Somite and axial development in vertebrates

CLEMENTINE HOFMANN

1. Introduction

The development of the axial skeleton starts with somitogenesis. This is an example of a segmentation process in vertebrates. An initially unsegmented body progressively becomes segmented, leading to the formation of the vertebral column, the most obviously segmented structure in vertebrates. In the mesoderm of all vertebrate embryos, the underlying fundamental patterning process generates periodic blocks of cells, the somites, which form on either side of the neural tube. Somites have been studied in embryos from many vertebrate classes using a variety of experimental methods. However, the majority of studies have been performed in avian and amphibian species, because their embryos are more amenable for experimental manipulations and microsurgery than mammalian embryos. For quite some time, the analysis of somitogenesis and axis formation in mammalian embryos was not the focus of attention, although mutations in mice affecting somites and vertebrae have long been described (reviewed in ref. 1). In the past few years, the zebrafish has emerged as an additional model system for the analysis of somite and axis formation, since the transparency of the embryos allows resulting morphological defects in mutants to be observed and analysed *in vivo*. The establishment of *in vitro* culture systems for postimplantation mammalian embryos (2, 3) has also enabled the study of somitogenesis in mammalian embryos. In addition, targeted mutagenesis in the mouse by embryonic stem-cell technology has led to a rapid further understanding of the underlying molecular processes.

In this review, the principles and mechanisms that appear to be common to all vertebrates will be emphasized, but with the focus on those species where most insights have been gained, i.e. particularly the chick and the mouse. The morphological basis of skeleton formation will be described, as will the genetic regulation underlying these processes and malformations of the skeleton, due to mutations. Understanding how somitogenesis is regulated at the genetic level is a fundamental question in vertebrate embryonic development.

2. Morphogenesis: development of the skeleton

2.1 Establishing the primary body axis: from primitive streak to somites

2.1.1 Formation of the paraxial mesoderm

The development of the vertebrate axis is closely linked to the formation of the primary body axis during gastrulation. In higher vertebrates, the paraxial mesoderm (also called segmental plates in chick embryos), where the somites arise, is laid down during the process of gastrulation. With the establishment of the primitive streak, the primary body axis, i.e. the future anteroposterior axis of the embryo, is fixed. Between about day 6.5 and 9.5 of mouse development or between 6 and 18 h of incubation of a chick embryo, epithelial cells from the primitive ectoderm (or the epiblast in chick embryos) delaminate and migrate through the primitive streak to generate mesoderm (reviewed in ref. 4; 5, 6) (see also Chapter 2). Thereafter, new mesoderm is generated from the tail bud until axis elongation ceases at about day 12.5 of mouse development (see ref. 7 for a review) and about stage 22 of chick development (8). Cells ingressing at the anterior end of the streak (dorsal marginal zone (7)), become paraxial mesoderm.

Immediately after gastrulation, two essential structures for the development of the axial skeleton can be identified: the notochord and the paraxial mesoderm. The paraxial mesoderm, which gives rise to the somites, appears on each side of the midline neural epithelium (i.e. neural folds) as the primitive streak regresses along the anteroposterior axis of the embryo (see Fig. 1). Progenitor cells for the paraxial mesoderm are continuously found in the regressing primitive streak, and then in the tail bud (9–11). The notochord arises concomitantly with the regression of the primitive streak and Hensen's node and is the most crucial structure for midline axis development in early embryogenesis. It not only promotes the anteroposterior axis formation, but also acts as an organizer and establishes the dorsoventral identity of axial structures. In vertebrates, the notochord is essential for dorsoventral patterning of both the neural tube and the paraxial mesoderm (12–15). In particular, the immediate proximity of a notochord induces the floor plate and ventralizes adjacent neural tissue (16). The notochord also plays a key role in dorsoventral patterning of the somite, in particular of the sclerotome (15, 17, 18), from which the cartilage, bone, and connective tissue of the axial skeleton arise, and which will be discussed below.

2.1.2 Segmentation of paraxial mesoderm into somites

The axial skeleton derives entirely from the paraxial mesoderm (19 and references therein, 20). The paraxial mesoderm is connected laterally to the intermediate mesoderm, and this itself is connected to the lateral mesoderm, consisting of somatopleura and splanchnopleura layers (Fig. 1). In the mouse and in the chick embryo, somites form from the anterior end of each segmental plate by budding off at the anterior end of the unsegmented paraxial mesoderm, also called presomitic

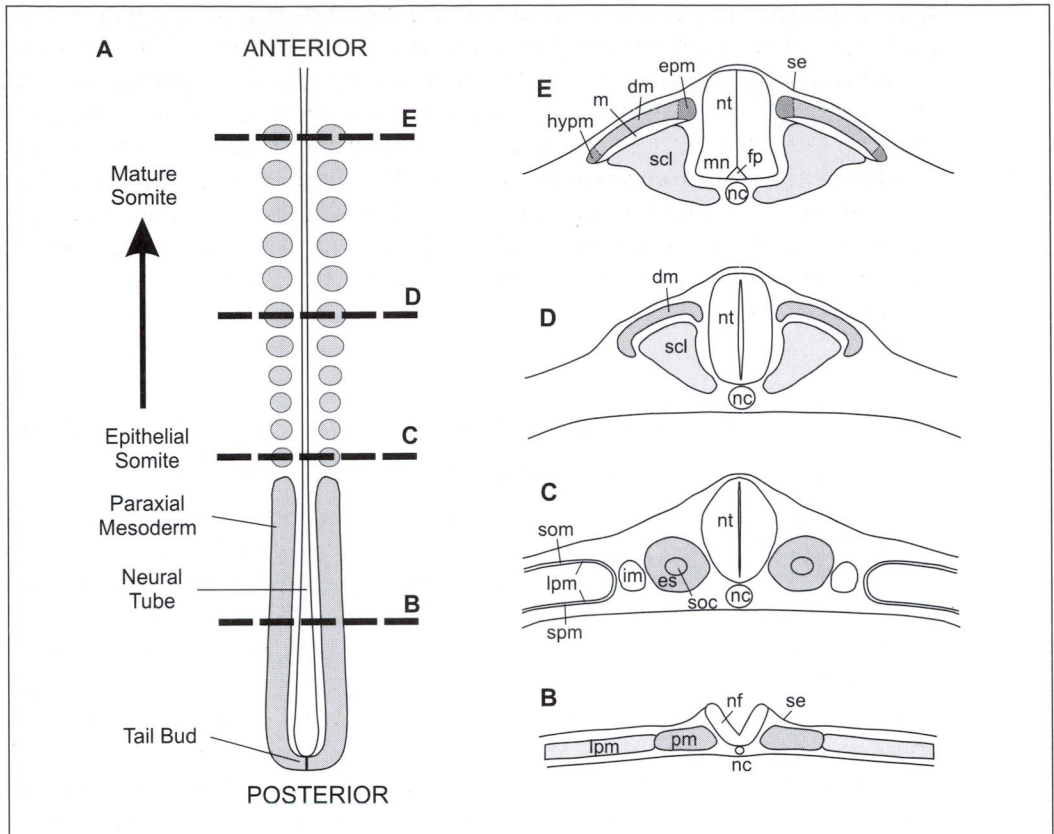

Fig. 1 Schematic drawing of a horizontal section illustrating differentiation of the paraxial mesoderm (A) and the relative maturity of somites at each level (in transverse sections, B–E). The lines in A indicate the levels of transverse sections schematically shown in B–E. As formation and segmentation of paraxial mesoderm into somites and their subsequent differentiation proceeds from anterior to posterior, there is an anterior to posterior gradient of maturity of developing somites within the embryo. While somites at an anterior position of the embryo have already started differentiation, the most posterior, nascent somites are still epithelial. Even further posteriorly, the paraxial mesoderm is still unsegmented, and at the very posterior end of the embryo, in the tail bud, gastrulation is still ongoing. (B) Transverse section through the unsegmented paraxial mesoderm. Paraxial mesoderm (pm) rods flank the neural folds (nf) and notochord (nc) and are laterally connected to lateral plate mesoderm (lpm). Dorsally, paraxial mesoderm is covered by surface ectoderm (se). (C) Transverse section through a level where epithelial somites (es) have formed. At this stage, somites are balls of epithelial cells with a central cavity, the somitocoele (soc), which is filled with mesenchymal cells. Epithelial somites flank the neural tube (nt) and connect laterally to intermediate mesoderm (im) which is connected to the splanchnopleuric (spm) and somatopleuric (som) layers of the lateral plate mesoderm (lpm). (D) Transverse section through somites which have just started differentiation. At their ventral side, somites undergo an epithelial–mesenchymal transition leading to the formation of the sclerotome (scl), while the dorsal part of somites retains the epithelial organization and forms dermomyotome (dm). (E) Transverse section showing a later stage of somite differentiation. Medial dermomyotome cells have migrated underneath the dermomyotome and form a myotome (m). This medial part of the dermomyotome gives rise to future epaxial muscle (epm). The very lateral region of the dermomyotome will give rise to hypaxial muscle (hypm). Sclerotome cells (scl) are further migrating towards the notochord; within the neural tube, a floor plate (fp) and motor neurons (mn) have been induced by the notochord.

mesoderm (in the chick embryo at an approximate rate of one pair every 100 min). During segmentation, the length of each segmental plate remains quite constant; cells are added both by mitosis within the plate and by recruitment of cells at the posterior end (21). Segmentation of the paraxial mesoderm seems to be independent of the notochord, since mouse embryos lacking a node and notochord form somites (22, 23).

Each somite is first constructed as a ball of radially arranged epithelial cells which line a small central lumen, the somitocoele (see Fig. 1). The lumen contains a cluster of mesenchymal cells; the outer side of the somite is surrounded by a basement membrane, and the somitic cells are polarized, their apical sides facing the somitocoele (24).

2.2 Constructing the vertebral column: differentiation of somites into dermomyotome and sclerotome

Several hours after their formation from the anterior end of the paraxial mesoderm, somites start to differentiate, first into sclerotome and dermomyotome, and subsequently the dermomyotome is patterned along the mediolateral axis. The cells in the ventromedial part of the somite lose their epithelial character and form, together with the luminal cells, the mesenchymal sclerotome that will give rise to axial bones and cartilage (20). The sclerotome cells migrate towards the notochord, surround it, and these cells, together with the notochord, give rise to the vertebral bodies and contribute to the intervertebral discs (20, 25), i.e. form the vertebral column. Sclerotome cells that migrate dorsally around the neural tube form the pedicles and laminae of the neural arches, which surround and protect the neural tube.

The dorsolateral cells of the somite initially retain their epithelial arrangement, providing the dermomyotome (see Fig. 1). This later subdivides into medial, central, and lateral compartments. The cells from the medial part, which are adjacent to the neural tube, give rise to the back muscles, which are referred to as epaxial muscles, whereas hypaxial muscles of the ventral body wall and the skeletal muscles of the limbs develop from the lateral part of the dermomyotome. The central, dorsal portion of the dermomyotome will form the dermis of the trunk (26, 27; reviewed in ref. 19).

The fate of sclerotome cells also depends on their relative position within the somite (25): cells located ventromedially contribute to intervertebral disks and vertebral bodies. Lateral sclerotome cells form the ribs; cells from the posterior halves of the segments form neural arches, suggesting that the anterior and posterior somite halves are determined to give rise to distinct structures (28). In the thoracic region, posterior sclerotome cells and somitocoele cells migrate ventrolaterally and also contribute to the intervertebral disks and ribs (29, 30).

Although we have learned a lot about genes involved in the differentiation of the distinct somitic regions in recent years (see below), the mechanisms underlying the disintegration of the basement membrane and the transition of epithelial somitic cells into mesenchymal cells are not yet known.

2.3 Differentiation and maturation: from sclerotomes to vertebrae, from cartilage to bone

Pattern formation determines not only the body plan of the early skeleton but also the shape of each individual skeletal element (see ref. 31). In these patterning processes, cell interactions and molecules that mediate intercellular communication are critical, in addition to 'regulator genes' and signalling molecules (see below).

The sclerotome will give rise to the vertebrae and ribs. After the initial de-epithelialization of somite cells at the ventromedial side of the somite, the medially located sclerotome cells migrate towards and around the notochord. The metameric structure of these cells along the anteroposterior axis is lost and the perinotochordal tube forms, but the lateral sclerotomes retain their segmented organization. The cells of the perinotochordal tube are loosely arranged and of mesenchymal character with no apparent segmentation, where soon centres of condensations can be observed. These will develop into intervertebral discs, and the loosely arranged intermediate zones will form vertebral bodies. The lateral parts of each sclerotome remain segmented and differentiate into anterior and posterior parts (32). Cell density within the lateral sclerotome is higher in the posterior half, from which the pedicle, neural arch, and ribs arise, and lower in the anterior half. When motor axons and neural crest cells emerge from the developing neural tube, they exclusively traverse the anterior halves of the adjacent sclerotomes, which will form the connective tissue surrounding the dorsal root ganglia. The anteroposterior sclerotomal subdivision exists in all vertebrate classes (33); this subdivision of the sclerotome was first observed by Remak (34) who proposed the concept of 'resegmentation', suggesting that on each side of the embryo the anterior half of one sclerotome merges with the posterior half of the preceding rostral sclerotome to form one vertebra. Experimental results from chick–quail transplantation studies and from manipulations of somite halves provided support for this concept (29, 30, 35–37), but the molecular mechanisms underlying sclerotome cell migration and the basis of cell condensation are not understood.

However, the crucial role of the notochord for the specification of ventral sclerotome has been demonstrated. Extirpation, transplantation, and *in vitro* culture studies all pointed to the fact that the ventral spinal cord and notochord are responsible for the 'induction' of cartilage differentiation in somites (reviewed in ref. 38). In fact, notochord ablation and transplantation experiments in the chick and a molecular analysis of mouse notochord mutants have demonstrated that the notochord is essential for the specification of ventral sclerotome derivatives (15, 17, 18, 39). Notochord ablation leads to the loss of ventral sclerotome derivatives, whereas transplantation of an ectopic notochord next to the unsegmented paraxial mesoderm induces ectopic sclerotome formation and leads to inhibition of dermomyotome development in the vicinity (see below). The concept that a specific component produced by ventral spinal cord or notochord is necessary for sclerotome and cartilage induction has long been proposed (see ref. 38) and has finally led to candidate molecule(s) (SHH), see below.

After long-term incubation of an ectopic notochord next to unsegmented paraxial mesoderm, ectopic cartilage eventually arises. Thus, notochord tissue is also important for chondrogenesis in normal vertebral development and seems to be sufficient to induce chondrogenesis. During vertebral development, the notochord and the ventral portion of the spinal cord induce chondrogenic differentiation in sclerotome cells (40; reviewed in ref. 41). Furthermore, enzymatic removal or alterations of perinotochordal material reduces the inductive capacity of the notochord (for a review see ref. 42), suggesting that extracellular matrix molecules produced by the notochord and spinal cord cells and surrounding the somite are critical in somite chondrogenesis (42, 43).

The vertebrae of all vertebrates form by chondrification of the sclerotomal mesoderm, starting in the mesenchymal condensations in the centre of the vertebral bodies and from two positions within the pedicles of the neural arches (for a review see ref. 44, and references therein). With the chondrification of the vertebral bodies, the notochord is deformed such that its diameter decreases at the site of a forming vertebral body, whereas it enlarges at the site of a developing intervertebral disk (20). There, the notochord cells participate in the formation of the nucleus pulposus of the intervertebral disk, while the annulus fibrosus develops at the periphery of the disk. Further differentiation into cartilage and the local control of bone development is similar to mechanisms described for other chondrification sites in the body (see, for example, ref. 31).

3. Patterning the vertebral column: signalling molecules and molecular responses

In this section, the molecular basis of the different steps of vertebral column morphogenesis outlined in the previous section will be described.

3.1 Growth factors, signalling molecules, and transcription factors in paraxial mesoderm formation

The axial skeleton is entirely derived from mesoderm, and its formation is closely linked to the organization of the primary body axis. Thus, genes involved in the development and patterning of mesoderm during gastrulation also mostly affect the paraxial mesoderm. Deficiency in these genes, however, often leads to severe defects and/or early embryonic lethality. This makes it difficult to analyse the specific function of these genes in vertebral column development. However, during the past few years considerable progress in understanding the genetic control of paraxial mesoderm development has been made, due to various techniques such as defined knock-outs based on the Cre/loxP system in mice or overexpression methods.

A number of genes encoding growth factors, signalling molecules, and their receptors, play an important role in early patterning of the axis. These include genes of the Wnt family (e.g. *Wnt3*, *Wnt5*), of the fibroblast growth factor (FGF) family, and of the transforming growth factor (TGF)-β superfamily (e.g. the secreted factor *nodal*). Whereas TGF-β like signals are required in the initiation of gastrulation, peptide growth factors of the Wnt family and of the FGF family play essential roles in specifying paraxial mesoderm.

Several members of the TGF-β superfamily are implicated in the onset of gastrulation and mesoderm formation. Bone morphogenetic protein-4 (BMP4) is essential for lateral ventral mesoderm formation in mouse (45), and the type I bone morphogenetic protein receptor (Bmpr) is required for gastrulation in mouse embryogenesis (46). The growth factor nodal, another member of the TGF-β superfamily, was shown to be disrupted by the 413.d retroviral insertion (47). This insertion causes a loss of *nodal* mRNA expression (48), and 413.d mutant embryos are unable to form a distinct primitive streak, or dorsal mesoderm and notochord. In wild-type embryos, *nodal* is expressed at the onset of gastrulation in proximal posterior regions of the embryonic ectoderm, later during gastrulation at the periphery of the node. Together with the analysis of several markers, the studies of the mutant phenotype suggest that the primary role of *nodal* is not in mesoderm induction, but rather in induction and/or maintenance of the primitive streak (48).

Wnt- and FGF signals appear to play essential roles in the specification of paraxial mesoderm. Three members of the Wnt gene family, *Wnt3a*, *Wnt5a*, and *Wnt5b*, are expressed in the murine primitive streak (49); *Wnt5a* and *Wnt5b* are expressed in posterior regions of the streak starting at day 6.5, *Wnt3a* starts to be expressed later at a stage when the primitive streak is extended (day 7.5) and is found throughout the streak. All three genes are then expressed in the regressing streak and in the tail bud. Inactivation of *Wnt3a* results in truncation of dorsal somitic mesoderm development posterior to the forelimb buds, leading to a lack of caudal somites and tail bud formation (49). These results suggest a necessity for *Wnt3a* in the formation of embryonic mesoderm, specifying paraxial mesoderm and generation of the tail bud. *Wnt5a*, which is expressed in a gradient at the posterior end of the growing embryo including the tail bud, and later in the distal-most aspect of several structures that extend from the body axis, seems also to be involved in primitive streak mesoderm specification (50). Yet, its inactivation does not affect gross somite formation (50), but rather seems to regulate a proliferation pathway common to many structures whose development requires extension from the primary body axis, such as the limb, the developing face, ears, and genitals (50). For *Wnt5b* as well as for the other members of the *Wnt* gene family expressed later during somite formation, the function has not yet been determined by genetic means.

Several members of the FGF gene family (*Fgf3*, *Fgf4*, *Fgf5*, and *Fgf8*) and two of the FGF receptor genes (*Fgfr1* and *Fgfr2*) are expressed during gastrulation and paraxial mesoderm specification (see also ref. 7, and references therein). Recently, two more of the 18 members of the *Fgf* family (*Fgf17* and *Fgf18*) have been found to be expressed in paraxial mesoderm (51). Mutational analysis of most of these *Fgfs* and their receptors has been reported. Inactivation of *Fgf4* affects the proliferation of the inner cell mass leading to embryonic death shortly after implantation at day 5 of development (52). So far, the Cre/loxP system has been used to circumvent this early lethality for studying the limb phenotype at later stages of development (53, 54), but there is no analysis yet for putative defects in the axis. For *Fgf5* mutant mice, which show increased hair length, no mesoderm defects have so far been reported, although *Fgf5* is expressed during gastrulation (55). Loss of *Fgf3* function leads to malformation of the caudal vertebrae. Mutant mice have a shortened, curly tail. *Fgf3* is shown to be expressed early in the anterior region of the streak and in embryonic mesoderm in cells emerging from the streak. Expression in the embryonic mesoderm is confined to the presomitic mesoderm and is found in the tail bud at least until day 11.5. The severe disorganization in the caudal vertebrae and the tail of mutant embryos was traced back to a defective organization of mesodermal cells in the tail bud at day 11.5 (56). For *Fgf8*, an allelic series of mutations was generated, and analysis of embryos carrying different combinations of the resulting alleles revealed requirements for *Fgf8* gene function during gastrulation, as well as cardiac, cranio-facial, and brain development (57). Further studies identified *Fgf8* as a gene essential for gastrulation (58). As *Fgf8* mutants fail to express *Fgf4*, this study shows that *Fgf8* and/or *Fgf4*, which both are coexpressed in the primitive streak of the gastrulating embryo, are required for cell migration away from the primitive streak, otherwise no embryonic mesoderm- or endoderm-derived tissues develop (58).

Deficiency in the *Fgfr1* gene results in an expansion of axial mesoderm at the expense of paraxial mesoderm, leading to a lack of somites and embryonic death before day 9.5 (59, 60). To further study FGFR1 function and to circumvent the early lethality of mutant embryos, two groups generated chimeras by injecting *Fgfr1*-deficient embryonic stem (ES) cells into wild-type blastocysts (61) or by aggregating mutant ES cells with tetraploid embryos (62). Ciruna and co-workers (62) focused on FGFR1 function in patterning of the mesoderm at gastrulation, and found that chimeric embryos formed secondary neural tubes which might be caused by a deficiency in the ability of epiblast cells to traverse the primitive streak. The second study analysed the role of *Fgfr1* gene after gastrulation. Embryos with a high contribution of mutant ES cells die during gastrulation, but embryos with a low contribution complete gastrulation and display malformations of posterior embry-onic structures such as limb-bud malformation, partial duplication of the neural tube, tail distortion, and spina bifida in the posterior portion of the spinal cord (61).

A null mutation encompassing both isoforms of *Fgfr2* results in lethality at or shortly after implantation at embryonic day 4.5–5.5, suggesting that *Fgfr2* gene function is required for early postimplantation development and that it contributes to the outgrowth, differentiation, and maintenance of the inner cell mass (63). Embryos with a homozygous hypomorphic allele die around 10.5 days of gestation, because mutants fail to form a functional placenta (64). Cre-mediated excision to generate mice lacking only the isoform IIIb of FGFR2 (and retaining expression of the IIIc form) resulted in mice that are viable until birth, but with severe defects in organs such as the limbs, lungs, and anterior pituitary gland (65). Similar results were obtained in a study where chimeras were raised from homozygous *Fgfr2* mutant ES cells and wild-type tetraploid embryos, suggesting that during early development FGFR2 might interact with FGF4, an absence of either one causes early lethality, whereas later, interactions between FGF10 and FGFR2 may be required for limb and lung development (66, 67).

Among the transcription factors that control gastrulation and/or mesoderm formation, the *Brachyury* gene (or *T* gene, for Tail) (68), a member of the T-box family of transcription factors (see ref. 69 for a review), plays a very important role (see Chapter 3). The *T* gene is initially expressed throughout the primitive streak and subsequently in the notochord and tail bud. In homozygous *Brachyury* mutants, the formation of mesoderm is defective and the notochord and tail bud are absent (70). Misexpression of *Brachyury* (*Xbra*) in *Xenopus* in prospective ectodermal tissue is sufficient to induce ectopic mesoderm (71). Mice heterozygous for *T* show very short tails and skeletal abnormalities. Taken together, the data suggest a requirement of an increasing amount of *T* gene product for the development of the notochord and formation of mesoderm along the anteroposterior axis (72). In the fish embryo, *Brachyury* (*no tail*) regulates the formation of notochord and tail, while another T-box gene, *VegT* (*spadetail*), regulates the formation of the trunk. The mechanism of how the formation of different regions along the anteroposterior axis is genetically controlled is still unclear (see ref. 69, and references therein). In the mouse, the *Tbx6* gene, which is highly related to the frog *VegT* gene, is strongly expressed in the paraxial mesoderm (73). *Tbx6*-deficient mice show a dramatic phenotype in which, at trunk and tail levels, the two stripes of paraxial mesoderm normally flanking the neural tube are replaced by two ectopic neural tubes (74), suggesting that, in the absence of *Tbx6*, the tail bud cells which normally give rise to paraxial mesoderm are unable to do so and instead acquire a neural tube cell fate. Recent work has investigated the extent of the similarities between phenotypes in *Tbx6*-, *T* gene-, and *Wnt3a*-deficient mice. Mutations in any of these genes lead to a loss of trunk and tail mesoderm, and ectopic neural structures appear (see ref. 75, and references therein). Interestingly, analysis of the *T*-gene promoter identifies *T* (*Brachyury*) as a direct transcriptional target of the Wnt signalling pathway, probably regulated through Lef1/Tcf1 binding sites (75). Moreover, *Lef1*$^{-/-}$*Tcf1*$^{-/-}$ compound mutant mice show a similar phenotype to *Wnt3a* null mutants (76). Yamaguchi and colleagues (75) suggest that *Wnt3a* might specify the fate of pluripotent primitive-streak cells to give rise to paraxial mesoderm daughter cells expressing high levels of *T*, which eventually contribute to trunk somites. Considering that neural development is a default state (from *Xenopus* studies), *Wnt3a* is required for paraxial mesoderm development and its absence leads to ectopic neural development (75, and references therein).

3.2 Segmentation of paraxial mesoderm into somites

The subdivision of paraxial mesoderm into somites is the mesodermal segmentation event of vertebrate development and comprises three distinct processes: pre-patterning of the mesoderm; boundary formation; and patterning within the somite itself. The various genes involved in these processes (Fig. 2) can be subdivided into four groups.

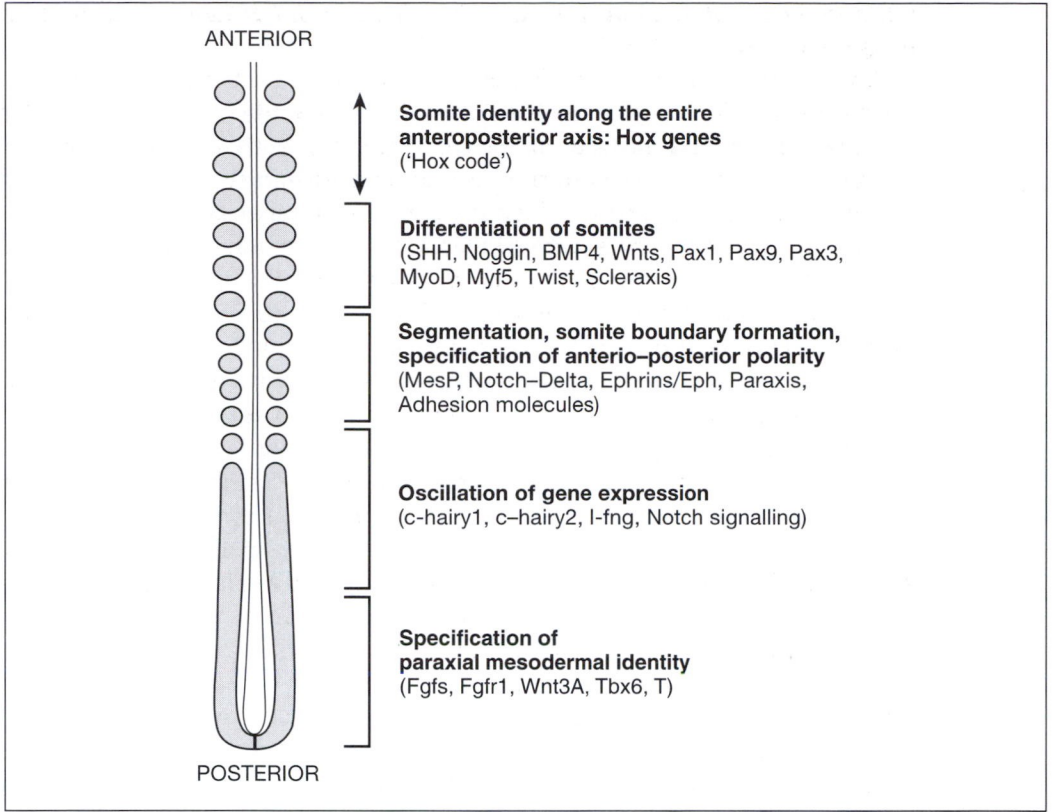

Fig. 2 Schematic representation summarizing patterning and differentiation events during somite and axial development. Beginning at the posterior end (bottom) which represents the 'youngest' developmental stage in the embryo, paraxial mesoderm is produced at the level of the primitive streak and then in the tail bud. Specification of paraxial mesoderm is controlled by T-box containing genes and by FGF and Wnt signalling. After the formation of paraxial mesoderm and organization as two stripes flanking the neural tube, the 'segmentation clock' drives oscillation of gene expression of several cycling genes and segmentation appears at the anterior end of paraxial mesoderm. Segmentation and somite boundary formation is then refined by genes of the Notch–Delta pathway, ephrins and paraxis. In addition, anterior and posterior polarity is specified by these genes and by genes of the Mesp family; adhesion molecules are essential for the epithelial and migratory properties of forming somites. Differentiation of somites along the dorsoventral axis is mainly controlled by members of the SHH, BMP, and Wnt families as well as by BMP-antagonists, e.g. Noggin. These signalling molecules regulate the expression of further differentiation markers in somitogenesis (e.g. Pax-genes, Myf-5, MyoD, see text). Along the anteroposterior axis, Hox genes specify somite identity.

Clock genes

Several indications have suggested for quite some time that the presomitic meso-derm is prepatterned into metameric subunits (see ref. 7). Recently, molecular evidence for segmental subdivision of the presomitic mesoderm in a defined temporal pattern has been obtained with the identification of *c-hairy1*, an avian homologue of the *Drosophila* segmentation gene *hairy* (77). *c-hairy1* is strongly expressed in presomitic mesoderm in cyclic waves with a periodicity that correlates precisely with the formation time of one somite (i.e. 90 min). To exclude that this expression wavefront simply reflects caudal–rostral movement of *c-hairy1*-expressing cells, labelling experiments with DiI (a lipophilic vital dye) were performed, which clearly indicated that the rostral progression of the *c-hairy1* wavefront occurs independently of cell movement and also independently of the presence of a caudal presomitic mesoderm (77). This autonomous expression of *c-hairy1* is consistent with clock models for somitogenesis, which state that prospective somites are determined almost concomitantly with paraxial mesoderm formation (reviewed in ref. 32). Thus, *c-hairy1* expression in the presomitic mesoderm might be driven by an underlying molecular clock linked to somitogenesis, called the 'segmentation clock' (77).

In zebrafish, the *her1* gene (78) was first reported to show a 'pair-rule' expression pattern (stripes in alternating somites), whereby the cells of the first band will give rise to the fifth somite, those of the second to the seventh somite, and so on, implicating that *her1* defines somite identity. However, *her1* is only distantly related to *c-hairy1* and is also very different to *Drosophila hairy* (77). Remarkably, recent detailed re-examination of the *her1* expression pattern also indicates an oscillating pattern (79), thus bringing into question the original description of its expression pattern.

A second avian hairy-related gene, *c-hairy2*, also cycles in the presomitic mesoderm and is closely related to the mammalian *HES1* gene, which is a downstream target of the Notch pathway (80) (see also Chapter 5). The *HES1* expression pattern in the mouse is closely related to *c-hairy1* in the chick (80), demonstrating oscillations also for the *HES1* gene in the mouse. *HES1*-deficient mice exhibit strong neurogenesis defects but do not show overt segmentation defects, and even maintain the dynamic expression of *lunatic fringe*, *l-fng*, another oscillating gene (see below). This indicates that the segmentation clock remains functional in the *HES1* mutant mouse. However, in the *Delta-like1* mutants (in which Notch signalling is thought to be altered), where *HES1* expression is absent, the segmentation clock is also affected and segmentation disrupted (see below), thus, in *Delta-like1* mutant mice the somitic defect is not due just to a lack of *HES1* expression.

Cell–cell signalling molecules

The Notch–Delta pathway is essential for cell–cell signalling, and is required for cell-fate determination in a large number of developmental processes in both *Drosophila* and *Caenorhabditis elegans* (81, and references therein). In vertebrates, a number of studies has shown that Notch–Delta intercellular signalling is involved in boundary formation and initial somitic segmentation (see also Chapter 5 for the role of Notch–Delta signalling in the regulation of neuron production). The analysis of mice, mutant for the Notch1 transmembrane receptor (82, 83), for its ligands Delta-like1 (Dll1) (84) and Delta-like3 (Dll3) (85), and for a transcription factor that mediates its cellular response (RBP-Jκ) (86), have indicated the importance of these proteins for refining the positioning of somitic boundaries as well as for defining the anterior and posterior identities of somites. In *Notch1* mutant embryos, somitogenesis is delayed and disorganized (82), and the mutant embryos die at day 11 of gestation. Across the midline of the mutants, asymmetries in segmentation are evident and undivided mesoderm is often present near the presomitic mesoderm on one side only of the embryonic axis (82). Normally, *Notch1* is expressed at high levels in a somite-sized domain at the anterior end of the presomitic mesoderm (in somite -I, after ref. 19), and is downregulated after the somite is formed (87). The specificity of Notch signalling might be derived from the various ligands that are expressed in a spatial- and temporal-specific manner (see below).

Delta-like1 (*Dll1*), one of the Notch receptor ligands, is expressed in the presomitic mesoderm and posterior halves of somites, and later in the myotomes (88). In *Dll1* null-mutant embryos, segmentation is perturbed, i.e. the segments are irregularly shaped and have no craniocaudal polarity, and somites show a lack of epithelialization (84). Caudal sclerotome halves do not condense; in addition, dorsal root ganglia and myotomes span somitic segment borders, suggesting that these borders are not maintained. The loss of segmental borders in the *Dll1*$^{-/-}$ embryos suggest that *Dll1* is necessary for compartmentalization of somites, and setting of anteroposterior identity is necessary for this, but sclerotome and myotome differentiation are independent of epithelialization and anteroposterior subdivision (84).

Mutation in the *Dll3* gene, another member of Notch ligands in vertebrates related to *Drosophila Delta*, is the molecular basis of the spontaneous mutant *pudgy* (*pu*) (85). A comparison of *Dll3* and *Dll1* expression demonstrates that these genes have distinct patterns of expression, but nevertheless together they may operate in many of the same processes, e.g. during somitogenesis *Dll3* and *Dll1* might coordinate the building of the intersomitic boundaries (89). *Dll3* transcripts are found in the primitive streak, in presumptive paraxial mesoderm, and *Dll3* is strongly expressed in the paraxial mesoderm and also persists in the tail bud. Expression of *Dll3* is observed in presomitic mesoderm and in nascent somites, in the most anterior region (89).

The *pudgy* mutants exhibit patterning defects at the earliest stages of somitogenesis, resulting in mice with severe vertebral and rib deformities (85). In *pu* embryos, the morphological borders between somites and the compartmental borders within somites are severely disrupted. Therefore, various genes expressed during somitogenesis show an abnormal pattern: *l-fng* is expressed diffusely, whereas in wild-type embryos it is expressed in defined bands in the presomitic mesoderm and forming somite. *Pax1*, normally expressed in every sclerotome compartment, shows no clear borders in *pu* embryos. In normal embryos, *Dll1* is expressed in the posterior compartment of each somite, *Dll3* in the anterior compartment. In *pu* mice, *Dll1* and *Dll3* show no such spatial segregation. Taken together, these abnormal expression patterns indicate that the critical early boundaries and compartment borders are not correctly formed in *pu* mice, which results in deformed vertebrae and fused and bifurcated ribs in adult animals, whereas the dermis and musculature reveal no evident histological defects (85).

RBP-Jκ, a transcription factor that is highly conserved in a number of species and whose *Drosophila* homologue is Suppressor of Hairless (*Su(H)*), has been reported to be a key downstream element in the Notch-receptor signalling pathway (86, and references therein). Mouse embryos deficient for *RBP-Jκ* revealed specific defects in neural and somitic development leading to early embryonic lethality before day 10.5 of gestation (86). Mutants show severe developmental delay compared to their littermates as early as day 8.5 of embryogenesis, and *RBP-J*κ deficiency results in defective somite formation, i.e. somites were poorly formed and irregularly arranged (86). Analysis of expression of *Mox1*, a homeobox-containing gene that is strongly expressed in the presomitic mesoderm and in the forming somites (90), but whose function remains to be elucidated, reveals that in *RBP-J*κ$^{-/-}$ embryos this somitic marker *Mox1* is still expressed. However, mutant embryos do not express myogenin, indicating a block of myogenesis in differentiating somites (86). Irregularly shaped somites form in *RBP-J*κ-deficient mice, suggesting that, similar to *Notch1*-deficient mice, the mutation results in uncoordinated somite formation (82, 86). (For a detailed comparison of mutants in the Notch–Delta pathway and the expression of relevant marker genes involved in somite boundary formation and anteroposterior somite polarity in these mutants see ref. 91.)

Presenilin-1 (PS1), a transmembrane protein that facilitates the transduction of the activated Notch signal might act upstream of the Notch–Delta pathway (92, 93). Inactivation of *PS1* leads to a markedly reduced expression of *Dll1* and *Notch1* in the presomitic mesoderm of early mutant embryos. *PS1*$^{-/-}$ embryos further exhibit abnormal patterning of the axial skeleton and spinal ganglia, a phenotype traced back to irregularly shaped somites along the entire length of the neural tube, defects in somite segmentation and differentiation. Moreover, somites are misaligned in mutants, and dorsal root ganglia are fused over multiple segments (93). These abnormal somite patterns are highly reminiscent of defects described in mice with inactivated *Notch1* or *Dll1* (see above), suggesting that *PS1* is essential for a functional *Notch1* or *Dll1* to establish or maintain somite borders and anteroposterior segment polarity (93; see also ref. 91 for a comparison of mutants in the Notch–Delta pathway).

The periodic expression pattern of *lunatic fringe* (*l-fng*), encoding a secreted protein and related to the *Drosophila fringe* gene, suggests a link between the autonomous oscillator that drives somite segmentation (the 'segmentation clock') and the Notch signalling pathway. In mouse and chick, *l-fng* is expressed in a dynamic pattern similar to that of *c-hairy1* and *c-hairy2*, thus, it seems to be regulated by the same segmentation clock (reviewed in ref. 94; see also ref. 80). In *Drosophila*, one component of the Notch signalling pathway is *fringe*, which is required for the activation of Notch during specification of the wing margin (95, 96). Also in vertebrates, *lunatic fringe* is required for boundary formation, as shown in mice homozygous for a targeted mutation of the gene (97, 98). The *lunatic fringe* mutants fail to form distinct boundaries between individual somites, show irregular somites in size and shape, and disruption of anteroposterior somite patterning. Moreover, for example, *Pax1* expression is still confined to the sclerotome as in wild-type embryos, but its expression is no longer segmented (97). It is striking that the oscillating genes identified so far are related to the Notch pathway. This leads to the possibility that one role of the segmentation clock might be to periodically activate Notch signalling to eventually generate the regular arrangement of somite boundaries (94).

Certain Eph family members (99) also show metameric expression in somitomeres and therefore may be candidate molecules for determining somite boundaries. Eph receptors represent the largest subfamily of receptor tyrosine kinases and fall into two subclasses, EphA and EphB (99, and references therein); the ligands of Eph receptors, known as 'ephrins' (Eph-receptor interacting proteins), also fall into two subclasses, A and B. Members of this signalling pathway are shown to have a role in intercellular signalling and patterning of the neural tube, but several Eph receptors and ephrins are also expressed in somites (100, and references therein). The mouse Eph-receptor gene *EphA4* is expressed at a low level in the presomitic mesoderm, and is strongly expressed in two discrete bands at the anterior half of newly forming somite -I and in the whole presumptive somite -II (101). Mice with a null mutation in the *EphA4* gene are viable and fertile, but have a gross motor dysfunction, which is evidenced by a loss of coordination of limb movement (102, 103). However, defects in somites and derivatives have not been reported.

Of the receptors examined in zebrafish, only *EphA4* expression is detected in the presomitic mesoderm and forming somites throughout somitogenesis; of the ligands, *ephrin-A-L1* and *ephrin-B2* are detected in somites as they are formed, and *ephrin-B2* transcripts are detected in three stripes in the paraxial mesoderm (100). Overexpression of dominant-negative forms of these genes by injecting mRNA into the early stages of zebrafish embryos results in the failure of somite segmentation from the presomitic mesoderm, leading to disrupted somite formation; whereas control injections with wild-type forms causes no disturbance of somitogenesis (100).

Transcription factors

In addition to the members of Notch–Delta and of the FGFR signalling pathways, the *Mesp* family of basic helix–loop–helix (bHLH) transcription factors have been implicated in establishing a segmental prepattern in the presomitic mesoderm. Murine *Mesp1* and *Mesp2* show a segmental expression pattern in the presomitic mesoderm (104, 105). *Mesp2*-deficient mice fail to initiate segmentation and have fused vertebral columns and dorsal root ganglia and impaired sclerotomal polarity (105). The defects can be rescued by *Mesp1* gene replacement in a dose-dependent manner (106). *Mesp1* is, in addition, expressed at the onset of gastrulation, and homozygous *Mesp1* null mice exhibit growth retardation after day 7.5 of development and die before 10.5 days post coitum (dpc) (106). The function of *Mesp1* during somitogenesis has not yet been clearly revealed because of early embryonic lethality and possible compensation by *Mesp2*.

Sawada *et al.* (79) identified two zebrafish *Mesp*-related genes: *mesp-a* and *mesp-b*. Their expression in the presomitic mesoderm is confined to the anterior halves of presumptive and forming somites. Ectopic expression of Mesp-b causes a loss of posterior identity within somites and leads to a segmentation defect, suggesting that Mesp-b is involved in specifying anterior identity within the presumptive somites (79).

Paraxis is also a bHLH transcription factor expressed in paraxial mesoderm and somites (107). Before the onset of somitogenesis, *paraxis* is weakly expressed in part of the primitive mesoderm, and later it is upregulated in the anterior part of paraxial mesoderm and is strongly expressed in a region equivalent to two somites in size (somites -I and -II) (107). Expression continues throughout the epithelial somite and, as somites mature, *paraxis* becomes restricted to the dermomyotome (see also ref. 108 for a review). *Paraxis* null-mutant mice are unable to form epithelia resulting in the failure of epithelialization of somites (109), and somite formation is thus disrupted. Nevertheless, in the absence of normal somites, the axial skeleton and skeletal muscle form but are abnormally patterned. Unexpectedly, the formation of epithelial somites was neither required for segmentation, nor for the establishment of somitic cell lineages (109; see also ref. 110). Furthermore, it is not possible to separate, on the one hand, the effect of the mutations of genes involved in epithelialization and segmentation and, on the other hand, specification of the anteroposterior polarity of the somite. In many of these mutants of genes expressed at the segmentation stage of somites, epithelialization is disturbed and, as a result, segmentation occurs later or segments are disrupted.

A very recent report shows that *paraxis*-deficient mice have defects in the axial skeleton and in peripheral nerves that are consistent with a failure in anteroposterior patterning of somites, but expression levels of *Mesp2* and of genes in the Notch pathway or downstream of it (such as *Dll1*, *l-fng*, *EphA4*) are not altered in the presomitic mesoderm of paraxis mutants (111). This indicates that paraxis is not required for Notch signalling in early somite development. However, expression of genes that are normally restricted to the posterior half of somites is present in a diffuse pattern, suggesting a loss of polarity. Thus, *paraxis* might have a role in maintaining somite polarity independently of Notch signalling (111).

Cell adhesion and extracellular matrix molecules

During formation of a somite, cells have to undergo profound changes in their cell adhesion and migratory properties as they switch between a migratory mesenchymal state and a polarized epithelial state, and this process has to be repeated (see above; see also ref. 7 and references therein). Therefore, molecules that mediate cell–cell interaction must be involved, as well as components that specifically establish contacts between cell and extracellular matrix plus matrix molecules themselves.

Cell adhesion   The expression of genes encoding both the calcium-dependent adhesion molecules (*N*-cadherin and cadherin 11) and the neural cell-adhesion molecule (NCAM), which is calcium-independent, is upregulated prior to and during somite formation. While the expression of *N-cadherin* is subsequently downregulated in the ventromedial portion of the somite (the presumptive sclerotome, which first starts to de-epithelialize), *NCAM* expression continues throughout the somite (112). *Cadherin 11* expression is complementary to *N-cadherin*; it is downregulated in the dermomyotome but maintained in the cranial and caudal edges of the sclerotomes (113). In mice with a targeted disruption of the *NCAM* gene, no defects in somite formation have been detected (114). However, a targeted mutation generating a soluble NCAM molecule causes dominant embryonic lethality and leads to a reduced number of irregular and small somites (115). Loss-of-function mutation in *N-cadherin* results in embryonic lethality around day 10 of gestation, and mutant embryos display a severe cell-adhesion defect in the heart (116). The somites of the mutants are small, irregularly shaped, and less cohesive than those of their wild-type littermates, and the epithelial organization of the somites is partially disrupted. The more severely affected embryos show undulated neural tubes and malformed somites (116). In *cadherin 11*-deficient mice only the brain phenotype and behavioural responses have been analysed; a putative somite phenotype has still to be assessed (117).

Cell-matrix interactions Cell-matrix contacts are mediated by adhesion receptors that make transmembrane connections, linking extracellular matrix and adjacent cells to the intracellular cytoskeleton. Laminin and fibronectin (among other extra-cellular matrix (ECM) components) are known to function in cell-substratum adhesion and cell migration and are present in the ECM of the paraxial mesoderm (7, and references therein; 118, and references therein). Laminin is expressed in basement membranes during early murine and avian embryogenesis (119, 120). Fibronectin is also found in early embryos and in somitogenesis (121), and has been implicated to play a role in the aggregation of paraxial mesoderm cells that precedes somite formation (122). Mouse embryos deficient for the fibronectin gene initiate gastrulation normally but have a shortened anteroposterior axis, show mesoderm defects during gastrulation, and fail to form notochord and somites (123). All characterized laminin variants are heterotrimeric molecules formed by the covalent bonding of one polypeptide from the α-, β-, and γ-laminin subunit families. Thus, many variant laminin trimers may potentially form depending on differential subunit gene expression (124). Null mutants have been generated for some of the various laminin genes, showing, for example, that the absence of laminin $\beta 2$ leads to the disruption of neuromuscular junction development and of kidney function (125). Mutation of the $\alpha 2$-subunit can result in an autosomal form of muscular dystrophy (126).

Laminin $\gamma 1$ is one of the earliest expressed laminin subunits, and is, together with the $\alpha 1$- and $\beta 1$-subunits, expressed in the preimplantation embryo (127). A targeted mutation of laminin $\gamma 1$ would alter the formation of all known basement membrane laminin isoforms (128). Homozygous mutant embryos do not survive beyond day 5.5 post coitum. The embryos lack basement membranes and, although the blastocysts expand, primitive endoderm cells remain in the inner cell mass. Thus, these results show that the laminin $\gamma 1$ subunit is necessary for laminin assembly and that laminin is essential for basement membrane organization (128).

Similar results are obtained upon inactivation of the receptor subunit $\beta 1$ integrin, the subunit which is essential for many of the various α-subunits to form an intact heterodimeric receptor for ECM molecules (see ref. 118 for a review). Homozygous $\beta 1$ integrin null embryos develop normally to the blastocyst stage, implant, and invade the uterine basement membrane, but die shortly thereafter (129, 130). The sets of receptors α3β1, α6β1, and α7β1 are all laminin receptors, and the three knock-out lines show different phenotypes (see ref. 131, and references therein). Moreover, several integrin heterodimers can serve as fibronectin receptors (e.g. α5β1), but none of the integrin null phenotypes is as severe as the fibronectin null phenotype, suggesting the possibilities of overlapping functions and/or compensations (131).

Inactivation of the ECM molecules and receptors relevant for somitogenesis show a very early, rather global, mesodermal or endodermal defect, and their specific role in somitogenesis has yet to be elucidated.

3.3 Signalling molecules in somite patterning and differentiation

3.3.1 Dorsoventral patterning

Shortly after their formation, somites of vertebrate embryos differentiate along the dorsoventral axis into dermomyotome, myotome, and sclerotome. The dermomyotome is later further patterned into epaxial muscle, dermis, and hypaxial muscle along the mediolateral axis. In the past few years, important signalling molecules have been identified and implicated in the patterning of a variety of embryonic structures, including somites. Somite patterning and differentiation depends on signalling molecules of the Hedgehog, Wnt, and TGF-β families, as well as on noggin, which all are released from surrounding tissues in a coordinated manner (see Fig. 3). As a result, the somites exhibit distinct polarities and subdivide into dorsoventral and mediolateral regions with particular patterns of cell differentiation and cell fates (see ref. 7 for a review, and references therein).

Sclerotome development

The formation of mesenchymal sclerotome cells at the ventral side of the somites (20) is followed by their migration towards and around the notochord to give rise to the vertebral bodies and ribs.

Sclerotome induction has been shown to be mediated by Sonic hedgehog (SHH), secreted from the notochord and floor plate of the neural tube (132–134). Sclerotome formation has long been suggested to be positively regulated by signals from notochord and ventral spinal cord. In chick embryos, grafting of an ectopic notochord or floor plate (ventral neural tube) to the unsegmented paraxial mesoderm induced sclerotome formation at the expense of myotomal differentiation (14, 15, 17). In contrast, when the notochord was removed, no sclerotome differentiation was observed, and myotomes fused at the midline (135, 136). These inductive effects of notochord and floor plate on sclerotome formation could be mimicked by the ectopic expression of *Shh* in chick embryos (133). Similarly, in explants of mouse presomitic mesoderm, sclerotome differentiation could be induced by *Shh*-expressing cells (132), and myotome differentiation was repressed (132–134). *Shh* is one of three vertebrate homologues of the *Drosophila* segment polarity gene *hedgehog* and is expressed in notochord and floor plate and in the posteriormost cells of the limb bud, the polarizing region (137–139). *Shh* is involved in various patterning processes in the embryo, such as sclerotome induction, motor neuron induction and patterning of the vertebrate limb and eye (reviewed in ref. 140). The vertebrate SHH protein (similar to the *Drosophila* hh) undergoes autoproteolytic cleavage to generate two fragments (134, 141, 142). The amino-terminal fragment (SHH-N), but not the carboxy-terminal fragment (SHH-C) is thought to have inducing capability and to be responsible for local and long-range signalling properties of the notochord.

Several *Pax* genes function as markers of different somite regions. *Pax* genes encode transcription factors and are characterized by a 384-bp paired-box sequence originally found in the *Drosophila* segmentation and segment polarity genes *paired* and *gooseberry* (see ref. 143 for a review, and references therein). *Pax1* and *Pax9* are expressed in sclerotomal cells (144–146), *Pax3* and *Pax7* are markers for dermo-myotome cells (135, 147, 148). The inductive SHH signal leads to changes in the expression pattern of *Pax* genes within the somite which are thought to be causally related to its dorsoventral patterning (for review, see ref. 44; and see below). *Pax1*, a marker for sclerotomal cells (144, 145), is induced ectopically in the somite upon grafting of notochord or *Shh*-expressing cells laterally to the neural tube, and the *Pax1* expression domain is expanded; moreover, *Pax1* expression is reduced or lost after removal of the notochord (15, 17, 18, 132, 133, 136, 149). Conversely, *Pax3*, which is initially expressed weakly throughout the segmental plate and epithelial somite (135, 147), becomes restricted to the dorsal–lateral somite, the dermomyotome. After ectopic *Shh* expression or transplantation of an ectopic notochord, *Pax3* expression was repressed or abolished (132, 133).

Furthermore, it has been shown that the surface ectoderm inhibits sclerotome formation (132) and, if co-cultured in contact with paraxial mesoderm, induces the expression of *Pax3*, *Pax7*, and *Sim1*, markers for dermomyotome (132). These results suggest that the ventralizing effect of notochord and floor plate and the dorsalizing signals from the neural tube and surface ectoderm act antagonistically and direct the fate of somite cells (see below).

However, the analysis of *Shh*-deficient mouse embryos has shown that initially *Pax1* is expressed in the sclerotomes of *Shh* null-mutant mice, but *Pax1* expression is not maintained during subsequent development (150). These results suggest that SHH might be an essential maintenance signal for sclerotome development, but that notochord and floor plate provide additional signals that are needed to initiate ventralization of the somites (see also ref. 7).

Dermomyotome development

While the sclerotome cells migrate ventromedially towards the vertebral column, the dorsal segment of the somite, which is located under the surface ectoderm, remains epithelial and forms the dermomyotome, the source of proliferating myoblasts and fibroblasts of the dorsal dermis (26, 151; reviewed in ref. 152). Cells from the dorso-medial part of the dermomyotome, which are adjacent to the neural tube, invaginate and migrate laterally underneath the dermomyotome to form the myotome (151; for review see refs 152, 153). These myotome cells rapidly become postmitotic myoblasts and constitute the first differentiated skeletal muscle in the embryo. The dorsomedial part of the dermomyotome contributes to the back muscles, which are referred to as epaxial muscles; cells coming from the lateral part of the dermomyotome will migrate to the limb to form skeletal muscles and the hypaxial muscles of ventral body wall; the dorsal portion of the dermomyotome gives rise to the dermis. For a long time, the exact origin of the myotomal precursor cells from the dermomyotome was not completely clear (for review see refs 7, 152), but recent work by Kalcheim and co-workers has shed light on the roots of myoblasts derived from the dermo-myotomal lips (154, 155; see 156 for a review). Although the muscle cells arising from medial and lateral compartments share a very similar phenotype, their development exhibits important differences: cells immediately adjacent to the midline differentiate first to form the epaxial muscles, myogenesis of lateral cells is delayed in comparison, and consequently the onset of expression of *MyoD* and *Myf5*, which are markers for muscle differentiation, is delayed in lateral, hypaxial precursors (157). *MyoD* and *Myf5* belong to the myogenic bHLH regulatory factors (MRFs), and individual members of this class of transcription factors are able to confer myogenic differentiation on non-muscle cell types. The bHLH myogenic proteins are skeletal muscle-specific and bind to DNA as dimers (for review see ref. 153); the four myogenic bHLH genes *MyoD*, *myogenin*, *Myf5*, and *Mrf4* are expressed in a specific spatiotemporal sequence and control muscle cell differentiation. In mice, *Myf5* is expressed earliest in the medial part of the dermomyotome. When the myotome subsequently forms, early myotomal cells continue to express *Myf5*, while *MyoD* is initiated later in cells of the lateral dermomyotome. In birds, the order of activation of these genes is reversed, with a short time separating the initiation of expression of the two genes (157; see also ref. 153).

Other dermomyotomal markers are *Pax3*, a member of the *Pax* gene family (see above) and c-*met*, encoding a tyrosine kinase receptor. *Pax3* is expressed in the dermomyotome, later becomes more concentrated laterally in the dermomyotome, and is a marker for muscle progenitor cells (135, 147, 158). The proto-oncogene c-*met* is expressed in muscle precursor cells migrating from the lateral dermomyotome. For example, at the limb level, *Pax3*- and c-*met*-expressing precursor cells, which migrate away from the lateral dermomyotome, do not initially express any member of the MRFs; differentiation in the limbs is initiated by *Myf5* expression, followed by expression of *MyoD* and myogenin (152, and references therein).

Myogenic bHLH regulatory factors can functionally substitute for each other. When either *MyoD* or *Myf5* alone are inactivated by gene targeting, no major muscle defect is observed, and the mice eventually produce skeletal musculature. However, in the absence of both genes, the double-mutant mice not only lack skeletal muscle but the precursor myoblast population is also absent (159). The identification of upstream regulators of the myogenic transcription factors remained elusive until evidence for a role of *Pax3* as an upstream regulator of *MyoD* in the mouse was reported (160, 161). The requirement for the activity of the myogenic transcription factors has been revealed by the analysis of *Pax3* (*splotch*), *Myf5*, and *splotch/Myf5* homozygous mutant mice (161). Remarkably, *splotch/Myf5* double homozygotes have no body muscles. *MyoD* does not rescue this double-mutant phenotype, and *MyoD* fails to be expressed in these mutants, suggesting either Pax3 or Myf5 activity is required for the initiation of *MyoD* expression and the onset of myogenesis. Maroto *et al.* (160) showed that signals from the overlying ectoderm can induce somitic expression of *Pax3* and *Pax7*, concomitant with the expression of *Myf5*, but prior to that of *MyoD*. Moreover, sufficiency of Pax3 to induce myogenesis was demonstrated by the ability of Pax3-transfected non-muscle cells to activate *MyoD* expression and to initiate myogenic differentiation (see ref. 153 for a review). Furthermore, the bHLH protein Twist should be mentioned, which is thought to be an inhibitory factor of myotomal differentiation (162). It is initially expressed throughout somitic mesoderm, is excluded from forming myotome, but continues to be expressed in sclerotome, lateral plate, and cranial neural crest. Its loss-of-function mutation results in disorganized and apoptotic somites (163) (see ref. 153 for characterization of further positive and negative regulators of myogenesis).

Several lines of evidence (e.g. rotation studies of epithelial somites (19)) have indicated for quite some time that the fate of somitic cells is not derived intrinsically, but requires signals emanating from the surrounding environment. Like sclerotome (see above), dermomyotome differentiation also requires inductive signals from the surrounding tissues. Muscle-inducing activities have been found in midline structures (notochord/neural tube) and in the dorsal surface ectoderm. If paraxial mesoderm is cultured in the presence of neural tube, myogenesis is significantly increased (e.g. 164, 165; and see ref. 153 and references therein). Combinations of neural tube/notochord with somite explants (166) or dorsal ectoderm with paraxial mesoderm in co-culture (167) was found to be sufficient to induce myogenesis. Furthermore, explanted young somites (I–III) required signals from the ventral midline (notochord or floor plate) and from the neural tube; whereas more mature somites (IV–VI) only required signals from the neural tube, and somites XII and older did not require extrinsic signals (166, 168). The observation that more mature somites no longer require the presence of notochord/floor plate suggests that prior exposure to a ventral midline signal makes the somites responsive to the muscle-inducing activity of the neural tube (168). Consistent with these *in vitro* assays, showing that axial structures also have dermomyotome-promoting activity, ablation of the dorsal neural tube caused failure of dermomyotome differentiation *in vivo*. However, when the whole neural tube was removed, a central fused dermomyotome formed in the operated region (169). This suggests that in the absence of the neural tube, surface ectoderm is sufficient to induce dermomyotome formation, whereas the presence of ventral neural tube antagonizes this activity. Thus, in somite dorsoventral patterning, competing diffusible signals from the dorsal neural tube and from the notochord/floor plate complex specify dermomyotome and sclerotome, respectively.

Different signals might also mediate the generation of medial versus lateral muscle fates. Experiments which remove the neural tube but leave the lateral somite in contact with the adjacent lateral mesoderm show that hypaxial muscle progenitors can still form and migrate to the limb and ventral body wall (170). Similarly, epaxial, but not hypaxial, muscle development is severely impaired in the mouse mutation *open brain* (*opb*), most likely because of the absence of dorsal neural tube signals (171).

In contrast, the lateral mesoderm is implicated as a source of inhibitory signals in the specification of myogenesis, and is able to delay the promoting effect of, for example, the ectoderm (167) on myogenic differentiation of co-cultured paraxial mesoderm. These signals for lateral somite specification result in a blockade of the myogenic programme, leaving the lateral somitic cells for a longer time as undifferentiated muscle progenitor cells (expressing *Pax3* gene, but not yet *MyoD*). Thus, *in vivo*, this mechanism could account for the delay observed in the onset of myogenesis between muscles of epaxial and hypaxial domains (172).

Among the signals regulating somite differentiation (Fig. 3), a number of secreted molecules are expressed in the axial and lateral structures. The stimulating signal for myogenesis of ventral structures is probably mediated by SHH (see above). Ectopic expression of SHH induced expansion of the *MyoD* expression domain in chick (133) and in zebrafish embryos (173), and SHH substituted for the activity of notochord

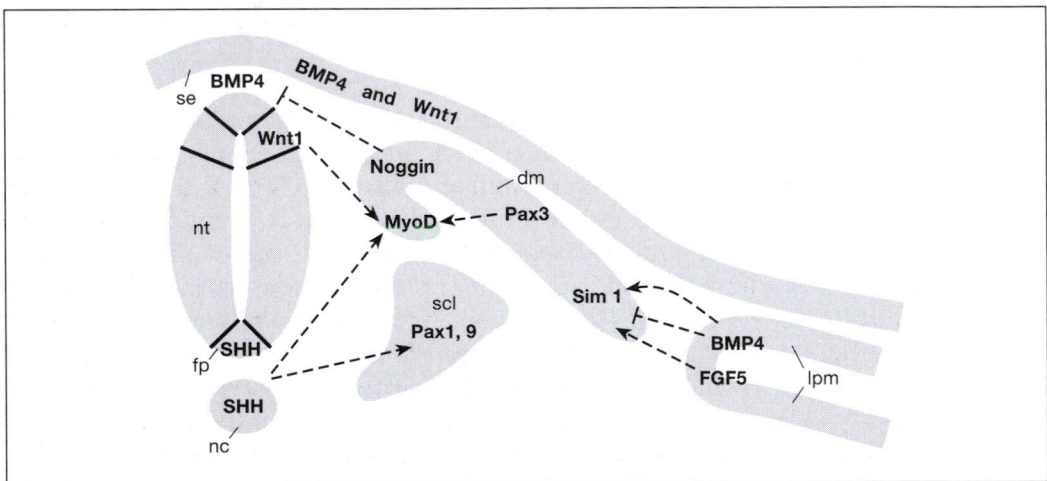

Fig. 3 Schematic transverse section through the midline of an embryo, which shows a differentiating somite and the signals that are involved in dorsoventral patterning of the somite leading to sclerotome and dermomyotome differentiation. SHH expressed in notochord and ventral neural tube (floor plate, fp) induces sclerotome (scl), and subsequently *Pax1* and *Pax9* are expressed. Various studies on myogenic induction suggest the following model: Wnt genes (e.g. *Wnt1*) expressed in the dorsal neural tube (nt) and surface ectoderm (se) and SHH expressed in notochord (nc) and ventral neural tube positively regulate the induction of myotome, leading to the expression of *MyoD* upon the onset of myogenesis. BMP4 expressed in the lateral plate mesoderm negatively regulates myogenesis and controls the specification of the lateral somitic lineages, the hypaxial muscles which start myogenesis later. FGF5 is thought to positively regulate this process. However, BMP4 is also expressed in the dorsal neural tube and its signalling may be controlled by BMP4 binding proteins, e.g. Noggin, which is highly expressed in the dorsomedial part of the dermomyotome (dm). These proteins have been shown to directly bind BMP4 and negatively regulate its activity. The role of Noggin, which is probably induced by Wnt1, may be to inhibit the activity of neural tube-derived BMP4. The role of BMP4 in the lateral plate mesoderm is to inhibit differentiation of the lateral migratory hypaxial myoblasts by activating the expression of the lateral marker *Sim1*. Thus, various combinatorial stimulatory and inhibitory signals regulate somite differentiation (see text).

and floor plate in explant cultures (174). Furthermore, it has been shown that a number of members of the Wnt family of secreted proteins (Wnt1, Wnt3a, and Wnt4) display a synergism with SHH to induce myogenic bHLH gene expression and to promote myogenesis (174, 175). These *Wnt* genes are expressed predominantly in the dorsal neural tube, while non-inducing *Wnt* genes (*Wnt7a* and *Wnt7b*) are localized primarily to the ventral neural tube. *Pax3* expression can also be induced by the same combination of SHH and Wnt proteins in somite explants, which demonstrates that similar signals might control the induction of upstream regulators of the MRFs (160). This combination of SHH and Wnt signals is needed in early phases of somitogenesis, whereas more mature somites require only Wnt signalling to activate *MyoD* expression (174). The requirement for both SHH and Wnt to activate myogenesis in nascent somites fits well with previous observations demonstrating that both the notochord/floor plate complex and dorsal neural tube are required for high-level activation of myogenic transcription factors (see above). Expression analysis of myogenic transcription factors in *Shh*-deficient mice indeed reveals a reduced expression of *Myf5* in the medial region of the myotome (150). However, skeletal muscle develops in *Shh* mutant mice, and expression of *MyoD* in the lateral myotome is unaffected, suggesting there is a possible redundancy in signals that activate myogenesis.

It was also shown *in vivo* by surgical ablation and grafting experiments in quail embryos that notochord is necessary to induce myogenic bHLH gene expression in dorsomedial myotomal cells, and that neural tube signals then contribute to the maintenance of high-level expression (176). Furthermore, myogenic bHLH gene expression is localized to dorsal medial cells of the somite by inhibitory signals produced by the lateral plate (172, 176). Removal of lateral plate mesoderm led to the lateral expansion of the *MyoD* expression domain in avian embryos (176); grafting lateral plate mesoderm between the neural tube and the unsegmented paraxial mesoderm resulted in downregulation of *Pax3* and concomitant activation of *MyoD* and *Myf5* in the lateral myotome (172). These findings indicate that signals from the lateral plate mesoderm are required to maintain myogenic cells in the lateral myotomal region as undifferentiated progenitors. Pourquié *et al.* (177) have demonstrated that expression of the bHLH transcription factor gene *cSim1*, an avian homologue of the *Drosophila single minded* gene, is restricted to cells of the lateral somitic region, and that this expression requires the presence of the lateral plate mesoderm. BMP4, a member of the transforming growth factor-β (TGF-β) family of genes, is highly expressed in chick lateral plate mesoderm and dorsal neural tube during somitogenesis (177). The authors showed that the lateral plate produces a diffusible signal, most likely BMP4, which is required for the activation of *cSim1* expression in the lateral somitic cells, and provide evidence that *Bmp4*-expressing cells, like the lateral plate, can induce *cSim1* expression when grafted between the neural tube and medial somite. Similarly positioned *Bmp4*-expressing cells can also inhibit the induction of myogenic bHLH gene expression within the medial somite. In addition, it was shown that axial tissue, e.g. a segment of neural tube, inhibited *cSim1* expression in the lateral somite domain when grafted laterally to the paraxial mesoderm. This suggests that lateral plate mesoderm activates and the neural tube represses *cSim1* expression in somites, thus, antagonistic signals of neural tube and lateral plate direct mediolateral somite patterning (177). The fact that *Bmp4* is also expressed in the dorsal neural tube complicates such a model, but recent findings have shown that there are proteins such as Noggin, Chordin, and Follistatin which inhibit BMP4 activity. These proteins are able to directly antagonize BMP4 by binding to it and preventing an interaction with its receptor(s) (178, 179). Thus, for example, Noggin, which is expressed initially widely within paraxial mesoderm and neural tube, but later becomes restricted to the dorsomedial part of the dermomyotome (180, 181), is likely to antagonize BMP4-induced repression of muscle fates within the developing medial dermomyotome (for review and further references see ref. 153). In addition, *follistatin* and *follistatin*-related (*flik*) genes are expressed in the medial dermomyotome and may inhibit BMP activity on medial myoblasts (182; for review see 153). When cells expressing Noggin are transplanted between the lateral mesoderm and the lateral edge of the somite, downregulation of lateral somitic markers (such as *cSim1*) and upregulation or expansion of the expression domain of medial somite differentiation markers (such as MyoD) are observed (180, 183). In addition, this ectopic expression of Noggin represses *Pax3* expression in the lateral somite and induces the formation of a lateral myotome (183). This indicates that Noggin is able to antagonize the lateralizing effect of BMP4. Furthermore, it was demonstrated that a neural tube-derived signal is required for Noggin expression in the medial somite, which is thought to be Wnt1. Cells expressing Wnt1 are able to substitute for neural tube in inducing Noggin expression (180) or to induce Noggin in co-cultured young (stage I–III) somites (183). BMP4 is also expressed within the dorsal neural tube, in a similar region to Wnt1, and BMP4-expressing cells, when placed in the lumen of the neural tube, are able to induce ectopic Wnt1 expression in the dorsal neural tube (181).Thus, BMP4 might induce Wnt1 expression in the neural tube, which induces Noggin in the medial dermomyotome. The main function of the medial dermomyotomal expression of Noggin may, in turn, be to counteract the potentially lateralizing BMP4 signal within the neural tube (see Fig. 3).

A possible model is that competing gradients of diffusible signals mediate cell-type specification in the developing myotome and dermomyotome. In addition to the activity of members of the BMP, HH, and Wnt protein families, other secreted molecules are quite certainly also involved in specifying positional cues within the dermomyotome. For example, FGF proteins are known for their ability to stimulate the proliferation of myoblasts and inhibit their differentiation *in vitro*, as well as to control aspects of early myotome development (184, 185). FGF5 is highly expressed in the lateral mesoderm during the time of specification of cell types within the dermomyotome and may play an active role in this process (186, 187). Neurotrophin-3 (NT-3), which is a signal also secreted by dorsal neural tube cells, is essential for differentiation of the dermal somite compartment, as demonstrated in chick embryos (188). Separating neural tube and dermomyotome by an impermeable membrane prevented dorsal dermis formation, but treatment of the membrane with NT-3 completely restored dermis differentiation. Moreover, NT-3-blocking antibodies disrupted early dermis differentiation in embryos, indicating that NT-3 is a neural tube signal required for dermatome differentiation (188).

In summary, various combinatorial stimulatory and inhibitory signals regulate somite differentiation, and the molecular mechanisms underlying this regulation are now emerging, but how these signals and factors are coordinated to control the pattern, growth, and differentiation of the segmental units of the vertebrate body has yet to be worked out in detail.

3.3.2 Anteroposterior patterning within the somite

Differences between the anterior and posterior halves of somites already exist in the early epithelial somite (189), which are thought to be necessary for generating the segmental organization of the peripheral nervous system (32, 190; and see Chapter 8). Craniocaudal polarity of the sclerotome is important for defining not only vertebral periodicity but also for the patterning of neural crest migration and spinal neuron development, as motor axons and neural crest cells migrate exclusively through the cranial half of the sclerotome (191, 192). Sclerotomal polarity has been shown to be generated before segmentation (193), and there are a number of genes known to be expressed either in the anterior or the posterior domain of a forming somite (see above). In addition, there are extracellular matrix glycoproteins (e.g. tenascin) expressed in the cranial sclerotome half, which are thought to guide neural cells (194).

However, the molecular mechanisms underlying anteroposterior regionalization have not been fully worked out. Genes that are involved in the establishment of segment polarity in the *Drosophila* embryo, such as *hedgehog*, do not appear to be involved in vertebrate somite segmentation (195). The establishment of anterior and posterior differences within segments is essential for the development of somites, and several genes expressed in the paraxial mesoderm of the vertebrate embryo are now known to regulate anteroposterior polarity within somites; these include members of the Notch–Delta pathway, *Mesp2* and *EphA4* (see above). There are four zebrafish mutations that lead to defects in somite boundary formation, and, in addition, anteroposterior regionalization within segments is disturbed (196). In *fss-* (*fused somites*), there is a loss of anterior segment markers, thus all segments appear posteriorized, whereas in *bea-* (*beamter*), *des-* (*deadly seven*), and *aei-* (*after eight*), anterior and posterior markers are expressed throughout each segment. The authors show that it is possible to rescue boundary formation in *fss-* animals through the ectopic expression of *EphA4*, an anterior segment marker in the paraxial mesoderm, indicating that correct anteroposterior patterning within segments may be a prerequisite for somite boundary formation (196).

3.4 *Hox* genes in anteroposterior specification along the axis

Although somites and the resulting vertebrae, the structural elements of the bony part of the vertebral column, resemble each other at different axial levels morphologically and in the expression of various genes that regulate the mechanisms to establish the basic form of somites (see above), it is obvious that vertebrae in different regions have distinct morphological features. Furthermore, vertebral morphology varies between regions not only in an individual organism but also between the vertebrae of different species (197; see also ref. 198 for a review). The positional specification of the paraxial mesoderm that leads to the formation of specific patterns in the different parts of the vertebral column requires the function of members of the *Hox* gene family (see ref. 199 for a review). The vertebrate *Hox* genes are related to the eight homeotic selector genes of *Drosophila*, clustered into two groups known, respectively, as the *Antennapedia complex* (ANT-C) and the *Bithorax complex* (BX-C), which are essential for specifying segment identity in insects. The homeobox genes have a common 183-bp motif that encodes a 61-amino-acid protein domain that folds into a helix–turn–helix motif capable of sequence-specific DNA binding (see, for example, ref. 200 for a review). Higher vertebrates have four clusters of *Hox* genes (*HoxA*, *HoxB*, *HoxC*, and *HoxD*), which are located on different chromosomes. Accordingly, the individual genes of the vertebrate *Hox* complexes are arranged in the same relative order as their homologues in *Drosophila*, and the activation of their expression along the anteroposterior body axis during development correlates with their ordered arrangement in the *Hox* complex, a phenomenon known as 'colinearity'. There are two types of colinearity, spatial and temporal, found in vertebrate embryos (reviewed in ref. 199). The spatial colinearity rule refers to the correlation between the anterior expression boundary of a *Hox* gene and its relative anteroposterior position in the complex: the rostral limit of expression of a *Hox* gene is more anterior than that of a 5′ neighbouring gene and more posterior than that of a 3′ neighbouring gene. The temporal colinearity rule refers to the correlation between the time of activation of a *Hox* gene and its position in the complex: transcriptional activation of a given *Hox* gene is earlier than that of a 5′ neighbouring gene and later than that of a 3′ neighbouring gene (see ref. 7 and references therein; 201). Thus, more 5′ genes are activated late and function in the posterior part of the embryo, genes at the extreme 3′ end of a *Hox* gene cluster are activated earliest and operate in the anterior part of the embryo.

Another interesting feature of the *Hox* clusters is that *Hox* genes vary in their response to retinoic acid, a molecule involved in the specification of limb and vertebral axes during vertebrate development (202, 203; for a review see ref. 201), depending on their location in the cluster. More 3′ genes respond to low concentrations of retinoic acid and are highest-sensitive, whereas 5′ genes have a low sensitivity and respond to high concentrations of retinoic acid (for reviews see refs 199, 201, 204).

Transcriptional regulation of *Hox* genes following colinearity rules is thought to be responsible for the establishment of positional information along the anteroposterior body axis and also along anteroposterior and proximodistal axes of the limbs. The resulting defined expression of characteristic combinations of *Hox* genes at a given axial level is thought to specify the identity of vertebral segments along the axis (205). On the basis of comparisons of expression patterns of *Hox* genes in normal and in experimentally manipulated mouse embryos and on the resulting morphological alterations, Kessel and Gruss (206) have proposed the '*Hox* code model'. This model postulates that the identity of a vertebral segment is specified by a particular combination of expressed *Hox* genes. If the *Hox* code is altered by the overexpression or inactivation of *Hox* genes or by modification of their expression by treatment with retinoic acid, the *Hox* code model predicts a homeotic transformation. Thus, this model is very useful in correlating phenotypes and altered *Hox* gene expression, although not all observed defects can be explained with simple alterations of the combinatorial expression of single *Hox* genes. This is because there is substantial overlap in expression domains among members of paralogous *Hox* groups, which may cause compensation to various degrees of loss-of-function phenotypes. Ectopic expression or targeted disruption of particular *Hox* genes results in homeotic transformations, i.e. the morphology of individual vertebrae is changed in a way that they acquire the identity of a vertebra anterior or posterior to their axial position. For example, in gain-of-function mutations where ectopic expression of *Hoxd-4* (207) or *Hoxa-7* (208) is performed, the occipital bones gain a more posterior phenotype that resembles cervical vertebrae; thus, a posterior transformation of part of the occipital bone into an additional cervical vertebra occurs. An example for a loss-of-function mutation is given by the work of Ramirez-Solis *et al.* (209), who showed in *Hoxb-4* mutant mice that the second cervical vertebra (axis) was transformed to a more anterior one and showed characteristics of the first cervical vertebra (atlas). Furthermore, the mutant mice showed defective morphogenesis of the sternum. Similar results are obtained in mice which are deficient for *Hoxd-3* (210). The mutants exhibit anterior transformations of the first and second cervical vertebrae in such a way that the atlas is transformed to an extension of the basioccipital bone of the skull, and the axis shows characteristics resembling an atlas (210). Targeted inactivation of *Hoxd-13* leads to a transformation of the fourth sacral vertebra into a vertebra resembling the third, and to abnormalities in the limbs (211).

Many single, double, and triple *Hox* mutants have been generated (for review see refs 7, 198, and references therein) to try to overcome compensation by paralogues, and to study further the basis of (multiple) *Hox* gene function. It has been shown, for example, that paralogous genes have redundant functions; mice with multiple mutations of *Hox* genes in the same paralogous group tend to show a more severe phenotype than mice with a mutation in a single gene of that paralogous group (see, for example, refs 212–214). An interesting approach is taken in the study of Suemori and Noguchi (215), who address the question of the function of a whole *Hox* cluster. Utilizing the Cre/loxP system, *HoxC* cluster null mice were generated lacking all nine *HoxC* genes. Mutants die at the time of birth, but develop to this stage with only minor transformations in the skeletal system and a defect in the respiratory tract, but no other apparent internal organ defect. Similarly, Medina-Martinez *et al.* (216) report an analysis of mice lacking all *HoxB* cluster genes except *Hoxb-13*, because it is separated from the others by 70 kb. Surprisingly, these mice, which are deficient for *Hoxb-1* to *Hoxb-9*, show no detectable abnormalities in the heterozygous state, and homozygous mutants also die at the perinatal stage. In addition, they also exhibit a series of single-segment anterior homeotic transformations (216) along the cervical and thoracic vertebral column, defects in sternum morphogenesis, and abnormalities in the IXth cranial nerve. No synergistic or new phenotypes were observed as compared to phenotypes of single *HoxB* gene mutants, although a higher penetrance of the sum of phenotypes was noticed. Expression of *Hoxb-13* was not affected by the absence of the other genes of that cluster, suggesting that its expression could be achieved independently of the colinear pattern of the cluster. The authors discuss the possibility that the combined loss of more clusters would result in a progressive anteriorization of segments, because combined mutations in paralogous group 4 members lead to an increase in the number of cervical vertebrae transformed toward the anterior first cervical vertebra (213), and a deficiency in the unique HOM-C cluster of *Tribolium* causes the development of antennae in all body segments (217). Other combinations of removing *Hox* clusters would be necessary to test such hypotheses (216).

As mentioned above, *Hox* genes are sensitive to retinoic acid treatment—the influence of retinoic acid on the expression of *Hox* genes came from pioneering studies in embryonic carcinoma cell cultures by Gruss and co-workers (see, for example, ref. 218). Murine F9 teratocarcinoma cells in monolayer culture differentiate into non-tumorigenic parietal endoderm cells in response to retinoic acid, and almost all their *Hox* genes are expressed in the course of the terminal differentiation of these cells (see ref. 201, and references therein). Boncinelli and co-workers performed a comprehensive analysis of *Hox* gene induction in teratocarcinoma cells by retinoic acid (204). A systematic study of *Hox* gene activation in human embryonal carcinoma cells by retinoic acid revealed that the rate of induction varies considerably for each gene, and reflects the localization of a particular gene within the *Hox* clusters (see also above): in general, *Hox* genes are activated by retinoic acid in a 3′ to 5′ direction (219, 220).

Treatment of mouse embryos *in utero* with retinoic acid resulted in homeotic transformations of the vertebrae (206, 221). Depending on the time of retinoic acid administration, anterior or/and posterior transformations were observed. In embryos treated early during gastrulation (7–7.5 dpc), at a time when *Hox* genes that are sensitive to activation by retinoic acid start to be expressed in paraxial mesoderm, posterior transformations were observed and, concomitantly, anterior shifts of *Hox* gene expression boundaries. This induced, ectopic, *Hox* gene expression in a more anterior region is in agreement with the morphologically posterior transformation observed in response to retinoic acid. Administration of retinoic acid later in development (8–9.5 dpc), when *Hox* genes that are less sensitive to or repressed by retinoic acid begin to be expressed, resulted in anterior transformations and in posterior shifts of *Hox* gene expression. These shifts can be explained by repression of *Hox* gene activation by retinoic acid in a region in which the gene would normally be expressed (206, 221). When retinoic acid was applied even later, at stages when somites differentiate (10–11.5 dpc), complex anterior and posterior transformations were observed, indicating that cells might be particularly sensitive to retinoic acid during epithelial-to-mesenchymal transition phases (221).

In *Drosophila*, maintenance of spatially restricted, stable expression of homeotic genes, once they are activated, needs the function of transcriptional regulators of the *trithorax* group genes (*trxG*) and the *Polycomb* group genes (*PcG*), which are assumed to regulate and maintain the active and repressed states of homeobox genes (see ref. 222 for a review): the *trithorax* group is responsible for sustaining the active state of homeotic gene expression, the *Polycomb* group encodes a stable repressor system. Recent findings indicate evolutionary conservation of this regulation, and an increasing number of mammalian homologues to *PcG* and *trxG* genes are being isolated (see ref. 223 for a review). Mutations in such genes also result in homeotic transformations and simultaneous changes of expression of several *Hox* genes. For example, animals deficient for *Bmi1* or *Zfp144* (also known as *Mel18*), both members of the *PcG* genes, exhibit, among other defects, posterior transformations along the entire axial skeleton and they die postnatally (224, 225). The presence of skeletal transformations also in heterozygous *Bmi1* mice suggests a gene-dose effect and, interestingly, overexpression of *Bmi1* results in the complementary phenotype, i.e. dose-dependent anterior transformations of vertebral identities (226). The vertebral transformations, also in the *Mel18* mutant, are correlated with shifts of *Hox* gene-expression boundaries by one or two segments (see overview in ref. 223). In the case of *Bmi1* overexpression, the anterior expression boundary of the *Hoxc-5* gene is shifted in the posterior direction, indicating that *Bmi1* is involved in the repression of *Hox* genes, and, as a consequence, anterior transformations of vertebrae in the cervical and thoracic regions are observed (226). Also targeted disruption of *Cbx2* (another *PcG* gene, also known as *M33*) causes retarded growth, homeotic axial transformations, sternal malformations, and postnatal lethality, but here *Hox* expression changes and skeletal transformations are more regionally restricted (227). The *eed* (homologue to the *Drosophila PcG* gene *esc*) null phenotype is more severe than any other murine *PcG* mutation; *eed* null-mutant mice fail to develop node, notochord, and somites, and die at gastrulation (228). Hypomorphic alleles of *eed* result in posterior homeotic transformations (229).

Gene targeting of the murine *trxG* homologue *Mll* locus resulted in bidirectional homeotic transformations and altered gene expression in heterozygous mutant mice, whereas homozygous deficiency resulted in embryonic lethality around day 11.5, and the expression of several *Hox* genes was completely abolished, consistent with a role for *Mll* to positively regulate and maintain active states of *Hox* gene expression (230).

In an interesting piece of recent work, the axial–skeletal transformations and altered *Hox* expression patterns of *Mll*-deficient and *Bmi1*-deficient mice were normalized when both *Mll* and *Bmi1* were deleted (231). This demonstrates the antagonistic role of both genes in regulating *Hox* gene expression and determining segmental identities.

Very recent findings need to be mentioned, because they give evidence that FGF signalling is involved in regulating the maturation of cells along the anteroposterior axis to provide axial identity and segment boundary position. This includes the regulation of segmentation clock control and also *Hox* gene activation (see ref. 232 for a review). Thus, FGF signalling plays a role in coordinating the segmentation process and spatiotemporal *Hox* gene activation (233, 234). The segmentation clock is thought to play a role in the periodic Notch signalling in the presomitic mesoderm, which results in the formation of the regular array of somitic boundaries (see above). The identity of each somite at the axial level is characterized by the combinatorial expression of *Hox* genes, as discussed before. But how the temporal periodicity of the clock oscillations is translated into the periodic order of segmental boundaries remains unknown. Dubrulle *et al.* (233) demonstrate that the determination of periodic segmentation occurs at the level of a 'determination front' whose position along the anteroposterior axis is controlled by FGF8, which is expressed in the caudal part of the segmental plate. Moving the position of the determination front by altering FGF signalling leads to a shift of the position of somitic boundaries. Furthermore, FGF8 treatment can increase the number of clock oscillations, and cells which experience a supernumerary oscillation are included into a differently numbered somite and, interestingly, exhibit *Hox* expression characteristic to a more posterior fate. Thus, *Hox* expression remains linked to the somitic number rather than to the original axial level of cells. These findings indicate a tight coordination between the segmentation clock leading to the positioning of somite boundaries and the *Hox*-dependent patterning process along the anteroposterior axis.

Acknowledgements

I am particularly grateful to Prof. C. Tickle for critical reading of the manuscript, many helpful suggestions and comments, and her encouragement and advice through all phases of writing. I would like to thank Prof. R. Balling, in whose department the article started and where the stimulating environment brought up many discussions on skeletal axis formation. I thank Prof. A. Gossler very much for critical comments on the manuscript and insightful discussions. Last, but not least, I greatly thank my husband, Dr B. Lutz, for critical reading of the manuscript, many discussions and suggestions, and his constant support.

References

1. Johnson, D. R. (1986) *The genetics of the skeleton*. Clarendon Press, Oxford.
2. Beddington, R. (1987) Isolation, culture and manipulation of post-implantation mouse embryos. In *Mammalian development: a practical approach* (ed. M. Monk), p. 43. IRL Press, Oxford.
3. Sturm, K. and Tam, P. P. L. (1993) Isolation and culture of whole postimplantation embryos and germ layer derivatives. *Methods in Enzymology*, **225**, 164.
4. Hogan, B., Beddington, R., Costantini, F., and Lacy, E. (1994) *Manipulating the mouse embryo. A laboratory manual*. Cold Spring Harbour Laboratory Press, Cold Spring Harbour, New York.
5. Spratt, N. T. J. (1955) Analysis of the organizer center of the early chick embryo. Localization of prospective notochord and somitic cells. *J. Exp. Zool.*, **128**, 121.
6. Nicolet, G. (1971) Avian gastrulation. Adv. Morphogen., **9**, 231.
7. Gossler, A. and Hrabe De Angelis, M. (1998) Somitogenesis. *Curr. Top. Dev. Biol.*, **38**, 225.
8. Hamburger, V. and Hamilton, H. L. (1951) A series of normal stages in the development of the chick embryo. *J. Morphol.*, **88**, 49.

9. Tam, P. P. L. and Beddington, R. S. P. (1987) The formation of mesodermal tissues in the mouse embryo during gastrulation and early organogenesis. *Development*, **99**, 109.

10. Selleck, M. A. and Stern, C. D. (1991) Fate mapping and cell lineage analysis of Hensen's node in the chick embryo. *Development*, **112**, 615.

11. Psychoyos, D. and Stern, C. D. (1996) Fates and migratory routes of primitive streak cells in the chick embryo. *Development*, **122**, 1523.

12. Placzek, M., Tessier-Lavigne, M., Jessell, T., and Dodd, J. (1990) Mesodermal control of neural cell identity: floor plate induction by the notochord. *Science*, **250**, 985.

13. van Straaten, H. W. M., Hekking, J. W. M., Wiertz-Hoessels, E. J. L. M., Thors, F., and Drukker, J. (1988) Effect of the notochord on the differentiation of a floor plate area in the neural tube of the chick embryo. *Anat. Embryol.*, **177**, 317.

14. Watterson, R. L., Fowler, I., and Fowler, B. J. (1954) The role of the neural tube and notochord in development of the axial skeleton of the chick. *Am. J. Anat.*, **95**, 337.

15. Pourquié, O., Coltey, M., Teillet, M. A., Ordahl, C., and LeDouarin, N. M. (1993) Control of dorsoventral patterning of somitic derivatives by notochord and floor plate. *Proc. Natl Sci. USA*, **90**, 5342.

16. Yamada, T., Pfaff, S. L., Edlund, T., and Jessell, T. M. (1993) Control of cell pattern in the neural tube: motor neuron induction by diffusible factors from notochord and floor plate. *Cell*, **73**, 673.

17. Brand-Saberi, B., Ebensperger, C., Wilting, J., Balling, R., and Christ, B. (1993) The ventralizing effect of the notochord on somite differentiation in chick embryos. *Anat. Embryol.*, **188**, 239.

18. Koseki, H., Wallin, J., Wilting, J., Mitzutani, Y., Ebensperger, C., Christ, B., and Balling, R. (1993) Pax-1 as a mediator of notochordal signals in the dorsoventral specification of vertebrae. *Development*, **119**, 649.

19. Christ, B. and Ordahl, C. P. (1995) Early stages of chick somite development. *Anat. Embryol.*, **191**, 381.

20. Christ, B. and Wilting, J. (1992) From somites to vertebral column. *Ann. Anat.*, **174**, 23.

21. Stern, C. D. and Bellairs, R. (1984) Mitotic activity during somite segmentation in the early chick embryo. *Anat. Embryol.*, **169**, 97.

22. Ang, S. L. and Rossant, J. (1994) HNF-3 beta is essential for node and notochord formation in mouse development. *Cell*, **78**, 561.

23. Weinstein, D. C., Ruiz i Altaba, A., Chen, W. S., Hoodless, P., Prezioso, V. R., Jessell, T. M., and Darnell, J. E. J. (1994) The winged-helix transcription factor HNF-3 beta is required for notochord development in the mouse embryo. *Cell*, **78**, 575.

24. Solursh, M., Drake, C., and Meier, S. (1979) The role of extracellular matrix in the formation of the sclerotome. *J. Embryol. Exp. Morphol.*, **54**, 75.

25. Verbout, A. J. (1985) The development of the vertebral column. *Adv. Anat. Embryol. Cell Biol.*, **90**, 1.

26. Christ, B., Jacob, M., and Jacob, H. J. (1983) On the origin and development of the ventro-lateral abdominal muscles in the avian embryo, an experimental and ultrastructural study. *Anat. Embryol.*, **160**, 87.

27. Ordahl, C. P. and Le Douarin, N. M. (1992) Two myogenic lineages within the developing somite. *Development*, **114**, 339.

28. Goldstein, R. S. and Kalcheim, C. (1992) Determination of epithelial half-somites in skeletal morphogenesis. *Development*, **116**, 441.

29. Huang, R., Zhi, Q., Wilting, J., and Christ, B. (1994) The fate of somitocoele cells in avian embryos. *Anat. Embryol.*, **190**, 243.

30. Huang, R., Zhi, Q., Neubüser, A., Müller, T. S., Brand-Saberi, B., Christ, B., and Wilting, J. (1996) Function of somite and somitocoele cells in the formation of the vertebral motion segment in avian embryos. *Acta Anat.*, **155**, 231.

31. Erlebacher, A., Filvaroff, E. H., Gitelman, S. E., and Derynck, R. (1995) Towards a molecular understanding of skeletal development. *Cell*, **80**, 371.

32. Keynes, R. J. and Stern, C. D. (1988) Mechanisms of vertebrate segmentation. *Development*, **103**, 413.

33. Keynes, R. J. and Stern, C. D. (1984) Segmentation in the vertebrate nervous system. *Nature*, **310**, 786.

34. Remak, R. (1855) *Untersuchungen über die Entwicklung der Wirbelthiere*. Reimer, Berlin.

35. Bagnall, K. M., Higgins, S. J., and Sanders, E. J. (1988) The contribution made by a single somite to the vertebral column: experimental evidence in support of resegmentation using the chick–quail chimaera model. *Development*, **103**, 69.

36. Bagnall, K. M., Higgins, S. J., and Sanders, E. J. (1989) The contribution made by cells from a single somite to tissues within a body segment and assessment of their integration with similar cells from adjacent segments. *Development*, **107**, 931.

37. Huang, R., Zhi, Q., Brand-Saberi, B., and Christ, B. (2000) New experimental evidence for somite resegmentation. *Anat. Embryol.*, **202**, 195.

38. Levitt, D. and Dorfman, A. (1974) Concepts and mechanisms of cartilage differentiation. *Curr. Top. Dev. Biol.*, **8**, 103.

39. Dietrich, S., Schubert, F. R., and Gruss, P. (1993) Altered Pax gene expression in murine notochord mutants: the notochord is required to initiate and maintain ventral identity in the somite. *Mech. Dev.*, **44**, 189.

40. Holtzer, H. and Detwiler, S. R. (1953) An experimental analysis of the development of the spinal column. III. Induction of skeletogenous cells. *J. Exp. Zool.*, **123**, 335.

41. Hall, B. K. (1977) Chondrogenesis of the somitic mesoderm. *Adv. Anat. Embryol. Cell Biol.*, **53**, 3.

42. Vasan, N. S. (1986) Somite chondrogenesis: extracellular matrix production and intracellular changes. In *Somites in developing embryos* (ed. R. Bellairs, D. A. Ede, and J. W. Lash), p. 237. Plenum Press, New York.

43. Grobstein, C. and Holtzer, H. (1955) *In vitro* studies of cartilage induction in mouse somite mesoderm. *J. Exp. Zool.*, **128**, 333.

44. Neubüser, A. and Balling, R. (1996) Axial skeleton. In *Handbook of experimental pharmacology* (ed. R. Kavlok and G. Daston), p. 77. Springer Verlag, Heidelberg.

45. Winnier, G., Blessing, M., Labosky, P. A., and Hogan, B. L. M. (1995) Bone morphogenetic protein-4 is required for mesoderm formation and patterning in the mouse. *Genes Dev.*, **9**, 2105.

46. Mishina, Y., Suzuki, A., Ueno, N., and Behringer, R. R. (1995) BMPr encodes a type I bone morphogenetic protein receptor that is essential for gastrulation during mouse embryogenesis. *Genes Dev.*, **9**, 3027.

47. Zhou, X., Sasaki, H., Lowe, L., Hogan, B. L. M., and Kuehn, M. R. (1993) Nodal is a novel TGF-beta-like gene expressed in the mouse node during gastrulation. *Nature*, **361**, 543.

48. Conlon, F. L., Lyons, K. M., Takaesu, N., Barth, K. S., Kispert, A., Herrmann, B., and Robertson, E. J. (1994) A primary requirement for *nodal* in the formation and maintenance of the primitive streak in the mouse. *Development*, **120**, 1919.

49. Takada, S., Stark, K. L., Shea, M. J., Vassileva, G., McMahon, J. A., and McMahon, A. P. (1994) *Wnt-3a* regulates somite and tailbud formation in the mouse embryo. *Genes Dev.*, **8**, 174.

50. Yamaguchi, T. P., Bradley, A., McMahon, A. P., and Jones, S. (1999a) A *Wnt5a* pathway underlies outgrowth of multiple structures in the vertebrate embryo. *Development*, **126**, 1211.

51. Maruoka, Y., Ohbayashi, N., Hoshikawa, M., Itoh, N., Hogan, B. L. M., and Furuta, Y. (1998) Comparison of the expression of three highly related genes, *Fgf8, Fgf17* and *Fgf18*, in the mouse embryo. *Mech. Dev.*, **74**, 175.

52. Feldman, B., Poueymirou, W., Papioannou, V. E., DeChiara, T. M., and Goldfarb, M. (1995) Requirement of FGF-4 for postimplantation mouse development. *Science*, **267**, 246.

53. Sun, X., Lewandowski, M., Meyers, E. N., Liu, Y.-H., Maxson, R. E. Jr, and Martin, G. R. (2000) Conditional inactivation of *Fgf4* reveals complexity of signalling during limb bud development. *Nature Genet.*, **25**, 83.

54. Moon, A. M., Boulet, A. M., and Capecchi, M. R. (2000) Normal limb development in conditional mutants of *Fgf4*. *Development*, **127**, 989.

55. Hébert, J. M., Rosenquist, T., Gotz, J., and Martin, G. R. (1994) FGF5 as a regulator of the hair growth cycle: evidence from targeted and spontaneous mutations. *Cell*, **78**, 1017.

56. Mansour, S. L., Goddard, J. M., and Capecchi, M. R. (1993) Mice homozygous for a targeted disruption of the proto-oncogene *int-2* have developmental defects in the tail and inner ear. *Development*, **117**, 13.

57. Meyers, E. N., Lewandowski, M., and Martin, G. R. (1998) An *Fgf8* mutant allelic series generated by Cre- and Flp-mediated recombination. *Nature Genet.*, **18**, 136.

58. Sun, X., Meyers, E. N., Lewandowski, M., and Martin, G. R. (1999) Targeted disruption of *Fgf8* causes failure of cell migration in the gastrulating mouse embryo. *Genes Dev.*, **13**, 1834.

59. Deng, C., Wynshaw-Boris, A., Shen, M. M., Daughtery, C., Ornitz, D. M., and Leder, P. (1994) Murine FGFR-1 is required for early postimplantation growth and axial organization. *Genes Dev.*, **8**, 3045.

60. Yamaguchi, T. P., Harplan, K., Henkeyer, M., and Rossant, J. (1994) *fgfr-1* is required for embryonic growth and mesodermal patterning during mouse gastrulation. *Genes Dev.*, **8**, 3032.

61. Deng, C., Bedford, M., Li, C., Xu, X., Yang, X., Dunmore, J., and Leder, P. (1997) Fibroblast growth factor receptor-1 (FGFR-1) is essential for normal neural tube and limb development. *Dev. Biol.*, **185**, 42.

62. Ciruna, B. G., Schwartz, L., Harpal, K., Yamaguchi, T. P., and Rossant, J. (1997) Chimeric analysis of *fibroblast growth factor receptor-1 (Fgfr1)* function: a role for FGFR1 in morphogenetic movement through the primitive streak. *Development*, **124**, 2829.

63. Arman, E., Haffner-Krausz, R., Chen, Y., Heath, J. K., and Lonai, P. (1998) Targeted disruption of fibroblast growth factor (FGF) receptor 2 suggests a role for FGF signaling in pregastrulation mammalian development. *Proc. Natl Acad. Sci. USA*, **95**, 5082.

64. Xu, X., Weinstein, M., Li, C., Naski, M., Cohen, R. I., Ornitz, D. M., Leder, P., and Deng, C. (1998) Fibroblast growth factor receptor 2 (FGFR2)-mediated reciprocal regulation loop between FGF8 and FGF10 is essential for limb induction. *Development*, **125**, 753.

65. De Moerlooze, L., Spencer-Dene, B., Revest, J.-M., Hajihosseini, M., Rosewell, I., and Dickson, C. (2000) An important role for the IIIb isoform of fibroblast growth factor receptor 2 (FGFR2) in mesenchymal-epithelial signalling during mouse organogenesis. *Development*, **127**, 483.

66. Arman, E., Haffner-Krausz, R., Gorivodsky, M., and Lonai, P. (1999) Fgfr2 is required for limb outgrowth and lung-branching morphogenesis. *Proc. Natl Acad. Sci. USA*, **96**, 11895.

67. Sekine, K., Ohuchi, H., Fujiwara, M., Yamasaki, M., Yoshizawa, T., Sato, T., Yagishita, N., Matsui, D., Koga, Y., Itoh, N., and Kato, S. (1999) Fgf10 is essential for limb and lung formation. *Nature Genet.*, **21**, 138.

68. Herrmann, B. G., Labeit, S., Poustka, A., King, T. R., and Lehrach, H. (1990) Cloning of the T gene required in mesoderm formation in the mouse. *Nature*, **343**, 617.

69. Smith, J. (1999) T-box genes. *Trends Genet.*, **15**, 154.

70. Wilkinson, D. G., Bhatt, S., and Herrmann, B. G. (1990) Expression pattern of the mouse T gene and its role in mesoderm formation. *Nature*, **343**, 657.

71. Cunliffe, V. and Smith, J. C. (1992) Ectopic mesoderm formation in *Xenopus* embryos caused by widespread expression of a *Brachyury* homologue. *Nature*, **358**, 427.

72. Scott, D., Kispert, A., and Herrmann, B. G. (1993) Rescue of the tail defect of *Brachyury* mice. *Genes Dev.*, **7**, 197.

73. Chapman, D. L., Agulnik, I., Hancock, S., Silver, L. M., and Papaioannou, V. E. (1996) *Tbx6*, a mouse T-box gene implicated in paraxial mesoderm formation at gastrulation. *Dev. Biol.*, **180**, 534.

74. Chapman, D. L. and Papaioannou, V. E. (1998) Three neural tubes in mouse embryos with mutations in the T-box gene *Tbx6*. *Nature*, **391**, 695.

75. Yamaguchi, T. P., Takada, S., Yoshikawa, Y., Wu, N., and McMahon, A. P. (1999) T (Brachyury) is a direct target of Wnt3a during paraxial mesoderm specification. *Genes Dev.*, **13**, 3185.

76. Galceran, J., Farinas, I., Depew, M. J., Clevers, H., and Grosschedl, R. (1999) $Wnt3a^{-/-}$-like phenotype and limb deficiency in $Lef1^{-/-}Tcf1^{-/-}$ mice. *Genes Dev.*, **13**, 709.

77. Palmeirim, I., Henrique, D., Ish-Horowicz, D., and Pourquié, O. (1997) Avian hairy gene expression identifies a molecular clock linked to vertebrate segmentation and somitogenesis. *Cell*, **91**, 639.

78. Müller, M., v. Weizsäcker, E., and Campos-Ortega, J. A. (1996) Expression domains of a zebrafish homologue of the *Drosophila* pair-rule gene *hairy* correspond to primordia of alternating somites. *Development*, **122**, 2071.

79. Sawada, A., Fritz, A., Jiang, Y., Yamamoto, A., Yamasu, K., Kuroiwa, A., Saga, Y., and Takeda, H. (2000) Zebrafish Mesp family genes, *mesp-a* and *mesp-b* are segmentally expressed in the presomitic mesoderm, and Mesp-b confers the anterior identity to the developing somites. *Development*, **127**, 1691.

80. Jouve, C., Palmeirim, I., Henrique, D., Beckers, J., Gossler, A., Ish-Horowicz, D., and Pourquié, O. (2000) Notch signalling is required for cyclic expression of the hairy-like gene *HES1* in the presomitic mesoderm. *Development*, **127**, 1421.

81. Greenwald, I. and Rubin, G. M. (1992) Making a difference: the role of cell–cell interactions in establishing separate identities for equivalent cells. *Cell*, **68**, 271.

82. Conlon, R. A., Reaume, A. G., and Rossant, J. (1995) *Notch1* is required for the coordinate segmentation of somites. *Development*, **121**, 1533.

83. Swiatek, P. J., Lindsell, C. E., del Amo, F. F., Weinmaster, G., and Gridley, T. (1994) Notch1 is essential for postimplantation development in mice. *Genes Dev.*, **8**, 707.

84. Hrabe de Angelis, M., McIntyre II, J., and Gossler, A. (1997) Maintenance of somite borders in mice requires the *Delta* homologue Dll1. *Nature*, **386**, 717.

85. Kusumi, K., Sun, E. S., Kerrebrock, A. W., Bronson, R. T., Chi, D.-C., Bulotsky, M. S., Spencer, J. B., Birren, B. W., Frankel, W. N., and Lander, E. S. (1998) The mouse pudgy mutation disrupts *Delta* homologue *Dll3* and initiation of early somite boundaries. *Nature Genet.*, **19**, 274.

86. Oka, C., Nakano, T., Wakeham, A., de la Pompa, J. L., Mori, C., Sakai, T., Okazaki, S., Kawaichi, M., Shiota, K., Mak, T. W., and Honjo, T. (1995) Disruption of the mouse RBP-Jκ gene results in early embryonic death. *Development*, **121**, 3291.

87. Reaume, A. G., Conlon, R. A., Zirngibl, R., Yamaguchi, T. P., and Rossant, J. (1992) Expression analysis of a *Notch* homologue in the mouse. *Dev. Biol.*, **154**, 377.

88. Bettenhausen, B., Hrabe de Angelis, M., Simon, D., Guenet, J. L., and Gossler, A. (1995) Transient and restricted expression during mouse embryogenesis of Dll1, a murine gene closely related to *Drosophila* Delta. *Development*, **121**, 2407.

89. Dunwoodie, S. L., Henrique, D., Harrison, S. M., and Beddington, R. S. P. (1997) Mouse *Dll3*: a novel divergent *Delta* gene which may complement the function of other *Delta* homologues during early pattern formation in the mouse embryo. *Development*, **124**, 3065.

90. Candia, A. F., Hu, J., Crosby, J., Lalley, P. A., Noden, D., Nadeau, J. H., and Wright, C. V. E. (1992) *Mox-1* and *Mox-2* define a novel homeobox gene subfamily and are differentially expressed during early mesodermal patterning in mouse embryos. *Development*, **116**, 1123.

91. del Barco Barrantes, I., Elia, A. J., Wünsch, K., Hrabe de Angelis, M., Mak, T. W., Rossant, J., Conlon, R. A., Gossler, A., and Luis de la Pompa, J. (1999) Interaction between Notch signalling and Lunatic fringe during somite boundary formation in the mouse. *Curr. Biol.*, **9**, 470.

92. Shen, J., Bronson, R. T., Chen, D. F., Xia, W., Selkoe, D. J., and Tonegawa, S. (1997) Skeletal and CNS defects in Presenilin-1-deficient mice. *Cell*, **89**, 629.

93. Wong, P. C., Zheng, H., Chen, H., Becher, M. W., Sirinathsinghji, D. J. S., Trumbauer, M. E., Chen, H. Y., Price, D. L., Van der Ploeg, L. H. T., and Sisodia, S. S. (1997) Presenilin 1 is required for *Notch1* and *Dll1* expression in the paraxial mesoderm. *Nature*, **387**, 288.

94. Pourquié, O. (2000) Vertebrate segmentation: is cycling the rule? *Curr. Opin. Cell Biol.*, **12**, 747.

95. Irvine, K. D. and Wieschaus, E. (1994) *fringe*, a boundary-specific signaling molecule, mediates interactions between dorsal and ventral cells during *Drosophila* wing development. *Cell*, **79**, 595.

96. Panin, V. M., Papayannopoulos, V., Wilson, R., and Irvine, K. D. (1997) Fringe modulates Notch-ligand interactions. *Nature*, **387**, 908.

97. Evrard, Y. A., Lun, Y., Aulehla, A., Gan, L., and Johnson, R. L. (1998) *lunatic fringe* is an essential mediator of somite segmentation and patterning. *Nature*, **394**, 377.

98. Zhang, N. and Gridley, T. (1998) Defects in somite formation in *lunatic fringe*-deficient mice. *Nature*, **394**, 374.

99. Orioli, D. and Klein, R. (1997) The Eph receptor family: axonal guidance by contact repulsion. *Trends Genet.*, **13**, 354.

100. Durbin, L., Brennan, C., Shiomi, K., Cooke, J., Barrios, A., Shanmugalingam, S., Guthrie, B., Lindberg, R., and Holder, N. (1998) Eph signaling is required for segmentation and differentiation of the somites. *Genes Dev.*, **12**, 3096.

101. Nieto, M. A., Gilardi-Hebenstreit, P., Charnay, P., and Wilkinson, D. G. (1992) A receptor protein tyrosine kinase implicated in the segmental patterning of the hindbrain and mesoderm. *Development*, **116**, 1137.

102. Dottori, M., Hartley, L., Galea, M., Paxinos, G., Polizzotto, M., Kilpatrick, T., Bartlett, P. F., Murphy, M., Köntgen, F., and Boyd, A. W. (1998) EphA4 (Sek1) receptor tyrosine kinase is required for the development of the corticospinal tract. *Proc. Natl Acad. Sci. USA*, **95**, 13248.

103. Helmbacher, F., Schneider-Maunoury, S., Topilko, P., Tiret, L., and Charnay, P. (2000) Targeting of the EphA4 tyrosine kinase receptor affects dorsal/ventral pathfinding of limb motor axons. *Development*, **127**, 3313.

104. Saga, Y., Hata, N., Kobayashi, S., Magnuson, T., Seldin, M. F., and Taketo, M. M. (1996) MesP1: a novel basic helix–loop–helix protein expressed in the nascent mesodermal cells during mouse gastrulation. *Development*, **122**, 2769.

105. Saga, Y., Hata, N., Koseki, H., and Taketo, M. M. (1997) Mesp2: a novel mouse gene expressed in the presegmented mesoderm and essential for segmentation initiation. *Genes Dev.*, **11**, 1827.

106. Saga, Y. (1998) Genetic rescue of segmentation defect in MesP2-deficient mice by *MesP1* gene replacement. *Mech. Dev.*, **75**, 53.

107. Burgess, R., Cserjesi, P., Ligon, K. L., and Olson, E. N. (1995) A basic helix–loop–helix protein expressed in paraxial mesoderm and developing somites. *Dev. Biol.*, **168**, 296.

108. Rawls, A., Wilson-Rawls, J., and Olson, E. N. (2000) Genetic regulation of somite formation. *Curr. Top. Dev. Biol.*, **47**, 131.

109. Burgess, R., Rawls, A., Brown, D., Bradley, A., and Olson, E. N. (1996) Requirement of the *paraxis* gene for somite formation and musculoskeletal patterning. *Nature*, **384**, 570.

110. Tam, P. P. L., Goldman, D., Camus, A., and Schoenwolf, G. C. (2000) Early events of somitogenesis in higher vertebrates: allocation of precursor cells during gastrulation and the organization of a meristic pattern in the paraxial mesoderm. *Curr. Top. Dev. Biol.*, **47**, 1.

111. Johnson, J., Rhee, J., Parsons, S. M., Brown, D., Olson, E. N., and Rawls, A. (2001) The anterior/posterior polarity of somites is disrupted in paraxis-deficient mice. *Dev. Biol.*, **229**, 176.

112. Duband, J.-L., Dufour, S., Hatta, K., Takeichi, M., Edelman, G. M., and Thiery, J. P. (1987) Adhesion molecules during somitogenesis in the avian embryo. *J. Cell Biol.*, **104**, 1361.

113. Kimura, Y., Matsunami, H., Inoue, T., Shimamura, K., Uchida, N., Ueno, T., Miyazaki, T., and Takeichi, M. (1995) Cadherin-11 expressed in association with mesenchymal morphogenesis in the head, somite, and limb bud of early mouse embryos. *Dev. Biol.*, **169**, 347.

114. Cremer, H., Lange, R., Christoph, A., Plomann, M., Vopper, G., Roes, J., Brown, R., Baldwin, S., Kraemer, P., Scheff, S., Barthels, D., Rajewsky, K., and Wille, W. (1994) Inactivation of the N-CAM gene in mice results in size reduction of the olfactory bulb and deficits in spatial learning. *Nature*, **367**, 455.

115. Rabinowitz, J. E., Rutishauser, U., and Magnuson, T. (1996) Targeted mutation of *Ncam* to produce a secreted molecule results in a dominant embryonic lethality. *Proc. Natl Acad. Sci. USA*, **93**, 6421.

116. Radice, G. L., Rayburn, H., Matsunami, H., Knudsen, K. A., Takeichi, M., and Hynes, R. O. (1997) Developmental defects in mouse embryos lacking N-cadherin. *Dev. Biol.*, **181**, 64.

117. Manabe, T., Togashi, H., Uchida, N., Suzuki, S. C., Hayakawa, Y., Yamamoto, M., Yoda, H., Miyakawa, T., Takeichi, M., and Chisaka, O. (2000) Loss of cadherin-11 adhesion receptor enhances plastic changes in hippocampal synapses and modifies behavioral responses. *Mol. Cell. Neurosci.*, **15**, 534.

118. Hynes, R. O. (1999) Cell adhesion: old and new questions. *Trends Cell Biol.*, **9**, M33.

119. Dziadek, M. and Timpl, R. (1985) Expression of nidogen and laminin in basement membranes during mouse embryogenesis and in teratocarcinoma cells. *Dev. Biol.*, **111**, 372.

120. Duband, J.-L. and Thiery, J. P. (1987) Distribution of laminin and collagens during avian neural crest development. *Development*, **101**, 461.

121. Ostrovsky, D., Cheney, C. M., Seitz, A. W., and Lash, J. W. (1983) Fibronectin distribution during somitogenesis in the chick embryo. *Cell Differ.*, **13**, 217.

122. Lash, J. W., Seitz, A. W., Cheney, C. M., and Ostrovsky, D. (1984) On the role of fibronectin during the compaction stage of somitogenesis in the chick embryo. *J. Exp. Zool.*, **232**, 197.

123. George, E. L., Georges-Labouesse, E. N., Patel-King, R. S., Rayburn, H., and Hynes, R. O. (1993) Defects in mesoderm, neural tube and vascular development in mouse embryos lacking fibronectin. *Development*, **119**, 1079.

124. Paulsson, M. (1996) Biosynthesis, tissue distribution and isolation of laminins. In *The laminins* (ed. P. Ekblom and R. Timpl), p. 217. Harwood Academic Publishers, Amsterdam.

125. Noakes, P. G., Gautam, M., Mudd, J., Sanes, J. R., and Merlie, J. P. (1995) Aberrant differentiation of neuromuscular junctions in mice lacking s-laminin/laminin beta2. *Nature*, **374**, 258.

126. Helbling-Leclerc, A., Zhang, X., Topaloglu, H., Cruaud, C., Tesson, F., Weissenbach, J., Tome, F. M., Schwartz, K., Fardeau, M., and Tryggvason, K. (1995) Mutations in the laminin alpha2-chain gene (LAMA2) cause merosin-deficient congenital muscular dystrophy. *Nature Genet.*, **11**, 216.

127. Shim, C., Kwon, H. B., and Kim, K. (1996) Differential expression of laminin chain-specific mRNA transcripts during mouse preimplantation embryo development. *Mol. Reprod. Dev.*, **44**, 44.

128. Smyth, N., Vatansever, H. S., Murray, P., Meyer, M., Frie, C., Paulsson, M., and Edgar, D. (1999) Absence of basement membranes after targeting the *LAMC1* gene results in embryonic lethality due to failure of endoderm differentiation. *J. Cell Biol.*, **144**, 151.

129. Fässler, R. and Meyer, M. (1995) Consequences of lack of β1 integrin gene expression in mice. *Genes Dev.*, **9**, 1869.

130. Stephens, L. E., Sutherland, A. E., Klimanskaya, I. V., Andrieux, A., Meneses, J., Pedersen, R. A., and Damsky, C. H. (1995) Deletion of β1 integrins in mice results in inner cell mass failure and peri-implantation lethality. *Genes Dev.*, **9**, 1883.

131. Hynes, R. O. (1996) Targeted mutations in cell adhesion genes: what have we learned from them? *Dev. Biol.*, **180**, 402.

132. Fan, C.-M. and Tessier-Lavigne, M. (1994) Patterning of mammalian somites by surface ectoderm and notochord: evidence for sclerotome induction by a hedgehog homolog. *Cell*, **79**, 1175.

133. Johnson, R. L., Laufer, E., Riddle, R. D., and Tabin, C. (1994) Ectopic expression of *Sonic hedgehog* alters dorsal–ventral patterning of somites. *Cell*, **79**, 1165.

134. Fan, C.-M., Porter, J. A., Chiang, C., Chang, D. T., Beachy, P. A., and Tessier-Lavigne, M. (1995) Long-range sclerotome induction by sonic hedgehog: direct role of the amino-terminal cleavage product and modulation by the cyclic AMP signaling pathway. *Cell*, **81**, 457.

135. Goulding, M., Lumsden, A., and Paquette, A. J. (1994) Regulation of *Pax-3* expression in the dermomyotome and its role in muscle development. *Development*, **120**, 957.

136. Monsoro-Burq, A.-H., Bontoux, M., Teillet, M.-A., and Le Douarin, N. M. (1994) Heterogeneity in the development of the vertebra. *Proc. Natl Acad. Sci. USA*, **91**, 10435.

137. Echelard, Y., Epstein, D. J., St-Jacques, B., Shen, L., Mohler, J., McMahon, J. A., and McMahon, A. P. (1993) *Sonic hedgehog*, a member of a family of putative signaling molecules, is implicated in the regulation of CNS polarity. *Cell*, **75**, 1417.

138. Riddle, R. D., Johnson, R. L., Laufer, E., and Tabin, C. (1993) *Sonic hedgehog* mediates the polarizing activity of the ZPA. *Cell*, **75**, 1401.

139. Roelink, H., Augsburger, A., Heemskerk, J., Korzh, V., Norlin, S., Ruiz i Altaba, A., Tanabe, Y., Placzek, M., Edlund, T., Jessell, T. M., and Dodd, J. (1994) Floor plate and motor neuron induction by *vhh-1*, a vertebrate homolog of *hedgehog* expressed by the notochord. *Cell*, **76**, 761.

140. Hammerschmidt, M., Brook, A., and McMahon, A. P. (1997) The world according to *hedgehog*. *Trends Genet.*, **13**, 14.

141. Porter, J. A., von Kessler, D. P., Ekker, S. C., Young, K. E., Lee, J. J., Moses, K., and Beachy, P. A. (1995) The product of *hedgehog* autoproteolytic cleavage active in local and long-range signalling. *Nature*, **374**, 363.

142. Roelink, H., Porter, J. A., Chiang, C., Tanabe, Y., Chang, D. T., Beachy, P. A., and Jessell, T. M. (1995) Floor plate and motor neuron induction by different concentrations of the amino-terminal cleavage product of sonic hedgehog autoproteolysis. *Cell*, **81**, 445.

143. Gruss, P. and Walther, C. (1992) Pax in development. *Cell*, **69**, 719.

144. Deutsch, U., Dressler, G. R., and Gruss, P. (1988) Pax 1, a member of a paired box homologous murine gene family, is expressed in segmented structures during development. *Cell*, **53**, 617.

145. Wallin, J., Wilting, J., Koseki, H., Fritsch, R., Christ, B., and Balling, R. (1994) The role of Pax-1 in axial skeleton development. *Development*, **120**, 1109.

146. Neubüser, A., Koseki, H., and Balling, R. (1995) Characterisation and developmental expression of *Pax9*, a paired-box-containing gene related to *Pax1*. *Dev. Biol.*, **170**, 701.

147. Williams, B. and Ordahl, C. P. (1994) *Pax-3* expression in segmental mesoderm marks early stages in myogenic cell specification. *Development*, **120**, 785.

148. Jostes, B., Walther, C., and Gruss, P. (1991) The murine paired box gene *Pax7* is expressed specifically during the development of the nervous and muscular system. *Mech. Dev.*, **33**, 27.

149. Ebensperger, C., Wilting, J., Brand-Saberi, B., Mizutani, Y., Christ, B., Balling, R., and Koseki, H. (1995) *Pax-1*, a regulator of sclerotome development is induced by notochord and floor plate signals in avian embryos. *Anat. Embryol.*, **191**, 297.

150. Chiang, C., Litingtung, Y., Lee, E., Young, K. E., Corden, J. L., Westphal, H., and Beachy, P. A. (1996) Cyclopia and defective axial patterning in mice lacking *Sonic hedgehog* gene function. *Nature*, **383**, 407.

151. Kaehn, K., Jacob, H. J., Christ, B., Hinrichsen, K., and Poelmann, R. E. (1988) The onset of myotome formation in the chick. *Anat. Embryol.*, **177**, 191.

152. Cossu, G., Tajbakhsh, S., and Buckingham, M. (1996) How is myogenesis initiated in the embryo? *Trends Genet.*, **12**, 218.

153. Currie, P. D. and Ingham, P. W. (1998) The generation and interpretation of positional information within the vertebrate myotome. *Mech. Dev.*, **73**, 3.

154. Cinnamon, Y., Kahane, N., and Kalcheim, C. (1999) Characterization of the early development of specific hypaxial muscles from the ventrolateral myotome. *Development*, **126**, 4305.

155. Cinnamon, Y., Kahane, N., and Kalcheim, C. (2001) The sub-lip domain—a distinct pathway for myotome precursors that demonstrate rostral–caudal migration. *Development*, **128**, 341.

156. Kalcheim, C., Cinnamon, Y., and Kahane, N. (1999) Myotome formation: a multistage process. *Cell Tissue Res.*, **296**, 161.

157. Pownall, M. E. and Emerson, C. P. (1992) Sequential activation of 3 myogenic regulatory genes during somite morphogenesis in quail embryos. *Dev. Biol.*, **151**, 67.

158. Bober, E., Franz, T., Arnold, H.-H., Gruss, P., and Tremlay, P. (1994) Pax-3 is required for the development of limb muscles: a possible role for the migration of dermomyotomal muscle progenitor cells. *Development*, **120**, 603.

159. Rudnicki, M. A., Schnegelsberg, P. N. J., Stead, R. H., Braun, T., Arnold, H.-H., and Jaenisch, R. (1993) MyoD or Myf-5 is required for the formation of skeletal muscle. *Cell*, **75**, 1351.

160. Maroto, M., Rehef, R., Münsterberg, A. E., Koester, S., Goulding, M., and Lassar, A. B. (1997) Ectopic *Pax-3* activates *MyoD* and *Myf-5* expression in embryonic mesoderm and neural tissue. *Cell*, **89**, 127.

161. Tajbakhsh, S., Rocancourt, D., Cossu, G., and Buckingham, M. (1997) Redefining the genetic hierarchies controlling skeletal myogenesis: *Pax-3* and *Myf-5* act upstream of *MyoD*. *Cell*, **89**, 127.

162. Hebrok, M., Wertz, K., and Füchtbauer, E. M. (1994) M-twist is an inhibitor of muscle differentiation. *Dev. Biol.*, **165**, 537.

163. Chen, Z. F. and Behringer, R. R. (1995) Twist is required in head mesenchyme for cranial neural tube morphogenesis. *Genes Dev.*, **9**, 686.

164. Avery, G., Chow, M., and Holtzer, H. (1956) An experimental analysis of the development of the spinal column. *J. Exp. Zool.*, **132**, 409.

165. Kenny-Mobbs, T. and Thorogood, P. (1987) Autonomy of differentiation in avian brachial somites and the influence of adjacent tissue. *Development*, **100**, 449.

166. Stern, H. M. and Hauschka, S. D. (1995) Neural tube and notochord promote *in vitro* myogenesis in single somite explants. *Dev. Biol.*, **167**, 87.

167. Cossu, G., Kelly, R., Tajbakhsh, S., Di Donna, S., Vivarelli, E., and Buckingham, M. (1996) Activation of different myogenic pathways: myf-5 is induced by the neural tube and MyoD by the dorsal ectoderm in mouse paraxial mesoderm. *Development*, **122**, 429.

168. Münsterberg, A. E. and Lassar, A. B. (1995) Combinatorial signals from the neural tube, floor plate and notochord induce myogenic bHLH gene expression in the somite. *Development*, **121**, 651.

169. Spence, M. S., Yip, J., and Erickson, C. A. (1996) The dorsal neural tube organizes the dermamyotome and induces axial myocytes in the avian embryo. *Development*, **122**, 231.

170. Rong, P. M., Teillet, M.-A., Ziller, C., and Le Douarin, N. (1992) The neural tube/ notochord complex is necessary for vertebral but not limb body wall striated muscle differentiation. *Development*, **115**, 657.

171. Spörle, R., Günther, T., Struwe, M., and Schughart, K. (1996) Severe defects in the formation of epaxial musculature in *open brain* (*opb*) mutant mouse embryos. *Development*, **122**, 79.

172. Pourquié, O., Coltey, M., Bréant, C., and Le Douarin, N. M. (1995) Control of somite patterning by signals from the lateral plate. *Proc. Natl Acad. Sci. USA*, **92**, 3219.

173. Hammerschmidt, M., Serbedzija, G. N., and McMahon, A. P. (1996) Genetic analysis of dorsoventral pattern formation in the zebrafish: requirement of a BMP-like ventralizing activity and its dorsal repressor. *Genes Dev.*, **10**, 2452.

174. Münsterberg, A. E., Kitajewski, J., Bumcrot, D. A., McMahon, A. P., and Lassar, A. B. (1995) Combinatorial signaling by Sonic hedgehog and Wnt family members induces myogenic bHLH gene expression in the somite. *Genes Dev.*, **9**, 2911.

175. Stern, H. M., Brown, A. M. C., and Hauschka, S. D. (1995) Myogenesis in paraxial mesoderm: preferential induction by dorsal neural tube and by cells expressing *Wnt-1*. *Development*, **121**, 3675.

176. Pownall, M. E., Strunk, K. E., and Emerson, C. P. (1996) Notochord signals control the transcriptional cascade of myogenic bHLH genes in somites of quail embryos. *Development*, **122**, 1475.

177. Pourquié, O., Fan, C.-M., Coltey, M., Hirsinger, E., Watanabe, Y., Bréant, C., Francis-West, P., Brickell, P., Tessier-Lavigne, M., and Le Douarin, N. M. (1996) Lateral and axial signals involved in avian somite patterning: a role for BMP4. *Cell*, **84**, 461.

178. Piccolo, S., Sasai, Y., Lu, B., and DeRobertis, E. M. (1996) Dorsoventral patterning in *Xenopus*: inhibition of ventral signals by direct binding of chordin to BMP-4. *Cell*, **86**, 589.

179. Zimmerman, L. B., De Jesus-Escobar, J. M., and Harland, R. M. (1996) The Spemann organizer signal noggin binds and inactivates bone morphogenetic protein 4. *Cell*, **86**, 599.

180. Hirsinger, E., Duprez, D., Jouve, C., Malapert, P., Cooke, J., and Pourquié, O. (1997) Noggin acts downstream of Wnt and Sonic Hedgehog to antagonize BMP4 in avian somite patterning. *Development*, **124**, 4605.

181. Marcelle, C., Stark, M. R., and Bronner-Fraser, M. (1997) Coordinate actions of BMPs, Wnts, Shh and Noggin mediate patterning of the dorsal somite. *Development*, **124**, 3955.

182. Amthor, H., Connolloy, D., Patel, K., Brand-Saberi, B., Wilkinson, D. G., Cooke, J., and Christ, B. (1996) The expression and regulation of follistatin and a follistatin-like gene during avian somite compartmentalization and myogenesis. *Dev. Biol.*, **178**, 343.

183. Reshef, R., Maroto, M., and Lassar, A. B. (1998) Regulation of dorsal somitic cell fates: BMPs and Noggin control the timing and pattern of myogenic regulator expression. *Genes Dev.*, **12**, 290.

184. Grass, S., Arnold, H.-H., and Braun, T. (1996) Alterations in somite patterning of myf-5-deficient mice—a possible role for FGF-4 and FGF-6. *Development*, **122**, 141.

185. Grothe, C., Brand-Saberi, B., Wilting, J., and Christ, B. (1996) Fibroblast growth-factor receptor-1 in skeletal and heart-muscle cells—expression during early avian development and regulation after notochord transplantation. *Dev. Dyn.*, **206**, 310.

186. Brunetti, A. and Goldfine, I. D. (1990) Role of myogenin in myoblast differentiation and its regulation by fibroblast growth factor. *J. Biol. Chem.*, **265**, 5960.

187. Haub, O. and Goldfarb, M. (1991) Expression of the fibroblast growth factor-5 gene in the mouse embryo. *Development*, **112**, 397.

188. Brill, G., Kahane, N., Carmeli, C., von Schack, D., Barde, Y. A., and Kalcheim, C. (1995) Epithelial–mesenchymal conversion of dermatome progenitors requires neural tube-derived signals: characterization of the role of Neurotrophin-3. *Development*, **121**, 2583.

189. Norris, W. E., Stern, C. D., and Keynes, R. J. (1989) Molecular differences between the rostral and caudal halves of the sclerotome in the chick embryo. *Development*, **105**, 541.

190. Stern, C. D., Jacques, K. F., Lim, T.-M., Fraser, S. E., and Keynes, R. J. (1991) Segmental lineage restrictions in the chick embryo spinal cord depend on the adjacent somites. *Development*, **113**, 239.

191. Rickmann, M., Fawcett, L. W., and Keynes, R. J. (1985) The migration of neural crest cells and the growth of motor axons through the rostral half of the chick somite. *J. Embryol. Exp. Morphol.*, **90**, 437.

192. Teillet, M.-A., Kalcheim, C., and Le Douarin, N. M. (1987) Formation of the dorsal root ganglion in the avian embryo: segmental origin and migratory behavior of neural crest progenitor cells. *Dev. Biol.*, **120**, 329.

193. Bronner-Fraser, M. and Stern, C. (1991) Effects of mesodermal tissues on avian neural crest cell migration. *Dev. Biol.*, **143**, 213.

194. Wehrle, B. and Chiquet, M. (1990) Tenascin is accumulated along developing peripheral nerves and allows neurite outgrowth *in vitro*. *Development*, **110**, 401.

195. Ingham, P. W. (1995) Signalling by hedgehog family proteins in *Drosophila* and vertebrate development. *Curr. Biol.*, **5**, 492.

196. Durbin, L., Sordino, P., Barrios, A., Gering, M., Thisse, C., Thisse, B., Brennan, C., Green, A., Wilson, S., and Holder, N. (2000) Anteroposterior patterning is required within segments for somite boundary formation in developing zebrafish. *Development*, **127**, 1703.

197. Kessel, M. (1991) Molecular coding of axial positions by *Hox* genes. *Semin. Dev. Biol.*, **2**, 367.

198. Burke, A. C. (2000) *Hox* genes and the global patterning of the somitic mesoderm. *Curr. Top. Dev. Biol.*, **47**, 155.

199. Krumlauf, R. (1994) *Hox* genes in vertebrate development. *Cell*, **78**, 191.

200. McGinnis, W. and Krumlauf, R. (1992) Homeobox genes and axial patterning. *Cell*, **68**, 283.

201. Hofmann, C. and Eichele, G. (1994) Retinoids in development. In *The retinoids: biology, chemistry and medicine* (ed. M. B. Sporn, A. B. Roberts, and D. S. Goodman), p. 387. Raven Press, New York.

202. Tickle, C., Alberts, B. M., Wolpert, L., and Lee, J. (1982) Local application of retinoic acid to the limb bud mimics the action of the polarizing region. *Nature*, **296**, 564.

203. Tickle, C. (1991) Retinoic acid and chick limb bud development. *Development* (Suppl. 1), 113.

204. Boncinelli, E., Simeone, A., Acampora, D., and Mavilio, F. (1991) HOX gene activation by retinoic acid. *Trends Genet.*, **7**, 329.

205. Holland, P. W. H. and Hogan, B. L. M. (1988) Expression of homeobox genes during mouse development: a review. *Genes Dev.*, **2**, 773.

206. Kessel, M. and Gruss, P. (1991) Homeotic transformations of murine vertebrae and concomitant alteration of Hox codes induced by retinoic acid. *Cell*, **67**, 89.

207. Lufkin, T., Mark, M., Hart, C. P., Dollé, P., LeMeur, M., and Chambon, P. (1992) Homeotic transformation of the occipital bones of the skull by ectopic expression of a homeobox gene. *Nature*, **359**, 835.

208. Kessel, M., Balling, R., and Gruss, P. (1990) Variations of cervical vertebrae after expression of a *Hox-1.1* transgene in mice. *Cell*, **61**, 301.

209. Ramirez-Solis, R., Zheng, H., Whiting, J., Krumlauf, R., and Bradley, A. (1993) *Hoxb-4* (*Hox-2.6*) mutant mice show homeotic transformation of a cervical vertebra and defects in the closure of the sternal rudiments. *Cell*, **73**, 279.

210. Condie, B. G. and Capecchi, M. R. (1993) Mice homozygous for a targeted disruption of *Hoxd-3* (*Hox-4.1*) exhibit anterior transformations of the first and second cervical vertebrae, the atlas and axis. *Development*, **119**, 579.

211. Dollé, P., Dierich, A., LeMeur, M., Schimmang, T., Schuhbaur, B., Chambon, P., and Duboule, D. (1993) Disruption of the *Hoxd-13* gene induces localized heterochrony leading to mice with neotenic limbs. *Cell*, **75**, 431.

212. Davis, A. P., Witte, D. P., Hsieh-Li, H. M., Potter, S. S., and Capecchi, M. R. (1995) Absence of radius and ulna in mice lacking *hoxa-11* and *hoxd-11*. *Nature*, **375**, 791.

213. Horan, G. S. B., Kovacs, E. N., Behringer, R. R., and Featherstone, M. S. (1995) Mutations in paralogous *Hox* genes result in overlapping homeotic transformations of the axial skeleton evidence for unique and redundant function. *Dev. Biol.*, **169**, 359.

214. Wahba, G. M., Hostikka, S. L., and Carpenter, E. M. (2001) The paralogous Hox genes Hoxa10 and Hoxd10 interact to pattern the mouse hindlimb peripheral nervous system and skeleton. *Dev. Biol.*, **231**, 87.

215. Suemori, H. and Noguchi, S. (2000) Hox C cluster genes are dispensable for overall body plan of mouse embryonic development. *Dev. Biol.*, **220**, 333.

216. Medina-Martinez, O., Bradley, A., and Ramirez-Solis, R. (2000) A large targeted deletion of *Hoxb1–Hoxb9* produces a series of single-segment anterior homeotic transformations. *Dev. Biol.*, **222**, 71.

217. Stuart, J. J., Brown, S. J., Beeman, R. W., and Denell, R. E. (1991) A deficiency of the homeotic complex in the beetle *Tribolium*. *Nature*, **350**, 72.

218. Breier, G., Bucan, M., Francke, U., Colberg-Poley, A. M., and Gruss, P. (1986) Sequential expression of murine homeobox genes during F9 EC cell differentiation. *EMBO J.*, **5**, 2209.

219. Simeone, A., Acampora, D., Arcioni, L., Andrews, P. W., Boncinelli, E., and Mavilio, F. (1990) Sequential activation of *HOX 2* homeobox genes by retinoic acid in human embryonal carcinoma cells. *Nature*, **346**, 763.

220. Simeone, A., Acampora, D., Nigro, V., Faiella, A., D'Esposito, M., Stornaiuolo, A., Mavilio, F., and Boncinelli, E. (1991) Differential regulation by retinoic acid of the homeobox genes of the four HOX loci in human embryonal carcinoma cells. *Mech. Dev.*, **33**, 215.

221. Kessel, M. (1992) Respecification of vertebral identities by retinoic acid. *Development*, **115**, 487.

222. Bienz, M. and Müller, J. (1995) Transcriptional silencing of homeotic genes in *Drosophila*. *BioEssays*, **17**, 775.

223. Schumacher, A. and Magnuson, T. (1997) Murine *Polycomb-* and *trithorax*-group genes regulate homeotic pathways and beyond. *Trends Genet.*, **13**, 167.

224. van der Lugt, N. M. T., Domen, J., Linders, K., van Roon, M., Robanus-Maandag, E., te Riele, H., van der Valk, M., Deschamps, J., Sofroniew, M., van Lohuizen, M., and Berns, A. (1994) Posterior transformation, neurological abnormalities, and severe hematopoietic defects in mice with a targeted deletion of the *bmi-1* proto-oncogene. *Genes Dev.*, **8**, 757.

225. Akasaka, T., Kanno, M., Balling, R., Mieza, M. A., Taniguchi, M., and Koseki, H. (1996) A role for *mel-18*, a Polycomb group-related vertebrate gene, during the anteroposterior specification of the axial skeleton. *Development*, **122**, 1513.

226. Alkema, M. J., van der Lugt, N. M. T., Bobeldijk, R. C., Berns, A., and van Lohuizen, M. (1995) Transformation of axial skeleton due to overexpression of *bmi-1* in transgenic mice. *Nature*, **374**, 724.

227. Coré, N., Bel, S., Gaunt, S. J., Aurrand-Lions, M., Pearce, J., Fisher, A., and Djabali, M. (1997) Altered cellular proliferation and mesoderm patterning in Polycomb-M33-deficient mice. *Development*, **124**, 721.

228. Faust, C., Schumacher, A., Holdener, B., and Magnuson, T. (1995) The *eed* mutation disrupts anterior mesoderm production in mice. *Development*, **121**, 273.

229. Schumacher, A., Faust, C., and Magnuson, T. (1996) Positional cloning of a global regulator of anteroposterior patterning in mice. *Nature*, **383**, 250.

230. Yu, B. D., Hess, J. L., Horning, S. E., Brown, G. A. J., and Korsmeyer, S. J. (1995) Altered *Hox* expression and segmental identity in *Mll*-mutant mice. *Nature*, **378**, 505.

231. Hanson, R. D., Hess, J. L., Yu, B. D., Ernst, P., van Lohuizen, M., Berns, A., van der Lugt, N. M. T., Shashikant, C. S., Ruddle, F. H., Seto, M., and Korsmeyer, S. J. (1999)

Mammalian *Trithorax* and *Polycomb*-group homologues are antagonistic regulators of homeotic development. *Proc. Natl Acad. Sci. USA*, **96**, 14372.

232. Vasiliauskas, D. and Stern, C. D. (2001) Patterning the embryonic axis: FGF signaling and how vertebrate embryos measure time. *Cell*, **106**, 133.

233. Dubrulle, J., McGrew, M. J., and Pourquié, O. (2001) FGF signaling controls somite boundary position and regulates segmentation clock control of spatiotemporal *Hox* gene activation. *Cell*, **106**, 219.

234. Zákány, J., Kmita, M., Alarcon, P., de la Pompa, J.-L., and Duboule, D. (2001) Localized and transient transcription of Hox genes suggests a link between patterning and the segmentation clock. *Cell*, **106**, 207.

5 | Vertebrate neurogenesis

KATE G. STOREY

1. Introduction

The adult nervous system contains the greatest diversity of cell types of any tissue in the vertebrate body, and it is this variety which largely underlies the complexity of the thinking, behaving tissue that is the brain. Many of the molecular mechanisms underlying the critical steps leading to the formation of the vertebrate nervous system have now been identified. This process begins in the early embryo when cells become competent to respond to neural inducing signals, and extends on from the formation of neural precursors and the regulated production of neurons and glia within the patterned neural tube to axon outgrowth and guidance, target recognition, and synapse formation (see Chapter 7). These activities take place within a dynamic structure that begins as a flat neural plate, which undergoes a series of morphogenetic movements and defined patterns of cell proliferation that transform it into a neural tube with distinct regional characteristics.

Here we focus on the cellular and molecular mechanisms leading to the generation of neurons. Different vertebrate embryos lend themselves to the investigation of distinct aspects of these processes. The large early amphibian embryo has proved an important system for ectopic expression studies of candidate neural inducing factors, while loss-of-function approaches have been more easily carried out in transgenic mice and in mutants generated by large-scale mutagenesis screens afforded by the zebrafish. Experiments in avian embryos readily combine cellular techniques, for example lineage tracing and the localization of tissue sources of signals, with molecular approaches such as ectopic expression studies mediated by retroviral infection and, more recently, misexpression of genes by electroporation. The identification of the essential mechanisms underlying neural development has thus been facilitated by the investigation of these processes in different vertebrate embryos. Further, many of the genes that control neurogenesis in the fruitfly *Drosophila* and the nematode worm *Caenorhabditis elegans* have turned out to have homologues that are involved in very similar molecular pathways during vertebrate neurogenesis. Indeed, the remarkably high level of structural and functional conservation of genes across species has provided a fast track to the identification of mechanisms controlling the development of the more complex and rather different vertebrate nervous system.

2. Tissue sources of neural inducing signals

Early in the 1920s Hilde Mangold placed the amphibian dorsal blastopore lip into the blastocoel cavity of a host embryo, and demonstrated that this region is a source of signals capable of inducing a new embryonic axis. The ectopic structure included a neural tube derived from ectodermal tissue that would otherwise have formed epidermis (1) (see Chapter 2). This key experiment identified the phenomenon of neural induction and instigated what became a long search for the biochemical basis of this process (see ref. 2). Grafting experiments have shown that cell populations equivalent to the dorsal blastopore lip or organizer region exist in all model vertebrates, and that these tissues are sources of neural inducing signals (Fig. 1) (reviewed in ref. 3) (see Chapter 2). Further, interspecies grafting strongly suggests that these signals are conserved between vertebrates (4, 5). However, neural tissue can form following surgical or genetic (deletion of *HNF3β*) ablation of the organizer (6; reviewed in ref. 7), indicating that it is not required to initiate neural induction. Indeed, recent experiments that exploit the responsiveness of the chick extraembryonic epiblast to neural inducing signals (8) (see Fig. 1) show that neural induction begins prior to gastrulation, and that in the chick these early signals are provided by a small group of 'posterior cells' located at the posterior end of the blastoderm which will later

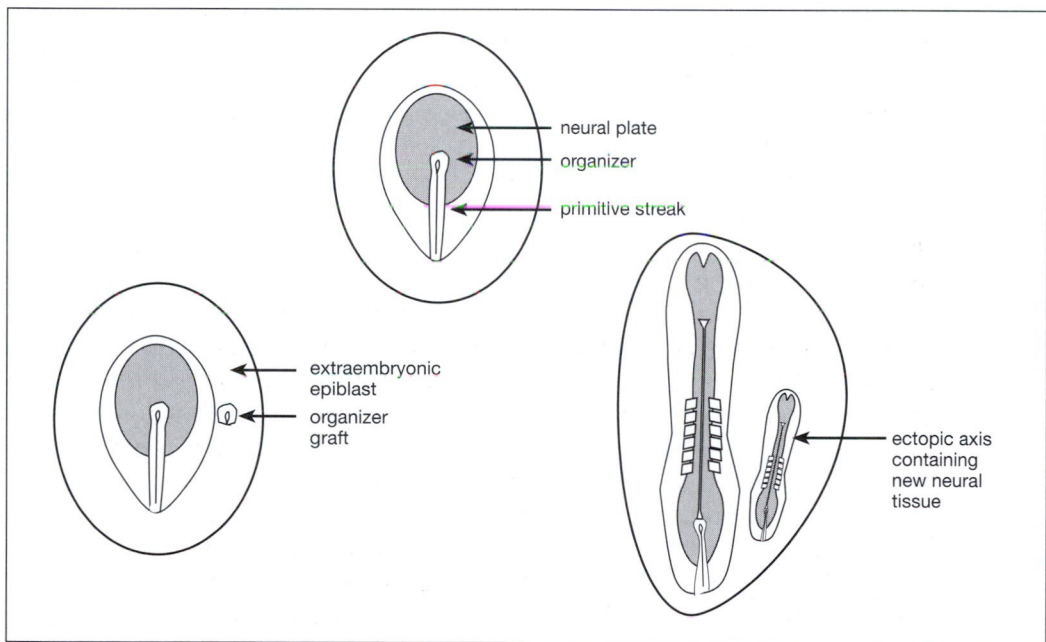

Fig. 1 The organizer is a source of neural inducing signals. Extraembryonic epiblast cells (which normally give rise to extraembryonic membranes) can be induced to form neural tissue in response to grafts of the organizer region (found in amniotes at the anterior tip of the primitive streak). Grafts placed close to the host generate more complete axes by recruiting host mesoderm cells. This chick assay has also been used to demonstrate neural inducing activity in organizer precursor cells located at the posterior margin of the early embryo.

contribute to the organizer (9). Early neural markers are also transiently induced by hypoblast cells that move beneath the early epiblast (10), perhaps priming this tissue for later, more localized, signalling from the developing organizer (see below).

In mammalian embryos, anterior visceral endoderm cells form a separate head organizer whose activity underlies, at least in part, the induction of the rostral (anterior) CNS or forebrain, and equivalent cell populations have now been identified in *Xenopus* spp. and zebrafish (11–13; reviewed in ref. 14) (see Chapter 6). However, a separate head signalling centre does not appear to be present in chick embryos, where the induction of rostral neural tissue is instead associated with signals from the hypoblast (10). These appear to act, in part, by directing cell movements away from the organizer, which provides both neural inducing and caudalizing signals (10, 15).

Signals provided by the mesendoderm underlying the established neural plate in the chick embryo also maintain neural identity (16). Neural plate explants cultured *in vitro* retain a neural character (17, 18). Thus these mesendoderm signals are not required for neural differentiation, and appear to act within the context of the early embryo to counter signals from other tissues that promote epidermal differentiation (16).

As the organizer ages it loses its neural inducing activity, but this is retained by the axial mesendoderm/notochord as it emerges from this region and moves beneath the neural plate, as observed in chick, frog, and mouse studies (19–21). As noted above, neural plate explants (e.g. chick (17, 18) and frog (22, 23)) derived prior to the emergence of axial mesoderm differentiate into neural tissue in isolation, thus it seems likely that neural inducing signals provided by the axial mesoderm (20, 24) serve to reinforce or stabilize earlier signals. There is also evidence in the frog that segmented paraxial mesoderm (somites) are a source of neural inducing signals that could play a similar role. However, this property is not conserved across species, as chick somites do not induce neural tissue (25).

Finally, there is evidence in chick and frog embryos that the differentiating neuro-epithelium itself emits neural inducing signals (see, for example, refs 16, 26). During development such homeogenetic signals may facilitate the spread of neural inducing/ maintenance and regionalizing signals within the prospective neural plate. Thus, signals from a number of different tissues collude in the induction and maintenance of the neural plate. This could reflect roles for many different signalling molecules in the assignment of neural cell fate, but it may also reveal functional redundancies within the embryo that ensure the formation of this essential tissue.

3. Neural competence

Neural induction depends not only on the provision of neural inducing molecules, but also on the competence of receiving cells to respond to such signals. This cell state has been considered as a cell autonomous property, which minimally requires the expression of appropriate receptors (27, 28) and must also depend on the presence of a signal transduction apparatus and the activity of transcription factors. Neural competence has been shown to be temporally and spatially regulated in several vertebrates and, in *Xenopus*, is temporally discrete from competence to respond to mesoderm inducing signals (29; reviewed in refs 28–31).

The molecular basis of neural competence is poorly understood, but has most recently been addressed in chick and frog embryos. In the chick, the ability to respond to organizer-derived neural inducing signals has been shown to correlate with the expression of the cell-surface antigen L5[220] which is progressively restricted to the neural plate (reviewed in ref. 32). L5[220] expression and neural competence in the extraembryonic ectoderm can be maintained by exposure to hepatocyte growth factor (HGF). Importantly, the ectopic application of HGF does not confer neural competence in regions outside the normal L5 domain and therefore cannot initiate the neural induction pathway, but acts only to prolong the responsiveness of cells to organizer-derived signals. In *Xenopus*, the dorsal ectoderm has a greater neural competence than ventral regions (33), a property that may be mediated by several genes that become restricted to the early dorsal ectoderm: these include the alpha isoform of protein kinase C (PKCα (34)) and the translation initiation factor eIF4AII (35). Overexpression of either of these genes increases neural competence as measured by a decreased requirement for neural inducing signals (35). A recent study has further identified a nucleotide exchange factor (*lfc*) that is expressed widely in the early ectoderm and which can upregulate PKCα and neural competence, although it is unclear how *lfc* activity is restricted to the dorsal ectoderm (36). Expression of the zinc-finger transcription factor *Opl* (37) and the HMG-containing factor *Sox2* (38) also begins prior to neural specification. While overexpression of these genes in non-neural ectoderm does not induce neural genes, it can sensitize dorsal ectoderm to neural inducing signals. However, *Sox2* expression in the chick embryo does not correlate with regions of neural competence, casting doubt on a conserved role for this gene (31). In conclusion, a hierarchy of gene activity underlying neural competence is beginning to emerge, and it is now important to establish how these genes interact and the extent to which pathways are conserved across species.

4. Candidate neural inducers and the default model of neural induction

All neural inducing molecules identified to date have been isolated in the frog. This embryo lends itself to the systematic screening of candidate signalling molecules; mRNAs can be ectopically expressed by injection into its large blastomeres and resulting embryos examined for effects on the developing nervous system. Further, since isolated animal-cap ectoderm derived from blastula stage embryos normally differentiates into epidermis, it is thus a useful tissue on which to test candidate neural inducing molecules (Fig. 2) (see also Chapter 3).

As endogenous neural inducing signals are provided by organizer-derived mesoderm, an important criterion for the identification of a neural inducing signal is that it

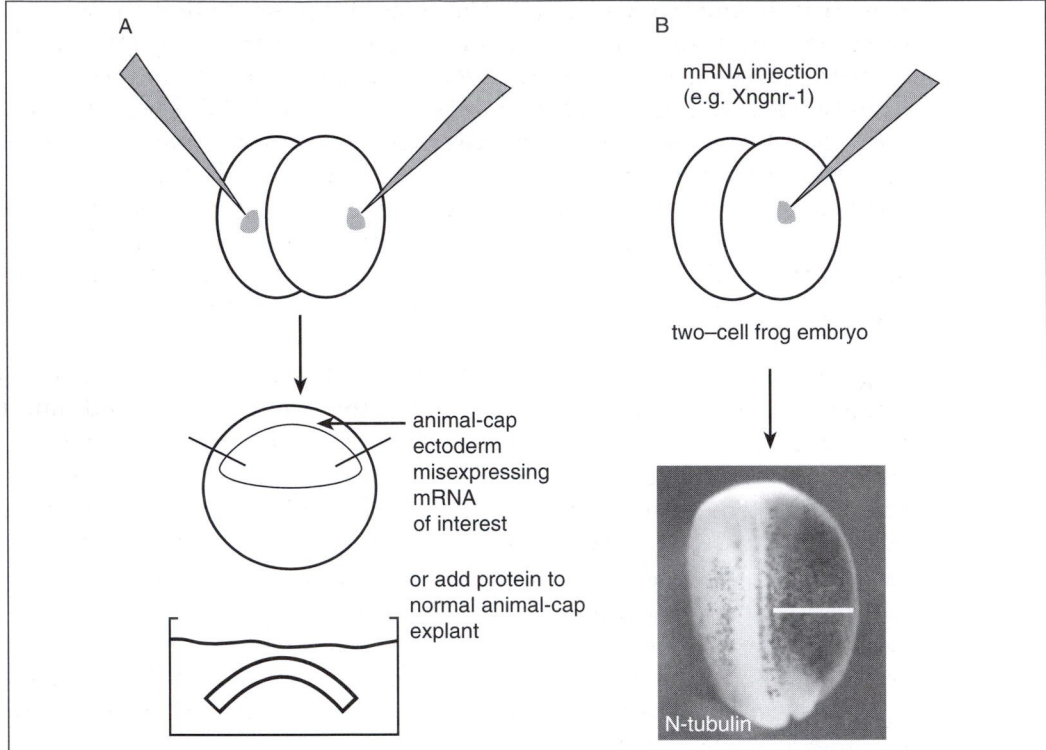

Fig. 2 Key gene function assays in the frog. mRNA injection into the two-cell-stage embryo. (A) Both cells are injected, the embryo cultured to blastoderm stage, and animal-cap ectoderm then removed and assessed for expression of neural genes (autoinduction assay). Animal-cap ectoderm can also be explanted from normal embryos and cultured with defined concentrations and different combinations of candidate neural inducers. (B) One cell is injected with test mRNA and the whole embryo cultured to late neural plate stage. This results in unilateral misexpression of the mRNA. In the case of *Xngnr-1* this leads to expansion of the neural plate (white line) and an increase in the number of *N*-tubulin-expressing neurons in comparison with the control non-injected half of the embryo. Anterior is up.

can induce neural tissue without also generating mesoderm. The first neural inducing molecule to be identified was Noggin, a secreted factor expressed in the organizer and isolated in a screen for molecules with the ability to dorsalize the frog embryo (39) (see Chapter 3). Addition of Noggin to blastula stage animal-cap explants induces neural tissue in the absence of muscle (40). Chordin was isolated in a screen for organizer-specific factors and has also been shown to induce neural tissue in the absence of mesoderm (41, 42) as have two further factors: Follistatin (43) and Xnr3 (44). Strikingly, all these molecules have been shown to act in a similar way, and the key to understanding the molecular basis of this activity has come with a reinterpretation of the significance of cell-dissociation experiments carried out in *Xenopus*.

Intact animal-cap ectoderm differentiates into epidermal tissue; however, if these cells are dissociated they come to express neural specific genes, a phenomenon known as 'autoneuralization' (45, 46). A similar effect is observed when a dominant-negative activin receptor (which blocks signalling by a number of TGF-β (transforming growth factor-β) molecules)) is expressed in the animal cap (47). It was quickly realized that the dilution or blockage of a TGF-β-like molecule is sufficient for this early neuralization step. This molecule was soon identified as BMP4 (bone morphogenetic protein), the addition of which to dissociated animal-cap cells reverses autoneuralization and induces epidermal differentiation (48). Conversely, expression of the dominant-negative BMP receptor (see Chapter 3 for an account of dominant-negative constructs) in the animal cap causes neuralization, as does expression of non-cleavable forms of BMPs or antisense BMP4 (49–51). All four neural inducers have now been shown to act by inhibiting BMP4 signalling, and Noggin, Chordin, and Follistatin do so by direct binding to BMPs (52–54) (see Chapter 6 for further BMP antagonists that act in head and tail formation). Antagonism of TGF-β signalling is also a molecular mechanism that operates in the fruitfly *Drosophila* to subdivide the dorsoventral axis, and as a consequence defines a neurogenic region within which the precursors of the central nervous system arise (55). Shortly after its identification, Chordin was found to be a homologue of the *Drosophila* gene *short gastrulation* (*sog*), which excludes the activity of decapentaplegic (*dpp*), a BMP4 homologue, from the neurogenic region. In vertebrates, antagonism of BMP signalling also mediates the establishment of the dorsoventral axis; Chordin and Noggin can induce dorsal fates in ventral mesoderm by antagonizing BMP signalling in this tissue (reviewed in ref. 56). Thus, within the early ectoderm, BMP antagonists inhibit the more ventral, epidermal cell fate and allow the dorsal, neural fate to proceed by 'default' (reviewed in refs 57—59).

The default model of neural induction represents a major breakthrough in this field, and is in the process of being evaluated using a number of different experimental approaches in other vertebrates (reviewed in detail in refs 7, 59, 60). An important test of the involvement of a molecule in a developmental process is the generation of mutant embryos lacking the gene encoding that molecule. Mice mutant for Noggin, Chordin, Follistatin, or BMP2, -4, or -7 and the BMP receptor (BMPR)-type 1 do not have an early neural phenotype, and nor do mice mutant for both Noggin and Chordin (61; reviewed in ref. 60). These findings may reflect functional redundancy between BMPs and BMP antagonists in the early embryo, which may also account for phenotypes observed in zebrafish mutant for Chordin and BMP2b/swirl which exhibit only changes in neural plate size (62–65). At present, therefore, there is no genetic evidence of a requirement for the attenuation of BMP signalling for neural induction.

Analysis of the role of BMP antagonists in the induction of neural tissue in the chick embryo has also generated some controversy. Extraembryonic epiblast competent to respond to neural inducing signals provided by the organizer does not form neural tissue when exposed to Chordin or Noggin. Further, treatment of primitive-streak stage embryos with BMP4 or -7 does not inhibit neural induction (66). These experiments show that BMP antagonists are not sufficient to induce neural tissue in the chick; instead, changes in neural plate size point to a later role for these factors in regulating the limits of this tissue, consistent with the expression of BMPs at the edges of the neural plate at primitive-streak stages (66). However, two recent reports provide evidence that neural induction begins prior to the formation of the chick organizer (9, 67). At this time *Bmp* transcripts are expressed in medial epiblast that will later contribute to the neural plate (66, 67). Using an *in vitro* approach to study the specification of explants of this very early epiblast, Wilson and colleagues have shown that the downregulation of *Bmp4* transcripts and the expression of early neural markers requires fibroblast growth-factor (FGF) signalling. Further, if FGF signalling is blocked in these explants, the addition of BMP antagonists can induce early neural genes (67). These findings thus suggest that repression of the BMP pathway could, after all, play an early role in neural induction in the chick.

One possibility is that the FGF-mediated repression of *Bmp* transcripts is the first step in neural induction in the chick; however, a number of findings indicate that a more complex network of interactions underlies the initiation of neural induction by FGF signalling. For instance, FGF8 is expressed in the chick 'posterior cells' and organizer and can mimic the neural inducing activity of posterior cells in the extraembryonic epiblast of primitive-streak stage embryos (9). However, FGF8 can induce early neural markers (*Sox3* and *ENRI*) even in the presence of excess BMP4, indicating that FGF signalling does not act merely by antagonizing BMP signalling (9, 68).

Several observations indicate that FGF signalling is also required for BMP antagonist-mediated neural induction in the frog (69–72). However, this again has been the subject of dispute, and other data suggests that FGF signalling is not required to induce anterior neural tissue in this embryo (73–75, 160). Wnt signalling has also recently been shown to initiate neural induction in the frog and acts, like FGF in the chick early medial epiblast, by repressing *Bmp4* transcription (77).

In conclusion, there is evidence in the frog and in the very early chick embryo that attenuation of BMP activity is an important step in neural induction, and that this can be achieved by signalling through different pathways as well as by direct binding of BMP ligand. This is perhaps not surprising given the many different tissue sources of neural inducing signals in the embryo. It is also likely that further novel neural inducing molecules remain to be discovered, which may help to account for neural induction by the chick organizer and by frog somites and for the phenomenon of homeogenetic neural induction.

It is striking that overexpression of BMP antagonists in frog animal-cap ectoderm leads to the induction of only early rostral (anterior) neural genes and that additional signals are required to induce caudal (posterior) genes (59) (see Chapter 6). This finding supports the activation/transformation hypothesis for neural induction put forward by Nieuwkoop (78) to explain his finding that explants of prospective neural plate implanted along the head to tail axis of the neural tube acquire first rostral and then caudal identities. These findings have led to the suggestion that initial neural inducing signals also convey rostral identity. However, recent experiments in the chick suggest that the organizer can induce a general neuralized cell state; addition of an organizer to extraembryonic ectoderm in which competence has been extended by HGF induces an ectopic neural tube that apparently lacks regional character (31). This experimental separation of neural and regional identities supports a hypothesis put forward by Waddington (79), who proposed the existence of general neuralized cell state that precedes the acquisition of regional identity. However, this phase must be early and transient, as it is now also clear in the chick that by late gastrulation the whole neural plate is specified as rostral and that the acquisition of caudal character requires exposure to further signals (17) (see Chapter 6).

5. Early responses to neural inducing signals and the neural precursor cell state

The misexpression of BMP-antagonists not only elicits the expression of early rostral neural genes in animal-cap ectoderm, but under some experimental conditions also leads to neuronal differentiation (40, 42). However, subsequent reports indicate that additional factors are required to generate neurons (80, 81). A number of transcription factors have now been identified that are rapidly expressed in response to neural inducing signals. These have been reviewed in detail in refs 82 and 83 and can be broadly subdivided into three groups; those enhanced by neural inducing signals but which fail to elicit neural specific gene expression; those that promote a neural precursor (proliferative) cell state, this also includes genes that repress neuronal differentiation; and, finally, genes that elicit neuronal differentiation (see Fig. 4).

The first group includes early pan-neural genes such as *Sox2* (31, 38, 84) (and possibly ENRI in the chick (9)). In the frog *Sox2* is induced within 1–2 h in response to Chordin; as discussed above, *Opl* and *Sox2* do not promote neural gene expression, and during normal development are enhanced rather than induced *de novo* by neural inducing signals (37, 38). Further, overexpression of a dominant-negative form of *Sox2* blocks neural differentiation (85). Thus, although *Sox2* cannot initiate neural induction, it is required for the maintenance of neural identity and subsequent differentiation.

The second group includes a related gene, *Sox1*, whose expression is restricted to the early neural plate and later to mitotically active cells in the ventricular zone of the differentiating murine neural tube, where it appears to define an early neural cell state (86). Expression of XBF-1, a winged helix transcription factor, may also help to regulate the neural precursor-cell state: while high concentrations of this gene suppress neuronal differentiation, lower levels promote neurogenesis (87). These observations are supported by the phenotype of BF-1 mutant mice in which neural progenitors undergo premature differentiation (88). However, BF-1 expression is restricted to the presumptive forebrain and other related genes may therefore play a similar role in different regions of the CNS (see ref. 89). *Xenopus* homologues of the *Drosophila iroquois* genes (see below) also induce the expression of neural transcripts in animal-cap ectoderm (90, 91) and, in the case of *Xiro3*, also suppress neuronal differentiation (91). Although the zinc-finger gene, *Zic2*, is expressed slightly later in the developing frog neural plate, it can also suppress neuronal differentiation (see below, and ref. 92). Suppression of neuronal differentiation may be more generally mediated by REST (RE1-silencing transcription factor) or NRSF (neuron-restrictive silencing factor), which bind to a conserved silencer element on a battery of structural neuronal genes including SCG10, human synapsin I, and the type-II sodium channel (93, 94). REST/NRSF is expressed widely in non-neural tissues, but is also found in the developing CNS where it is required for the repression of neuronal gene expression in undifferentiated neural tissue (95).

The third group consists of a growing number of transcription factors that are expressed in response to Noggin and other BMP antagonists, and which can act to repress epidermal differentiation and promote neuronal differentiation in the *Xenopus* embryo (*XlPOU2* (96, 97); *Zicr-1* (38), geminin (98), *XBF2* (99), *SoxD* (100), also see *Gli3* (92)). The former activity suggests a requirement for transcriptional repression of genes downstream of BMP signalling that reinforce the direct antagonism of BMP4 and loss of *Bmp4* transcripts. Unlike the genes described so far, this group of transcription factors also elicit neuronal differentiation in overexpression assays. This may seem contradictory, given that the misexpression of BMP antagonists does not generate neurons (see above). One explanation may be that BMP antagonists also induce inhibitors of neuronal differentiation, or that these may be present in the animal cap and are overcome when transcription factors in this group are misexpressed at high levels. In most cases, these factors have been shown to elicit neuronal differentiation via the proneural gene homologue *neurogenin-related* (*ngnr-1*) (see below). However, expression of this third group of genes is more widespread than that of *ngnr-1*, indicating that these effects are unlikely to be direct and that further factors regulate the activity of early-response genes (see below).

It is likely that these three groups of early genes act in a broadly sequential fashion (see Fig. 4), but they may also carry out complex regulatory interactions. The first two help to specify neural fate, defining and maintaining the neural precursor cell state, while the third acts to promote the formation of neurons. This early activity may be analogous to that mediated by the prepatterning genes in the fly, which define where neural precursors can arise via positive regulation of proneural gene expression. Indeed, it is important to understand the genetic cassette that regulates *Drosophila* neurogenesis in order to understand the roles and regulation of the many vertebrate proneural gene homologues

6. Assignment of neural cell fate in *Drosophila*

In the fly, the proneural genes (basic helix–loop–helix (bHLH) transcription factors, which encode four genes of the *achaete-scute* complex (AS-C) and *atonal* and *amos*, act to confer neural potential on ectoderm cells. Within the neurogenic region of the embryo, defined by *sog* antagonism of *dpp* signalling (see above), clusters of cells expressing AS-C genes prefigure the formation of neural precursors (reviewed in ref. 101). The position of proneural clusters are themselves defined by the activity of the pair-rule genes (reviewed in ref. 102), and in the adult peripheral nervous system by other prepatterning genes such as those of the *Iroquois* complex (103). Only one cell in each proneural cluster within the neurogenic region will become a neural precursor, while neighbouring cells will become epidermal unless they are chosen in a subsequent cluster. Selection of the neural precursor is brought about by a mechanism known as 'lateral inhibition'. This process is mediated by the neurogenic genes, principally, those encoding the transmembrane receptor Notch and its membrane-bound ligand Delta. The classical definition of lateral inhibition stipulates that all cells have the same developmental potential and express equal levels of Notch and Delta (104, 105). At random, one cell in a cluster comes to express higher levels of Delta than its immediate neighbours and therefore increases Notch signalling in adjacent cells. Notch signalling leads to the cell autonomous downregulation of Delta via repression of the proneural genes, which are direct positive transcriptional regulators of Delta (Fig. 3). In this way, intercellular signalling quickly leads to the selection of a single cell that inhibits its neighbours from following the same developmental pathway (reviewed in ref. 105). High proneural gene expression also leads to the expression of neural specific genes in the neural precursor cell, which now delaminates from the ectodermal cell layer and undergoes a stereotyped series of asymmetrical cell divisions that generate ganglion mother cells, which in turn divide to produce neurons (reviewed in ref. 101).

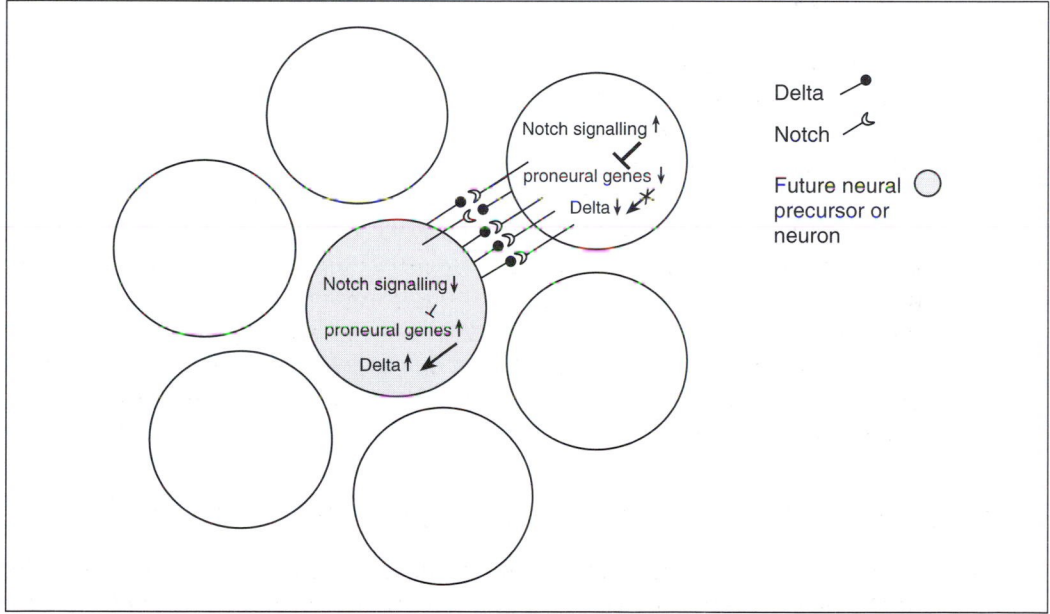

Fig. 3 Lateral inhibition is mediated by Notch–Delta signalling. In this model all cells initially express Notch and Delta; however, one cell comes to express slightly more Delta (shaded cell) (see text for details) and this leads to an increase in Notch signalling in neighbouring cells (for simplicity only two interacting cells are depicted). Notch signalling represses the expression of proneural genes which are positive regulators of Delta. Thus, cells receiving increased levels of Delta come to express less Delta themselves, resulting in a decrease in Notch signalling within the central cell. A feedback loop is thus established, which leads to the generation of two distinct cell states.

A number of observations, however, indicate that ectoderm cells in the neurogenic region are not equipotential, but are prepatterned with respect to neural cell fate (reviewed by ref. 106). This bias may be mediated by *achaete-scute* gene activity and/or by as yet unidentified signals that regulate a Notch-independent selection process. Indeed, there is also evidence for such a mechanism in the vertebrate embryo (see below).

7. Vertebrate proneural gene homologues

Proneural gene homologues have been identified in nearly all model vertebrates. Moreover, their roles have been studied in a number of sites where neurogenesis takes place, including the retina, neural crest, sensory placodes, the olfactory neuro-epithelium, and the *Xenopus* neural plate, which is characterized by precociously differentiating primary neurons (reviewed in refs 107–109). Here we will focus on events taking place within the CNS, although comparisons will be drawn with proneural gene activity in the peripheral nervous system (PNS). At the outset it is important to define the cellular context in which vertebrate proneural genes function. As outlined above, cells in the neurogenic region of *Drosophila* can become epidermal or neural, and neural precursors are selected by the coordinate activity of the pro-neural and neurogenic genes. In vertebrates, all cells in the neural plate will become neural, and therefore proneural gene activity is not thought to be required to select individual neural precursors. Instead, vertebrate proneural genes act to regulate the steps leading to neuronal differentiation. One exception to this may be the chick caudal neural plate, which is uniquely interspersed with epidermal and neural precursors (110, 111). The heterogeneous expression pattern of the proneural gene homologue *cash4* in caudal neural plate, together with functional studies (in the fly and frog) showing that *cash4* can promote neural cell fates (112), raises the possibility that this gene acts in the chick, like the *Drosophila* proneural genes, to promote the formation of individual neural precursors.

To date, three proneural gene homologues have been identified, which are expressed in the early *Xenopus* neural plate: *xash3* and the *atonal*-related genes *ATH3* and *ngnr-1*. Overexpression of *xash3* in the embryo leads to an expansion of the neural plate at the expense of neural crest and epidermal tissue (113, 114). Depending on the dose of injected mRNA, *xash3* can also elicit ectopic neuronal differentiation within the neural plate (115). However, this gene comes to be expressed in only a small group of neural precursors and is therefore unlikely to play a central role in promoting neuronal differentiation. However, *ngnr-1* and *ATH3* expression do prefigure primary neurogenesis in the *Xenopus* posterior neural plate, demarcating the medial, intermediate, and lateral territories where primary motor neurons, inter-neurons, and sensory neurons, respectively, will later differentiate (116, 117). Ectopic expression of *ngnr-1* or *ATH3* induces ectopic primary neurogenesis; however, *ATH3* is expressed only weakly at these early stages and appears to play a more prominent role later in neuronal differentiation in the anterior CNS. Ectopic *ngnr-1* induces the expression of another related bHLH transcription factor, *NeuroD*. During normal development, *ngnr-1* expression prefigures that of *NeuroD*, which is found later in a subset of *ngnr-1*-expressing cells. The ectopic expression of *NeuroD* also elicits neuronal differentiation, even in ventral epidermal tissue (118), but does not elicit *ngnr-1* expression, thus defining a unidirectional cascade of bHLH transcription factors (*ngnr-1–NeuroD*) that promotes neuronal differentiation (see ref. 107).

In the mouse, loss-of-function experiments have shown that *Mash-1* is required for the differentiation of autonomic, olfactory, and retinal neurons (119–121). This proneural gene homologue is also upstream of similar cascades of bHLH transcription factors that mediate neurogenesis; for instance, in the olfactory neuroepithelium Mash1 and Ngn1 are expressed during different phases of neuronal progenitor development and *NeuroD* is expressed still later in postmitotic cells (122). In *mash-1* mutant mice, the olfactory epithelium lacks both *ngn-1* and *NeuroD* expression defining the cascade *Mash-1–ngn-1–NeuroD* (122; and see ref. 123). Such cascades of bHLH and later of other classes of transcription factors have been shown to mediate neurogenesis at many different sites in a range of model vertebrates (reviewed in refs 108, 109).

8. The regulation of neuron production by Notch/Delta signalling

The ability of bHLH transcription factors in these cascades to promote neuronal differentiation has been shown to be differentially sensitive to inhibition provided by Notch signalling during the process of lateral inhibition, which serves to single out individual cells as neuronal precursors. In general, earlier expressed proneural gene homologues, such as *xash3*, *ngnr-1*, *Mash-1*, and *ngn-1*, are more sensitive to inhibition via Notch signalling than later expressed genes such as *NeuroD* or combinations of genes *xash3* or *ngnr-1* plus *MyT-1* (124). Other transcription factors, expressed in the frog after *ngnr-1* and prior to *NeuroD* in neuronal precursors, such as *MyT-1* and *Xcoe2*, may stabilize the neuronal cell fate by allowing cells to escape lateral inhibition (124, 125) (also see *NeuroM* (126)). It has been proposed that during evolution the addition of protein domains responsive to inhibitory signals has provided a mechanism for increasing cell number and delaying neuronal differentiation, so that specified neurons can migrate to new positions and differentiate at appropriate times (107). Thus, genes sensitive to inhibition are thought of as neural determination genes, while those insensitive to such inhibition are considered neural differentiation genes (see Fig. 4).

A number of vertebrate homologues of *Notch* and *Delta* have been identified, these include *Notch1* and *-3* which are expressed widely in the ventricular zone of the developing CNS in most model vertebrates, while the ligands *Delta-1* and *-3*, *Serrate-1* (*Jagged-1* in mammals), and the zebrafish *Delta B* gene are found in postmitotic prospective neurons (see ref. 127 for zebrafish *Delta A* and *D* genes; reviewed in ref.

128). Overexpression of an activated form of mammalian Notch-3 or Xotch-1 reduces the number of neurons produced (129, 130). The same phenotype is also generated by ectopic expression of the *Delta* ligand, while, conversely, overexpression of a dominant-negative form of *Delta-1*, which depletes Notch signalling, leads to the excessive production of neurons, as demonstrated in zebrafish, *Xenopus*, and chick embryos (127, 131–134). Thus, Notch-signalling induced by Delta-expressing prospective neuronal precursors inhibits proneural gene expression in neighbouring cells, and thereby regulates the production of postmitotic cells within the vertebrate neuroepithelium (see Fig. 3 and 4).

In the fly, inhibition by Notch signalling is mediated by the induction of the direct transcriptional repressors of proneural genes, the *HLH* genes of the Enhancer of Split (E(spl)) family (135). Vertebrate homologues of these genes have been identified (reviewed in ref. 109) and recently shown to be effectors of Notch signalling (136, 137). Mice mutant for *HES-1* (*Hairy/Espl-1*) exhibit precocious expression of *Mash-1* and premature neuronal differentiation (138, 139). Proneural gene activity may also be suppressed by homologues of the HLH protein extramacrochaete (emc) which antagonizes AS-C activity in the fly, although emc is not directly involved in lateral inhibition (140; reviewed in ref. 141). Vertebrate emc homologues, the *Id* genes, are expressed in the ventricular layer of the neuroepithelium, consistent with a role as inhibitors of neuronal differentiation (142, 143). This activity has recently been demonstrated in a double *Id1/Id3* knockout mouse which exhibits precocious neurogenesis and upregulation of later expressed neuronal differentiation genes, such as *NeuroD1* and *MATH1*, -2, and -3 (144).

In the fly, Notch signalling is itself regulated by Numb, a protein that acts by binding to the cytoplasmic domain of Notch, thereby preventing signal transduction and translocation to the nucleus. In vertebrates, a number of Numb and Numb-like homologues have been identified, which, along with the Notch protein, have striking and controversial patterns of expression in neural precursors and their postmitotic progeny (145–148). Recently, avian Numb has been shown to be localized to the basal cortex of mitotic neuroepithelial cells. These cells can divide asymmetrically to produce a basal postmitotic neuron, while retaining a mitotic cell at the apical (inner) surface of the neural tube. Ectopic expression experiments have demonstrated that c-Numb represses Notch signalling in nascent neurons, consistent with the effects of Notch signalling described above (148). The regulation of such asymmetric cell divisions in the ventricular layer of the developing neuroepithelium is a very active area of research and is reviewed in detail elsewhere (161).

Finally, there is evidence for a prepattern underlying the initial selection of neural precursors in *Xenopus* CNS and in mouse sensory placodes (116, 149, 150). For example, neural precursors still form in the correct spatial pattern in mice in which a downstream effector of the Notch-signalling pathway (RBPJ$_k$) is disrupted (149), as well as in embryos expressing a dominant-negative form of Kuzbanian (a protease required for Notch processing and function ((151), after ref. 149). This suggests that lateral inhibition mediated by Notch/Delta signalling may not be the only selection mechanism that operates during vertebrate neurogenesis.

9. Neurogenesis and regional identity

It is important to emphasize that different proneural genes operate at different sites of neurogenesis, where they may carry out similar activities. For example, *ngn-1* and -2 are expressed in overlapping domains in the CNS, but in the PNS these genes are expressed in different sensory placodes where they appear to play unique roles in the determination of specific populations of neural precursors (149, 150). Further, while *ngn-1* and -2 promote the formation of the sensory nervous system, *mash-1* carries out a similar role in the autonomic nervous system (119, 120). This suggests that these proneural genes may convey regional as well as neural identity, depending on the cellular context in which they are expressed. A clue to understanding how proneural genes mediate these different aspects of neurogenesis has come from the creation of chimeric proteins which combine the HLH domain of the *scute* gene with elements of the basic region of the *atonal* gene in the fly (152). These

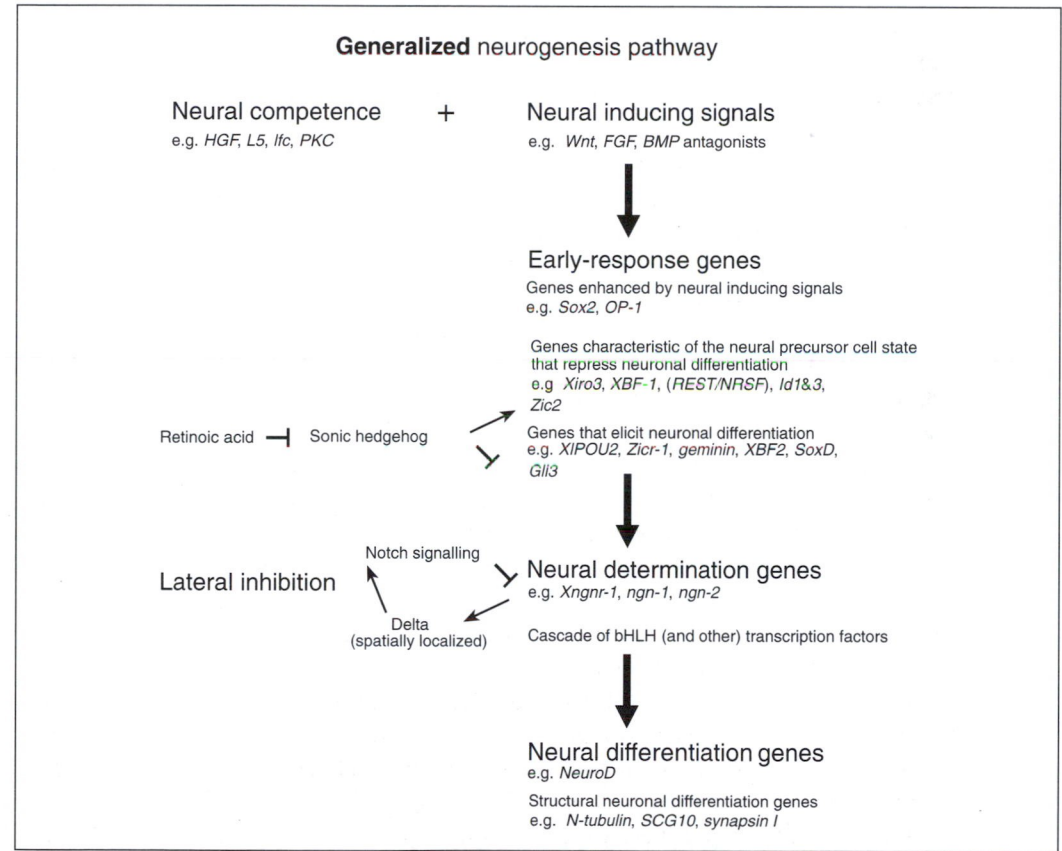

Fig. 4 Generalized neurogenesis pathway. Examples of genes involved in each key step are provided. Note that these examples are taken from different organisms and normally act in distinct developmental contexts, and thus do *not* represent a series of direct hierarchical interactions (see text for details).

experiments show that *scute* expression, which normally leads to the formation of external sensory organs, can elicit chordotonal organ formation (normally induced by *atonal* gene expression) on inclusion of a small part of the *atonal* basic domain. This region appears not to contact DNA but it may interact with cofactors, which provide regional/cell type specification. A single amino acid change in the basic domain of the vertebrate proneural gene *ATH3* also appears to separate the neural promoting activity of this gene from its ability to induce anterior regional character (117). These findings emphasize that, while specific proneural genes are expressed in particular neural lineages, the activity of these genes depends on the presence of cofactors. Indeed, *mash-1* has also been shown to mediate different steps of neurogenesis depending on the cellular context in which it is expressed; while this gene is required for the formation of neuronal progenitors in the olfactory neuroepithelium (119), in the autonomic nervous system it is involved in the terminal differentiation of noradrenergic neurons (see, for example, ref. 120; see also ref. 153). Thus, it is possible to see how small changes in the protein or in the regulatory regions of these genes may allow the generation of increasingly diverse cell types. Of course, other classes of regulatory genes (such as the POU domain, *PHD* and *LIM* genes) are also involved in this process, and a detailed account of the determination of neuronal phenotype is provided by Anderson and Jan (108) (see also ref. 154).

Finally, it is striking that, in nearly all vertebrates, neuronal differentiation commences in the caudal regions of the neural plate, and the onset of this process appears to be linked to the acquisition of caudal (posterior) neural character. Caudalizing signals such as retinoic acid (a derivative of vitamin A) also stimulate neuronal differentiation (see, for example, refs 38, 155, 156; see also ref. 76) and blocking the retinoic acid pathway disrupts both the acquisition of caudal character and neurogenesis in the frog (155, 157). This small molecule can also drive neuronal differentiation in murine-derived P19 cells, an early consequence of which is the induction of the proneural gene *mash-1* (see, for example, ref. 158). A recent study has further shown that retinoic acid acts via induction of *ngnr-1*, providing an important link between this pathway and the established cascade of transcription factors that lead to neuronal differentiation in the frog (159). These authors further show that retinoic acid and the secreted protein sonic hedgehog (shh) have opposing effects on the neurogenesis pathway, and provide evidence that retinoic acid activates primary neurogenesis by repressing *shh* in the frog. Shh signalling induces *Zic2* (a transcription factor which can repress *ngnr-1*) and also inhibits neuronal differentiation mediated by Gli3 (92). Shh activity therefore helps to define the initial pattern of *ngnr-1* expression. These findings further illustrate the close relationship between neuronal differentiation and patterning events in both rostrocaudal (anteroposterior) and dorsoventral axes (see Chapter 6 for details on anteroposterior patterning).

10. Themes and future perspectives

There are two major themes that emerge in this field: it is clear that both positive and negative factors combine to sculpt the gene-expression profile of ectoderm cells that will ultimately differentiate into neurons, and that at almost every level it is difficult to separate neural specification from regionalization/patterning events (see Fig. 4). Attenuation of BMP signalling by secreted BMP antagonists (as well as via FGF and Wnt signalling) and the activation of transcriptional repressors serve to inhibit epidermal differentiation, whilst simultaneously inducing early neural specific genes. Neuronal differentiation is further regulated by inhibition delivered by Notch signalling and by neural silencer genes that inhibit the inappropriate expression of structural neuronal genes. Attenuation of BMP signalling is also part of a global dorsoventral patterning mechanism that affects both the ectoderm and mesoderm. In the ectoderm, loss of BMP signalling results in the expression of only early rostral neural genes, further factors are required both to caudalize the neuroepithelium as well as to elicit neuronal differentiation via the proneural genes. These genes can also convey regional character, the activity of which depends on the cellular context in which they are expressed.

Key areas for future research include understanding the molecular basis of neural competence and the identification of new molecules with neural inducing activity. In particular, we are just beginning to understand events downstream of neural inducing signals and prior to the onset of neuronal differentiation. The genes involved in these early steps must help to coordinate cell cycle and cell movement with the assignment of neural fate and the acquisition of regional character, and future research aims to unravel the gene interactions that integrate these different aspects of neurogenesis. Finally, it is important to ascertain the extent to which the pathways underlying neurogenesis are conserved across vertebrate species, as this will allow us to identify fundamental molecular mechanisms controlling this essential process.

Acknowledgements

I am grateful to Ruth Diez del Corral, Nancy Papalopulu, Domingos Henrique, Claudio Stern, and Andrea Streit for discussions and helpful comments on all or parts of the manuscript. The photograph of the *Xngnr-1*-misexpressing embryo in Fig. 2 was kindly provided by Nancy Papalopulu. This work is supported by the Medical Research Council, the Wellcome Trust, and the Human Frontiers Science Programme. KGS is an MRC Senior Research Fellow.

References

1. Spemann, H. and Mangold, H. (1924) Uber Induktion von Embryoanlagen durch Implantation artfremder Organisatoren. *Wilh. Roux. Arch. EntwMech. Organ.*, **100**, 599.
2. Hamburger, V. (1988) *The heritage of experimental embryology, Hans Spemann and the organizer*, p. 191. Oxford University Press, New York.
3. Streit, A., Thery, C., and Stern, C. D. (1994) Of mice and frogs. *Trends Genet.*, **10**, 181.
4. Waddington, C. H. (1932) Experiments on the development of chick and duck embryos cultivated *in vitro. Phil. Trans. Roy. Soc. London, B*, **221**, 179.

5. Kintner, C. R. and Dodd, J. (1991) Hensen's node induces neural tissue in *Xenopus* ectoderm. Implications for the action of the organizer in neural induction. *Development,* **113**, 1495.

6. Klingensmith, J., Ang, S. L., Bachiller, D., and Rossant, J. (1999) Neural induction and patterning in the mouse in the absence of the node and its derivatives. *Dev. Biol.,* **216**, 535.

7. Harland, R. (2000) Neural induction. *Curr. Opin. Genet. Dev.,* **10**, 357.

8. Storey, K. G., Crossley, J. M., De Robertis, E. M., Norris, W. E., and Stern, C. D. (1992) Neural induction and regionalisation in the chick embryo. *Development,* **114**, 729.

9. Streit, A., Berliner, A. J., Papanayotou, C., Sirulnik, A., and Stern, C. D. (2000) Initiation of neural induction by FGF signalling before gastrulation. *Nature,* **406**, 74.

10. Foley, A. C., Skromne, I., and Stern, C. D. (2000) Reconciling different models of forebrain induction and patterning: a dual role for the hypoblast. *Development,* **127**, 3839.

11. Thomas, P. and Beddington, R. (1996) Anterior primitive endoderm may be responsible for patterning the anterior neural plate in the mouse embryo. *Curr. Biol.,* **6**, 1487.

12. Tam, P. P. and Steiner, K. A. (1999) Anterior patterning by synergistic activity of the early gastrula organizer and the anterior germ layer tissues of the mouse embryo. *Development,* **126**, 5171.

13. Ho, C. Y., Houart, C., Wilson, S. W., and Stainier, D. Y. (1999) A role for the extra-embryonic yolk syncytial layer in patterning the zebrafish embryo suggested by properties of the hex gene. *Curr. Biol.,* **9**, 1131.

14. Beddington, R. S. P. and Robertson, E. (1998) Anterior patterning in mouse. *Trends Genet.,* **14**, 277.

15. Knoetgen, H., Viebahn, C., and Kessel, M. (1999) Head induction in the chick by primitive endoderm of mammalian, but not avian origin. *Development,* **126**, 815.

16. Pera, E., Stein, S., and Kessel, M. (1999) Ectodermal patterning in the avian embryo: epidermis versus neural plate. *Development,* **126**, 63.

17. Muhr, J., Graziano, E., Wilson, S., Jessell, T. M., and Edlund, T. (1999) Convergent inductive signals specify midbrain, hindbrain, and spinal cord identity in gastrula stage chick embryos. *Neuron,* **23**, 689.

18. Darnell, D. K., Stark, M. R., and Schoenwolf, G. C. (1999) Timing and cell interactions underlying neural induction in the chick embryo. *Development,* **126**, 2505.

19. Rowan, A. M., Stern, C. D., and Storey, K. G. (1999) Axial mesendoderm refines rostrocaudal pattern in the chick nervous system. *Development,* **126**, 2921.

20. Hemmati Brivanlou, A., Stewart, R. N., and Harland, R. M. (1990) Region-specific neural induction of an engrailed protein by anterior notochord in *Xenopus*. *Science,* **250**, 800.

21. Ang, S. L. and Rossant, J. (1993) Anterior mesendoderm induces mouse Engrailed genes in explant cultures. *Development,* **118**, 139.

22. Sive, H. L., Hattori, K., and Weintraub, H. (1989) Progressive determination during formation of the anteroposterior axis in *Xenopus laevis*. *Cell,* **58**, 171.

23. Doniach, T. (1992) Induction of anteroposterior neural pattern in *Xenopus* by planar signals. *Dev. Suppl.,* **5**, 183.

24. Barnett, M. W., Old, R. W., and Jones, E. A. (1998) Neural induction and patterning by fibroblast growth factor, notochord and somite tissue in *Xenopus*. *Dev. Growth Differ.,* **40**, 47.

25. Storey, K. G., Selleck, M. A., and Stern, C. D. (1995) Neural induction and regionalisation by different sub-populations of cells in Hensen's node. *Development,* **121**, 417.

26. Nakamura, OaT, S. (1978) *Organiser: a milestone of a half century from Spemann*. Elsevier/North-Holland Biomedical, Amsterdam.

27. Servetnick, M. and Grainger, R. M. (1991) Homeogenetic neural induction in *Xenopus*. *Dev. Biol.*, **147**, 73.
28. Gurdon, J. B. (1987) Embryonic induction—molecular prospects. *Development*, **99**, 285.
29. Grainger, R. M. and Gurdon, J. B. (1989) Loss of competence in amphibian induction can take place in single non-dividing cells. *Proc. Natl Acad. Sci. USA*, **86**, 1900.
30. Woodside, G. L. (1937) The influences of host age on induction in the chick embryo. *J. Exp. Zool.*, **75**, 259.
31. Streit, A., Sockanathan, S., Perez, L., *et al.* (1997) Preventing the loss of competence for neural induction: HGF/SF, L5 and Sox-2. *Development*, **124**, 1191.
32. Streit, A. C. and Stern, C. D. (1997) Competence for neural induction: HGF/SF, HGF1/MSP and the c-Met receptor. *Ciba Found. Symp.*, **212**, 155.
33. Sharpe, C. R., Fritz, A., De-Robertis, E. M., and Gurdon, J. B. (1987) A homeobox-containing marker of posterior neural differentiation shows the importance of pre-determination in neural induction. *Cell*, **50**, 749.
34. Otte, A. P. and Moon, R. T. (1992) Protein kinase C isozymes have distinct roles in neural induction and competence in *Xenopus*. *Cell*, **68**, 1021.
35. Morgan, R. and Sargent, M. G. (1997) The role in neural patterning of translation initiation factor eIF4AII; induction of neural fold genes. *Development*, **124**, 2751.
36. Morgan, R., Hooiveld, M. H. W., and Durston, A. J. (1999) A novel guanine exchange factor increases the competence of early ectoderm to respond to neural induction. *Mech. Dev.*, **88**, 67.
37. Kuo, J. S., Patel, M., Gamse, J., *et al.* (1998) Opl: a zinc finger protein that regulates neural determination and patterning in *Xenopus*. *Development*, **125**, 2867.
38. Mizuseki, K., Kishi, M., Matsui, M., Nakanishi, S., and Sasai, Y. (1998) *Xenopus* Zic-related-1 and Sox-2, two factors induced by chordin, have distinct activities in the initiation of neural induction. *Development*, **125**, 579.
39. Smith, W. C. and Harland, R. M. (1992) Expression cloning of noggin, a new dorsalizing factor localized to the Spemann organizer in *Xenopus* embryos. *Cell*, **70**, 829.
40. Lamb, T. M., Knecht, A. K., Smith, W. S., *et al.* (1993) Neural induction by the secreted polypeptide noggin. *Science*, **262**, 713.
41. Sasai, Y., Lu, B., Steinbeisser, H., Geissert, D., Gont, L. K., and De Robertis, E. M. (1994) *Xenopus* chordin: a novel dorsalizing factor activated by organizer-specific homeobox genes. *Cell*, **79**, 779.
42. Sasai, Y., Lu, B., Steinbeisser, H., and De Robertis, E. M. (1995) Regulation of neural induction by the Chd and Bmp-4 antagonistic patterning signals in *Xenopus*. *Nature*, **376**, 333.
43. Hemmati Brivanlou, A., Kelly, O. G., and Melton, D. A. (1994) Follistatin, an antagonist of activin, is expressed in the Spemann organizer and displays direct neuralizing activity. *Cell*, **77**, 283.
44. Hansen, C. S., Marion, C. D., Steele, K., George S., and Smith, W. C. (1997) Direct neural induction and selective inhibition of mesoderm and epidermis inducers by Xnr3. *Development*, **124**, 483.
45. Godsave, S. F. and Slack, J. M. (1989) Clonal analysis of mesoderm induction in *Xenopus laevis*. *Dev. Biol.*, **134**, 486.
46. Grunz, H. and Tacke, L. (1989) Neural differentiation of *Xenopus laevis* ectoderm takes place after disaggregation and delayed reaggregation without inducer. *Cell. Differ. Dev.*, **28**, 211.
47. Hemmati Brivanlou, A. and Melton, D. A. (1994) Inhibition of activin receptor signaling promotes neuralization in *Xenopus*. *Cell*, **77**, 273.

48. Wilson, P. A. and Hemmati-Brivanlou, A. (1995) Induction of epidermis and inhibition of neural fate by Bmp-4. *Nature*, **376**, 331.

49. Hawley, S. H., Wunnenberg-Stapleton, K., Hashimoto, C., *et al.* (1995) Disruption of BMP signals in embryonic *Xenopus* ectoderm leads to direct neural induction. *Genes Dev.*, **9**, 2923.

50. Sasai, Y., Lu, B., Steinbeisser, H., and De Robertis, E. M. (1995) Regulation of neural induction by the Chd and Bmp-4 antagonistic patterning signals in *Xenopus*. *Nature*, **376**, 333. [Published errata appear in *Nature* (1995), **377**, 757 and *Nature* (1995), **378**, 419.]

51. Xu, R. H., Kim, J., Taira, M., Zhan, S., Sredni, D., and Kung, H. F. (1995) A dominant negative bone morphogenetic protein 4 receptor causes neuralization in *Xenopus* ectoderm. *Biochem. Biophys. Res. Commun.*, **212**, 212.

52. Piccolo, S., Sasai, Y., Lu, B., and De Robertis, E. M. (1996) Dorsoventral patterning in *Xenopus*: inhibition of ventral signals by direct binding of chordin to BMP-4. *Cell*, **86**, 589.

53. Zimmerman, L. B., Jesus Escobar, J. M. D., and Harland, R. M. (1996) The Spemann organizer signal noggin binds and inactivates bone morphogenetic protein 4. *Cell*, **86**, 599.

54. Fainsod, A., Deissler, K., Yelin, R., *et al.* (1997) The dorsalizing and neural inducing gene follistatin is an antagonist of BMP-4. *Mech. Dev.*, **63**, 39.

55. Ferguson, E. L. and Anderson, K. V. (1992) Localised enhancement and repression of activity of the TGF-beta family member, decapentaplegic, is necessary for dorsal-ventral patterning in the *Drosophila* embryo. *Development*, **114**, 583.

56. Dale, L. and Jones, C. M. (1999) BMP signalling in early *Xenopus* development. *BioEssays*, **21**, 751.

57. Sasai, Y. and De Robertis, E. M. (1997) Ectodermal patterning in vertebrate embryos. *Dev. Biol.*, **182**, 5.

58. Wilson, P. A. and Hemmati Brivanlou, A. (1997) Vertebrate neural induction: inducers, inhibitors, and a new synthesis. *Neuron*, **18**, 699.

59. Harland, R. (1997) Neural induction in *Xenopus*. In *Molecular and cellular approaches to neural development* (ed. W. M. Cowan, J. T. M., and S. L. Zipursky), p. 1. Oxford University Press, New York.

60. Streit, A. and Stern, C. D. (1999) More to neural induction than inhibition of BMPs. In *Cell lineage and fate determination* (ed. S E. Moody), p. 437. Academic Press, San Diego, CA.

61. Bachiller D., Klingensmith J., Kemp C., *et al.* (2000) The organizer factors Chordin and Noggin are required for mouse forebrain development. *Nature*, **403**, 658.

62. Schulte Merker, S., Lee, K. J., McMahon, A. P., and Hammerschmidt, M. (1997) The zebrafish organizer requires chordino. *Nature*, **387**, 862.

63. Kishimoto, Y., Lee, K. H., Zon, L., Hammerschmidt, M., and Schulte Merker, S. (1997) The molecular nature of zebrafish swirl: BMP2 function is essential during early dorsoventral patterning. *Development*, **124**, 4457.

64. Nguyen, V. H., Trout, J., Connors, S. A., Andermann, P., Weinberg, E., and Mullins, M. C. (2000) Dorsal and intermediate neuronal cell types of the spinal cord are established by a BMP signaling pathway. *Development*, **127**, 1209.

65. Barth, K. A., Kishimoto, Y., Rohr, K. B., Seydler, C., Schulte-Merker, S., and Wilson, S. W. (1999) Bmp activity establishes a gradient of positional information throughout the entire neural plate. *Development*, **126**, 4977.

66. Streit, A., Lee, K. J., Woo, I., Roberts, C., Jessell, T. M., and Stern, C. D. (1998) Chordin regulates primitive streak development and the stability of induced neural cells, but is not sufficient for neural induction in the chick embryo. *Development*, **125**, 507.

67. Wilson, S. I., Graziano, E., Harland, R., Jessell, T. M., and Edlund, T. (2000) An early requirement for FGF signalling in the acquisition of neural cell fate in the chick embryo. *Curr. Biol.*, **10**, 421.

68. Streit, A. and Stern, C. D. (1999) Establishment and maintenance of the border of the neural plate in the chick: involvement of FGF and BMP activity. *Mech. Dev.*, **82**, 51.

69. LaBonne, C. and Whitman, M. (1997) Localization of MAP kinase activity in early *Xenopus* embryos: implications for endogenous FGF signaling. *Dev. Biol.*, **183**, 9.

70. Launay, C., Fromentoux, V., Shi, D. L., and Boucaut, J. C. (1996) A truncated FGF receptor blocks neural induction by endogenous *Xenopus* inducers. *Development*, **122**, 869.

71. Sasai, Y., Lu, B., Picolo, S., and De Robertis, E. M. (1996) Endoderm induction by the organizer-secreted factors chordin and noggin in *Xenopus* animal caps. *EMBO J.*, **15**, 4547.

72. Xu, R. H., Kim, J., Taira, M., Sredni, D., and Kung, H. (1997) Studies on the role of fibroblast growth factor signaling in neurogenesis using conjugated/aged animal caps and dorsal ectoderm-grafted embryos. *J. Neurosci.*, **17**, 6892.

73. Kroll, K. L. and Amaya, E. (1996) Transgenic *Xenopus* embryos from sperm nuclear transplantations reveal FGF signalling requirements during gastrulation. *Development*, **122**, 3173.

74. Schulte Merker, S. and Smith, J. C. (1995) Mesoderm formation in response to Brachyury requires FGF signalling. *Curr. Biol.*, **5**, 62.

75. Hongo, I., Kengaku, M., and Okamoto, H. (1999) FGF signaling and the anterior neural induction in *Xenopus*. *Dev. Biol.*, **216**, 561.

76. Hardcastle, Z., Chalmers, A. D., and Papalopulu, N. (2000) FGF-8 stimulates neuronal differentiation through FGFR-4a and interferes with mesoderm induction in xenopus embryos. *Curr. Biol.*, **10**, 1511.

77. Baker, J. C., Beddington, R. S., and Harland, R. M. (1999) Wnt signaling in *Xenopus* embryos inhibits bmp4 expression and activates neural development. *Genes Dev.*, **13**, 3149.

78. Nieuwkoop, P. D. (1952) Activation and organisation of the central nervous system in amphibians. III Synthesis of a new working hypothesis. *J. Exp. Zool.*, **120**, 83.

79. Waddington, C. H. (1940) *Organisers and genes*. Cambridge University Press, London.

80. Goldstone, K. and Sharpe, C. R. (1998) The expression of XIF3 in undifferentiated anterior neuroectoderm, but not in primary neurons, is induced by the neuralizing agent noggin. *Int. J. Dev. Biol.*, **42**, 757.

81. Messenger, N. J., Rowe, S. J., and Warner, A. E. (1999) The neurotransmitter noradrenaline drives noggin-expressing ectoderm cells to activate *N*-tubulin and become neurons. *Dev. Biol.*, **205**, 224.

82. Sasai, Y. (1998) Identifying the missing links: genes that connect neural induction and primary neurogenesis in vertebrate embryos. *Neuron*, **21**, 455.

83. Chitnis, A. B. (1999) Control of neurogenesis—lessons from frogs, fish and flies. *Curr. Opin. Neurobiol.*, **9**, 18.

84. Rex, M., Orme, A., Uwanogho, D., *et al.* (1997) Dynamic expression of chicken Sox2 and Sox3 genes in ectoderm induced to form neural tissue. *Dev. Dyn.*, **209**, 323.

85. Kishi, M., Mizuseki, K., Sasai, N., *et al.* (2000) Requirement of Sox2-mediated signaling for differentiation of early *Xenopus* neuroectoderm. *Development*, **127**, 791.

86. Pevny, L. H., Sockanathan, S., Placzek, M., and Lovell Badge, R. (1998) A role for SOX1 in neural determination. *Development*, **125**, 1967.

87. Bourguignon, C., Li, J., and Papalopulu, N. (1998) XBF-1, a winged helix transcription factor with dual activity, has a role in positioning neurogenesis in *Xenopus* competent ectoderm. *Development*, **125**, 4889.

88. Xuan, S., Baptista, C. A., Balas, G., Tao, W., Soares, V. C., and Lai, E. (1995) Winged helix transcription factor BF-1 is essential for the development of the cerebral hemispheres. *Neuron*, **14**, 1141.

89. Zuber, M. E., Perron, M., Philpott, A., Bang, A., and Harris, W. A. (1999) Giant eyes in *Xenopus laevis* by overexpression of XOptx2. *Cell*, **98**, 341.

90. Gomez Skarmeta, J. L., Glavic, A., de la Calle Mustienes, E., Modolell, J., and Mayor, R. (1998) Xiro, a *Xenopus* homolog of the *Drosophila* Iroquois complex genes, controls development at the neural plate. *EMBO J.*, **17**, 181.

91. Bellefroid, E. J., Kobbe, A., Gruss, P., Pieler, T., Gurdon, J. B., and Papalopulu, N. (1998) Xiro3 encodes a *Xenopus* homolog of the *Drosophila* Iroquois genes and functions in neural specification. *EMBO J.*, **17**, 191.

92. Brewster, R., Lee, J., and Ruiz i Altaba, A. (1998) Gli/Zic factors pattern the neural plate by defining domains of cell differentiation. *Nature*, **393**, 579.

93. Schoenherr, C. J. and Anderson, D. J. (1995) The neuron-restrictive silencer factor (NRSF): a coordinate repressor of multiple neuron-specific genes. *Science*, **267**, 1360.

94. Chong, J. A., Tapia Ramirez, J., Kim, S., *et al.* (1995) REST: a mammalian silencer protein that restricts sodium channel gene expression to neurons. *Cell*, **80**, 949.

95. Chen, Z. F., Paquette, A. J., and Anderson, D. J. (1998) NRSF/REST is required *in vivo* for repression of multiple neuronal target genes during embryogenesis. *Nature Genet.*, **20**, 136. [See Comments.]

96. Witta, S. E., Agarwal, V. R., and Sato, S. M. (1995) XIPOU 2, a noggin-inducible gene, has direct neuralizing activity. *Development*, **121**, 721.

97. Matsuo-Takasaki, M., Lim, J. K., and Sato, S. M. (1999) The POU domain gene, XlPOU 2 is an essential downstream determinant of neural induction. *Mech. Dev.*, **89**, 75.

98. Kroll, K. L., Salic, A. N., Evans, L. M., and Kirschner, M. W. (1998) Geminin, a neuralizing molecule that demarcates the future neural plate at the onset of gastrulation. *Development*, **125**, 3247.

99. Mariani, F. V. and Harland, R. M. (1998) XBF-2 is a transcriptional repressor that converts ectoderm into neural tissue. *Development*, **125**, 5019.

100. Mizuseki, K., Kishi, M., Shiota, K., Nakanishi, S., and Sasai, Y. (1998) SoxD: an essential mediator of induction of anterior neural tissues in *Xenopus* embryos. *Neuron*, **21**, 77.

101. Campos-Ortega, J. A. (1993) Early neurogenesis in *Drosophila melanogaster*. In *The development of* Drosophila melanogaster (ed. Bate, C. M. and Martinez-Arias, A.), Vol. 1, p. 1091. Cold Spring Harbor Laboratory Press, New York.

102. Skeath, J. B. (1999) At the nexus between pattern formation and cell type specification: the generation of individual neuroblast cell fates in the *Drosophila* embryonic central nervous system. *BioEssays*, **21**, 922.

103. Gomez Skarmeta, J. L., Diez Del Corral, R., De La Calle Mustienes, E., Ferres Marco, D., and Modolell, J. (1996) Araucan and caupolican, two members of the novel Iroquois complex, encode homeoproteins that control proneural and vein-forming genes. *Cell*, **85**, 95.

104. Artavanis Tsakonas, S., Matsuno, K., and Fortini, M. E. (1995) Notch signaling. *Science*, **268**, 225.

105. Simpson, P. (1997) Notch signalling in development: on equivalence groups and asymmetric developmental potential. *Curr. Opin. Genet. Dev.*, **7**, 537.

106. Simpson, P. (1997) Notch signaling in development. *Perspect. Dev. Neurobiol.*, **4**, 297.

107. Lee, J. E. (1997) Basic helix–loop–helix genes in neural development. *Curr. Opin. Neurobiol.*, **7**, 13.

108. Anderson, D. J. and Jan, N. J. (1997) The determination of the neuronal phenotype. In *Molecular and cellular approaches to neural development* (ed. W. M. Cowan, T. M. Jessell, and S. L. Zipursky), p. 26. Oxford University Press, New York.

109. Kageyama, R. and Nakanishi, S. (1997) Helix–loop–helix factors in growth and differentiation of the vertebrate nervous system. *Curr. Opin. Genet. Dev.*, **7**, 659.

110. Selleck, M. A. and Bronner-Fraser, M. (1995) Origins of the avian neural crest: the role of neural plate-epidermal interactions. *Development*, **121**, 525.

111. Brown, J. M. S. and Storey, K. G. (2000) A region of the vertebrate neural plate in which neighbouring cells can adopt neural or epidermal cell fates. *Curr. Biol.*, **10**, 869.

112. Henrique, D., Tyler, D., Kintner, C., *et al.* (1997) Cash4, a novel achaete-scute homologue induced by Hensen's node during generation of the posterior nervous system. *Genes Dev.*, **11**, 603.

113. Turner, D. L. and Weintraub, H. (1994) Expression of achaete-scute homolog 3 in *Xenopus* embryos converts ectodermal cells to a neural fate. *Genes Dev.*, **8**, 1434.

114. Zimmerman, K., Shih, J., Bars, J., Collazo, A., and Anderson, D. J. (1993) XASH-3, a novel *Xenopus* achaete-scute homolog, provides an early marker of planar neural induction and position along the mediolateral axis of the neural plate. *Development*, **119**, 221.

115. Chitnis, A. and Kintner, C. (1996) Sensitivity of proneural genes to lateral inhibition affects the pattern of primary neurons in *Xenopus* embryos. *Development*, **122**, 2295.

116. Ma, Q., Kintner, C., and Anderson, D. J. (1996) Identification of neurogenin, a vertebrate neuronal determination gene. *Cell*, **87**, 43.

117. Takebayashi, K., Takahashi, S., Yokota, C., *et al.* (1997) Conversion of ectoderm into a neural fate by ATH-3, a vertebrate basic helix–loop–helix gene homologous to *Drosophila* proneural gene atonal. *EMBO J.*, **16**, 384.

118. Lee, J. E., Hollenberg, S. M., Snider, L., Turner, D. L., Lipnick, N., and Weintraub, H. (1995) Conversion of *Xenopus* ectoderm into neurons by Neuro D., a basic helix–loop–helix protein. *Science*, **268**, 836.

119. Guillemot, F., Lo, L. C., Johnson, J. E., Auerbach, A., Anderson, D. J., and Joyner, A. L. (1993) Mammalian achaete-scute homolog 1 is required for the early development of olfactory and autonomic neurons. *Cell*, **75**, 463.

120. Sommer, L., Shah, N., Rao, M., and Anderson, D. J. (1995) The cellular function of MASH1 in autonomic neurogenesis. *Neuron*, **15**, 1245.

121. Tomita, K., Nakanishi, S., Guillemot, F., and Kageyama, R. (1996) Mash1 promotes neuronal differentiation in the retina. *Genes Cells*, **1**, 765.

122. Cau, E., Gradwohl, G., Fode, C., and Guillemot, F. (1997) Mash1 activates a cascade of bHLH regulators in olfactory neuron progenitors. *Development*, **124**, 1611.

123. Torii, M., Matsuzaki, F., Osumi, N., *et al.* (1999) Transcription factors Mash-1 and Prox-1 delineate early steps in differentiation of neural stem cells in the developing central nervous system. *Development*, **126**, 443.

124. Bellefroid, E. J., Bourguignon, C., Hollemann, T., *et al.* (1996) X-MyT1, a *Xenopus* C2HC-type zinc finger protein with a regulatory function in neuronal differentiation. *Cell*, **87**, 1191.

125. Dubois, L., Bally Cuif, L., Crozatier, M., Moreau, J., Paquereau L., and Vincent A. (1998) XCoe2, a transcription factor of the Col/Olf-1/EBF family involved in the specification of primary neurons in *Xenopus*. *Curr. Biol.*, **8**, 199.

126. Roztocil, T., Matter Sadzinski, L., Alliod, C., Ballivet, M., and Matter, J. M. (1997) NeuroM, a neural helix–loop–helix transcription factor, defines a new transition stage in neurogenesis. *Development*, **124**, 3263.

127. Haddon, C., Smithers, L., Schneider Maunoury, S., Coche, T., Henrique, D., and Lewis, J. (1998) Multiple delta genes and lateral inhibition in zebrafish primary neurogenesis. *Development*, **125**, 359.

128. Lewis, J. (1996) Neurogenic genes and vertebrate neurogenesis. *Curr. Opin. Neurobiol.*, **6**, 3.

129. Lardelli, M., Williams, R., Mitsiadis, T., and Lendahl, U. (1996) Expression of the Notch 3 intracellular domain in mouse central nervous system progenitor cells is lethal and leads to disturbed neural tube development. *Mech. Dev.*, **59**, 177.

130. Coffman, C. R., Skoglund, P., Harris, W. A., and Kintner, C. R. (1993) Expression of an extracellular deletion of Xotch diverts cell fate in *Xenopus* embryos. *Cell*, **73**, 659.

131. Chitnis, A., Henrique, D., Lewis, J., Ish Horowicz, D., and Kintner, C. (1995) Primary neurogenesis in *Xenopus* embryos regulated by a homologue of the *Drosophila* neurogenic gene Delta. *Nature*, **375**, 761. [See Comments]

132. Dorsky, R. I., Chang, W. S., Rapaport, D. H., and Harris, W. A. (1997) Regulation of neuronal diversity in the *Xenopus* retina by Delta signalling. *Nature*, **385**, 67.

133. Henrique, D., Hirsinger, E., Adam, J., *et al.* (1997) Maintenance of neuroepithelial progenitor cells by Delta-Notch signalling in the embryonic chick retina. *Curr. Biol.*, **7**, 661.

134. Appel, B. and Eisen, J. S. (1998) Regulation of neuronal specification in the zebrafish spinal cord by Delta function. *Development*, **125**, 371.

135. Van Doren, M., Bailey, A. M., Esnayra, J., Ede, K., and Posakony, J. W. (1994) Negative regulation of proneural gene activity: Hairy is a direct transcriptional repressor of achaete. *Genes Dev.*, **8**, 2729.

136. de la Pompa, J. L., Wakeham, A., Correia, K. M., *et al.* (1997) Conservation of the Notch signalling pathway in mammalian neurogenesis. *Development*, **124**, 1139.

137. Ohtsuka, T., Ishibashi, M., Gradwohl, G., Nakanishi, S., Guillemot, F., and Kageyama, R. (1999) Hes1 and Hes5 as notch effectors in mammalian neuronal differentiation. *EMBO J.*, **18**, 2196.

138. Ishibashi, M., Ang, S. L., Shiota, K., Nakanishi, S., Kageyama, R., and Guillemot, F. (1995) Targeted disruption of mammalian hairy and Enhancer of split homolog-1 (HES-1) leads to up-regulation of neural helix–loop–helix factors, premature neurogenesis, and severe neural tube defects. *Genes Dev.*, **9**, 3136.

139. Tomita, K., Ishibashi, M., Nakahara, K., *et al.* (1996) Mammalian hairy and Enhancer of split homolog 1 regulates differentiation of retinal neurons and is essential for eye morphogenesis. *Neuron*, **16**, 723.

140. Van Doren, M., Ellis, H. M., and Posakony, J. W. (1991) The *Drosophila* extramacrochaetae protein antagonizes sequence-specific DNA binding by daughterless/achaete-scute protein complexes. *Development.*, **113**, 245.

141. Modolell, J. (1997) Patterning of the adult peripheral nervous system of *Drosophila*. *Perspect. Dev. Neurobiol.*, **4**, 285.

142. Duncan, M., DiCicco Bloom, E. M., Xiang, X., Benezra, R., and Chada, K. (1992) The gene for the helix–loop–helix protein, Id, is specifically expressed in neural precursors. *Dev. Biol.*, **154**, 1.

143. Jen, Y., Manova, K., and Benezra, R. (1997) Each member of the Id gene family exhibits a unique expression pattern in mouse gastrulation and neurogenesis. *Dev. Dyn.*, **208**, 92.

144. Lyden, D., Young, A. Z., Zagzag, D., Yan, W., Gerald, W., O'Reilly, ?. ?., Bader, B. L. Hymes, R. O., Zhuang, Y., Manova, K., and Benezra, R. (1999) Id1 and Id3 are required for neurogenesis, angiogenesis and vascularization of tumour xenografts. *Nature*, **401**, 670.

145. Chenn, A. and McConnell, S. K. (1995) Cleavage orientation and the asymmetric inheritance of Notch1 immunoreactivity in mammalian neurogenesis. *Cell*, **82**, 631.

146. Zhong, W., Feder, J. N., Jiang, M. M., Jan, L. Y., and Jan, Y. N. (1996) Asymmetric localization of a mammalian numb homolog during mouse cortical neurogenesis. *Neuron*, **17**, 43.

147. Zhong, W., Jiang, M. M., Weinmaster, G., Jan, L. Y., and Jan, Y. N. (1997) Differential expression of mammalian Numb, Numblike and Notch1 suggests distinct roles during mouse cortical neurogenesis. *Development*, **124**, 1887.

148. Wakamatsu, Y., Maynard, T. M., Jones, S. U., and Weston, J. A. (1999) NUMB localizes in the basal cortex of mitotic avian neuroepithelial cells and modulates neuronal differentiation by binding to NOTCH-1. *Neuron*, **23**, 71.

149. Ma, Q., Chen, Z., del Barco Barrantes, I., de la Pompa, J. L., and Anderson, D. J. (1998) neurogenin1 is essential for the determination of neuronal precursors for proximal cranial sensory ganglia. *Neuron*, **20**, 469.

150. Fode, C., Gradwohl, G., Morin, X., *et al.* (1998) The bHLH protein NEUROGENIN 2 is a determination factor for epibranchial placode-derived sensory neurons. *Neuron*, **20**, 483.

151. Pan, D. and Rubin, G. M. (1997) Kuzbanian controls proteolytic processing of Notch and mediates lateral inhibition during *Drosophila* and vertebrate neurogenesis. *Cell*, **90**, 271.

152. Chien, C. T., Hsiao, C. D., Jan, L. Y., and Jan, Y. N. (1996) Neuronal type information encoded in the basic-helix–loop–helix domain of proneural genes. *Proc. Natl Acad. Sci. USA*, **93**, 13239.

153. Lo, L., Tiveron, M. C., and Anderson, D. J. (1998) MASH1 activates expression of the paired homeodomain transcription factor Phox2a, and couples pan-neuronal and subtype-specific components of autonomic neuronal identity. *Development*, **125**, 609.

154. Edlund, T. and Jessell, T. M. (1999) Progression from extrinsic to intrinsic signaling in cell fate specification: a view from the nervous system. *Cell*, **96**, 211.

155. Papalopulu, N. and Kintner, C. (1996) A posteriorising factor, retinoic acid, reveals that anteroposterior patterning controls the timing of neuronal differentiation in *Xenopus* neuroectoderm. *Development*, **122**, 3409.

156. Sharpe, C. R. and Goldstone, K. (1997) Retinoid receptors promote primary neurogenesis in *Xenopus*. *Development*, **124**, 515.

157. Blumberg, B., Kang, H., Bolado, J., Jr, *et al.* (1998) BXR, an embryonic orphan nuclear receptor activated by a novel class of endogenous benzoate metabolites. *Genes Dev.*, **12**, 1269.

158. Johnson, J. E., Zimmerman, K., Saito, T., and Anderson, D. J. (1992) Induction and repression of mammalian achaete-scute homologue (MASH) gene expression during neuronal differentiation of P19 embryonal carcinoma cells. *Development*, **114**, 75.

159. Franco, P. G., Paganelli, A. R., Lopez, S. L., and Carrasco, A. E. (1999) Functional association of retinoic acid and hedgehog signalling in *Xenopus* primary neurogenesis. *Development*, **126**, 4257.

160. Ribisi, S. Jr., Mariani, F. V., Aamar, E., Lamb, T. M., Frank, D. and Harland, R. M. (2000) Ras-mediated FGF signalling is required for the formation of posterior but not anterior neural tissue in Xenopus laevis. *Dev. Biol.* **227**, 183–96.

161. Knoblich, J. A. (2001) Asymmetric cell division during animal development. Nat. Rev. Mol. Cell. Biol. **2**, 11–20.

6 | Anteroposterior regionalization of the vertebrate nervous system

ANTHONY GRAHAM and IVOR MASON

1. Introduction

The complex vertebrate nervous system is constructed in a stepwise fashion during development. Specification of the neural territory at the dorsal surface of the embryo occurs early in development: at the gastrula stage (see Chapter 5). The subdivision of the neural primordium into the four broad territories, which prefigure the gross anatomy of the adult brain (forebrain, midbrain, hindbrain, and spinal cord) occurs both concomitant with and following neural induction. As embryogenesis proceeds, these domains are further partitioned into smaller units, which define the placement of neuronal populations and pioneering of the first axonal pathways. Thereby, the ground plan of the brain's neuronal architecture is established. Thus these early patterning events are critical to the development of the adult nervous system, as they provide the scaffolding upon which the elaborate cytoarchitecture of the brain and spinal cord is realized. In this chapter we discuss the current view of how the nervous system becomes patterned along its long axis (anteroposterior axis). Such a review is timely as we are now beginning to understand both the developmental processes and the molecular effectors that underlie the anteroposterior regionalization of the nervous system.

2. Initial anteroposterior regionalization of the neural primordium

Experimental studies over many decades have firmly established the role of the organizer tissue (dorsal lip in amphibians, node in amniotes, and shield in zebrafish) in inducing the formation of neural tissue that is patterned along the anteroposterior axis (see Chapter 2). In amphibia, it has been shown that grafting the dorsal blastopore lip of an early-stage gastrula embryo to the ventral side of a host embryo, results in that embryo forming a second complete regionally patterned nervous system. Similarly, experiments in amniotes in which Hensen's node was grafted also resulted in the formation of a duplicated, regionalized nervous system. However, it was unclear how anteroposterior identity was actually imparted to the forming nervous system, and a number of models were put forward to explain this phenomenon. One prominent model proposed by Mangold suggested that neural tissue of a different regional character resulted from distinct organizer signals causing the induction of the head, trunk, and tail (1). By contrast, another model put forward by Nieuwkoop suggested that neural tissue is specified initially as forebrain, and that the other neural territories arise as a result of transforming factors which elicit the formation of more posterior fates (2). Although, there is experimental evidence to support each of these models, most recent studies seem to suggest that Nieuwkoop's analysis is closer to reality.

Studies in a number of vertebrate species have shown that the first step in anteroposterior patterning of the nervous system is the specification of forebrain identity. However, it is currently unclear as to whether all neural tissue is induced with forebrain character, or whether this is secondarily imposed upon truly naïve neuroectoderm. Prospective posterior neural tissue, even from regions subsequently fated to give rise to the spinal cord, initially expresses markers characteristic of the forebrain in both *Xenopus* and chick (3, 4). Although, other studies suggest that, while anterior markers (*Otx2*) are expressed throughout the avian epiblast prior to neural induction, certain posterior markers (*Gbx2*) are already localized to caudal epiblast (5–7). Thus, neural induction may occur in epiblast that is already crudely regionalized along its anteroposterior axis.

Regardless of whether or not induced neural tissue is truly naïve, it has recently become clear that anterior identity is either induced or maintained by signal(s) from the cells that underlie the anterior neural plate, which, at least in part, are the organizer-derived head mesendodermal cells (8–10). However, studies in both mouse and *Xenopus* have also suggested that initially the source of this signal is from non-organizer-derived endodermal cells, which express the homeobox gene *Hesx1*, (11, 12) and that it is only later that this activity is transferred to the head mesendodermal cells (11). Interestingly, these anterior cells do not have direct neural inducing activity themselves but rather act to confer or maintain anterior character on neural plate cells. In the mouse embryo, the endodermal cells are located in the anterior of the extraembryonic visceral endoderm (AVE) and express a number of molecular markers prior to the onset of gastrulation (*Otx2*, *Lim1*, *Gsc*, and *Cerberus-related-1* (13–16)). AVE can ectopically induce anterior markers in chicks and mice (10, 17) and its ablation affects forebrain patterning. The extent of the influence of the AVE and anterior mesendoderm is revealed by transgenic mice lacking Otx2 or Lim1 function, in which anterior brain structures are deleted as far posterior as rhombomere 3 of the hindbrain (14, 18–20). We currently understand rather little about the mechanism by which AVE or anterior mesendodermal cells signal to their overlying targets, but there is increasing evidence that part of their role involves inhibition of Wnt signalling (21–24)

In keeping with Nieuwkoop's model, it seems that specification of the more caudal neural territories involves the imposition of a posterior character on neural tissue that has initially been assigned a forebrain fate, and direct experimental support for this view has recently been forthcoming. Experiments in *Xenopus* have shown that direct neural inducers cause the formation of anterior neural tissue (25–27). In addition, chick caudal epiblast cells, fated to give rise to hindbrain and spinal cord, are first specified as forebrain (28): if caudal avian epiblast cells from early gastrulating embryos are explanted into a dish they will form forebrain tissue. However, as development proceeds the caudal epiblast cells no longer form forebrain tissue when placed in culture, but produce midbrain and hindbrain. This transformation of specification is due to the influence of posterior tissue, and it has been shown, for example, that forebrain tissue when recombined with posterior axial tissue gave rise to midbrain and hindbrain (29). More specifically, detailed studies in zebrafish and chicks have pinpointed the paraxial mesoderm as the source of these signals (28, 30). Transplantation experiments in zebrafish have shown that non-axial mesodermal cells can posteriorize forebrain progenitors (30). Similarly, work in chicks has demonstrated that if presumptive forebrain cells are grown with paraxial mesoderm then they adopt a caudal fate (28). This latter study also showed that the transforming ability of the paraxial mesoderm changes as the embryo grows. Epiblast cells specified as forebrain will form midbrain tissue when recombined with young paraxial mesoderm, yet if recombined with slightly older paraxial mesoderm they will form hindbrain. These epiblast cells can be further induced to form spinal cord when exposed to paraxial mesoderm from distinctly older embryos.

In contrast to the induction of anterior tissue, we do have some idea of the molecules that are involved in posteriorizing the nervous system, with many studies implicating fibroblast growth factors (FGFs) and retinoic acid in this process (29–34) (see also Chapter 5). These factors can certainly promote posteriorization of neural tissue, but they may not act alone, and there is emerging evidence for the existence of an as yet unidentified paraxial mesoderm caudalizing factor that acts in concert with these signals. Presumptive forebrain tissue will not acquire posterior traits when exposed only to FGFs, but will do so when this is coupled with treatment with paraxial mesoderm caudalizing factor (28). Similarly, in recombination experiments, if FGF signalling is blocked, then paraxial mesoderm cannot induce midbrain or hindbrain formation in presumptive forebrain tissue. While these results suggest that posterior fates are due to the activity of FGFs and a paraxial mesoderm caudalizing factor, there clearly must be other factors if separate midbrain versus hindbrain identities are to be imparted. Indeed, the development of midbrain characters seems to be a result of the influence of these posteriorizing factors acting in concert with the anterior head mesendodermal signal (28). In contrast, the specification of more caudal regions of the nervous system, the posterior hindbrain and spinal cord, does not involve FGFs. Rather the assignment of identity at these axial levels seems to involve retinoic acid, itself produced by paraxial mesoderm (35), again acting alongside a second paraxial mesoderm caudalizing factor (28, 34). Thus treatment of prospective forebrain tissue with retinoic acid alone does not generate spinal cord, but it does if combined with exposure to paraxial mesoderm caudalizing signal. Similarly, interfering with retinoid signalling, either in *Xenopus* embryos or in tissue recombinations between paraxial mesoderm and epiblast cells, perturbs the specification of caudal hindbrain and spinal cord identities (28, 34).

While the identities of some of the additional paraxial mesoderm-derived, posteriorizing activities remain elusive, we can infer something of the nature of their action from 'Keller explant' studies in *Xenopus*. These have shown that anteroposterior pattern, including assignment of midbrain and hindbrain identity, can occur in the absence of adjacent paraxial mesoderm tissue (36). These data can be reconciled if the influence of paraxial mesoderm is mediated in the plane of the neuroepithelium, which is not unexpected since the influence of the paraxial mesoderm signal apparently extends rostral to the postotic anterior limit of this tissue.

Once large territories, forebrain, midbrain, hindbrain, and spinal cord, have been specified they becomes further regionalized, and this is largely due to the action of specific local cues.

3. Patterning the forebrain

Current evidence suggests that all neural tissue is induced with forebrain or prosencephalic character, and that the extent of the forebrain anlage is determined by the anterior limit of the influence of secondary, posteriorizing factors (see above). The forebrain then becomes subdivided into the telencephalon and diencephalon, which are readily distinguishable morphologically and each of which forms the primordium for a number of anterior brain structures. Derivatives of the telencephalon include the cortex, neocortex, and hippocampus from the dorsally located pallium, the striatum and septum from the ventrally positioned subpallium, and the olfactory bulbs. While diencephalic derivatives include the pretectum, dorsal thalamus, ventral thalamus, and hypothalamus, which differ in their origin along the anteroposterior axis. Lateral diencephalic outgrowths form the optic lobes (which ultimately give rise to the retinas), a ventral outpouching forms the infundibulum which contributes to the pituitary, while a dorsal outgrowth forms the epiphysis. With reference to the anteroposterior axis, the telencephalon and diencephalic hypothalamus are the most anterior forebrain derivatives with the former developing dorsally and the latter ventrally, probably from the alar and basal plates, respectively.

A key question is how do the telencephalon and diencephalon become further subdivided in order to produce their diverse derivatives? Older 'columnar models' of His and Herrick (see, for example, ref. 37) have been superseded in recent years by neuromeric models (Fig. 1) that draw upon lessons learned from hindbrain organization and development (see later). Figdor and Stern (38) reported the morphological subdivision of the avian diencephalon into four neuromeres (D1–D4), the boundaries of which coincided with the position of axon tracts. Alternate neuromeres had high or low levels of acetylcholinesterase (AChE)-like immunoreactivity and cell-labelling experiments suggested that each was a lineage-restricted compartment. No evidence was found for such morphological divisions within the telencephalon. By contrast, in

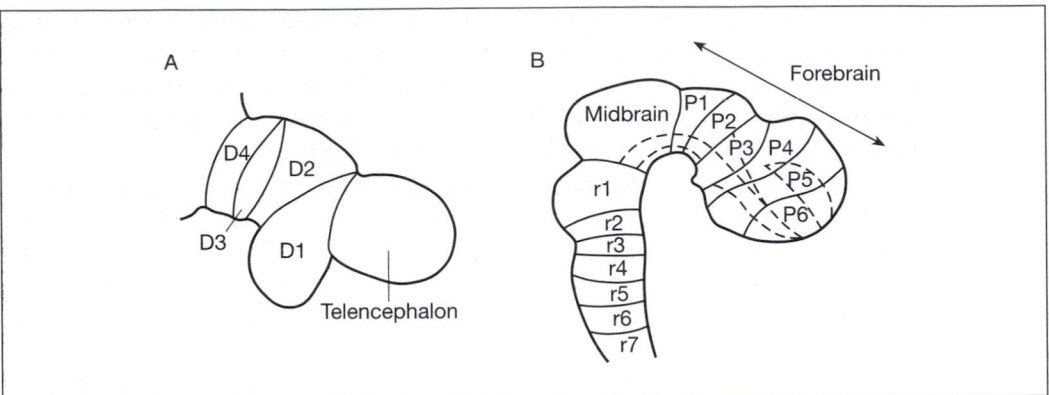

Fig. 1 Neuromeric subdivisions of the developing forebrain. (A) Subdivision of the diencephalon into four neuromeres: D1–D4. (After Figdor and Stern (38).) (B) Subdivision of the entire forebrain into six prosomeres: P1–P6. (After Puelles and Rubenstein (39). Some of the further transverse subdivisions are indicated by the dotted lines.

the same year, others proposed that the entire forebrain was subdivided into six 'prosomeres' (P1–P6) (39). This subdivision was based upon reported patterns of gene expression (*Em1/2, Dl1/2, Nk2.1/2.2, Pax-6, BF1/2, Tbr1,* and *Wnt3I*) in embryonic rodent forebrains, although subsequent work has shown that such gene-expression domains are largely conserved in other vertebrates (40, 41). In this model, the pretectum derives from P1, the dorsal thalamus from P2, the ventral thalamus from P3, and the telencephalon and hypothalamus from dorsal and ventral P4–6, respectively. Each prosomere is further subdivided into different territories along its dorsoventral axis (39).

It is now becoming clear, particularly from studies in zebrafish, that the boundaries of gene expression that define or further subdivide prosomeres serve to guide pioneering axons within the developing forebrain (42). However, the status of the prosomere or even the diencephalic neuromere, as defined by Figdor and Stern, as lineage-restricted compartments is uncertain. While marking cells with lipophilic dyes in avian embryos appeared to demonstrate a lack of cell movement across neuromere boundaries, other studies fail to detect such restrictions. The use of retroviruses as lineage markers in rodent embryos showed that sibling cells could colonize the entire extent of the diencephalon (43) and fail to respect prosomeric boundaries in the prospective cortical region (44). Likewise, cell-labelling experiments show tangential migration of postmitotic cells throughout the telencephalon without respect for prosomere interfaces (45, 46). While lineage restriction is unclear, there is some evidence that axial identity is maintained. Transplantation experiments suggest that, within the telencephalon, cells are completely plastic in their ability to mix and differentiate when relocated in the dorsoventral plane, but that they autonomously maintain their anteroposterior identity (47–49).

If prosomeric subdivisions do indeed underpin the patterning of the forebrain along its anteroposterior axis, we currently have few clues to the mechanism by which they are generated. However, organizer properties have been attributed to the most anterior cells of the neural plate at stages before the appearance of prosomeric patterns of gene expression. It also seems likely that a second signalling centre, the zona limitans intrathalamica (ZLI), forms subsequent to prosomeres at the P2/3 boundary (D2/3 boundary of Figdor and Stern (38)).

The fundamental patterning role of the most anterior neural plate cells was identified in zebrafish when the most anterior row of six to eight cells (row 1) was removed from embryos by aspiration, resulting in the loss of telencephalic patterning. Transplantation of these cells to more caudal positions within the neural plate resulted in the induction of ectopic gene expression characteristic of the telencephalon (50). The equivalent tissue in higher vertebrates is very likely an anterior thickening of the neural plate: the anterior neural ridge (ANR). Removal of the ANR from explants of mouse forebrain tissue results in loss of the anterior neural marker *BF1*. *BF1* expression can be induced by FGF8; *Fgf8* is expressed in the ANR and, subsequently, the anterior neuropore (51). Row 1 cells also express *Fgf8*, but FGF8 is unable to rescue telencephalic development in zebrafish lacking row 1, indicating that it does not account for all of their patterning properties (50). It is noteworthy that, in avian embryos, the onset of *Fgf8* and *BF1* expression occurs simultaneously, suggesting that *Fgf8* might normally serve to maintain rather than induce *BF1* (E. Bell, H. Shamim, A. Lumsden, and I. Mason, unpublished data). However, mice hypomorphic for FGF8 lack anterior forebrain structures, e.g. olfactory bulbs, indicating a crucial role for this molecule in forebrain patterning (52).

By contrast, the zona limitans intrathalamica (ZLI) is much less studied. However, drawing upon lessons learned from the function of the isthmus (midbrain–hindbrain boundary; see later), the association of several secreted signalling proteins with this interprosomeric boundary strongly suggests an anteroposterior axial patterning function. Most intriguingly, the ZLI is the one region in which *shh* expression extends dorsally across the entire neural tube. In addition, *Wnt*, *fringe*, and, dorsally, *Fgf8* transcripts are also associated with this boundary region (53). However, the patterning function(s) of the ZLI remain to be elucidated.

Differential vertical signalling from underlying prechordal mesoderm and noto-chord may also contribute to anteroposterior subdivision of the ventral forebrain. The anterior limit of the notochord approximates to the prospective position of the infundibulum, with prechordal mesoderm underlying more anterior regions. Prechordal mesoderm differs from notochord in its pattern of gene expression: *Bmp7*, *gsc*, *otx2*, and *ntl* are all specifically expressed in prechordal mesoderm, while *shh* is transient (54–56). Most strikingly, the loss of anterior ventral forebrain derivatives in the *Cyclops* zebrafish is due to mutation in a nodal-related gene expressed in the prechordal mesoderm (57–59).

4. Patterning the midbrain

The midbrain of the embryo gives rise dorsally to the colliculi structures of the mammalian brain (tectum in avian embryos), and within the ventral tegmentum are located the oculomotor nuclei and the dopaminergic neurons that are lost in Parkinson's disease.

Whereas segmentation underpins the development of the hindbrain, and possibly also the forebrain, anteroposterior patterning of the midbrain is regulated by an organizing centre located at the midbrain–hindbrain junction or isthmus. The isthmus is the best-characterized organizer region of the developing nervous system. Its properties were first demonstrated in a series of tissue-grafting experiments performed in the chicken embryo. These studies were initiated by Nakamura, Alvarado-Mallart, Sotello, and their colleagues, but were subsequently extended and refined by these and other groups. In essence, it was found that isthmic tissue when transplanted into anterior midbrain respecified that structure to develop as posterior midbrain tissue. More dramatically, transplantation into posterior diencephalon respecified that tissue to develop as an ectopic midbrain (reviewed in refs 60–62). Thus, the isthmus was identified as an organizing centre imparting posterior character to the midbrain but capable of ectopically inducing midbrain fate in other regions of the developing brain.

Clearly, identification of the factor or factors produced by the isthmus and responsible for its patterning activities was of considerable interest. Several groups reported that FGF8 was expressed specifically at the isthmus in mouse embryos (63, 64), thus it became a good candidate for the isthmic patterning activity. Studies in which the FGF8 protein was expressed ectopically within the chick embryo brain showed that it could induce patterns of gene expression characteristic of posterior midbrain within either anterior midbrain (Fig. 2) or posterior forebrain (P1 or P2). Gene-targeted mutagenesis in mice had previously shown that many of these genes (*Pax2*, *En1*, *Wnt1*) were themselves required for normal development and patterning of the midbrain and anterior hindbrain (see ref. 65 for discussion and references therein). FGF8 was also found to be mitogenic for midbrain tissue; the highest mitotic index within the developing midbrain normally occurs adjacent to the isthmus. Embryos, when allowed to develop further, showed morphological respecification of anterior midbrain to a posterior character. Most significantly, FGF8 alone was sufficient to redirect posterior forebrain to a midbrain fate (65–68). These results have been further supported by the phenotypes of *acerebellar* zebrafish which lack functional FGF8 protein and transgenic mice that are hypomorphic for *Fgf8* (69, 70). Thus, *Fgf8* alone is sufficient to account for all the patterning properties previously ascribed to the isthmus in tissue-grafting studies. However, temporal analyses have shown that *Fgf8* expression is detected only after the onset of expression of some of the posterior midbrain genes that it induces when expressed ectopically, e.g. *En1* and *Pax2*. This has prompted a suggestion that the normal function of FGF8 at the isthmus is not to induce their expression but to both maintain and refine it to the posterior midbrain (65).

In addition to respecification of anterior midbrain to a posterior character, ectopic cerebellum-like structures were induced locally around an FGF8-coated bead placed in the midbrain (68). This was somewhat surprising since the cerebellum is a derivative of the anterior hindbrain (see below). However, recent molecular analyses

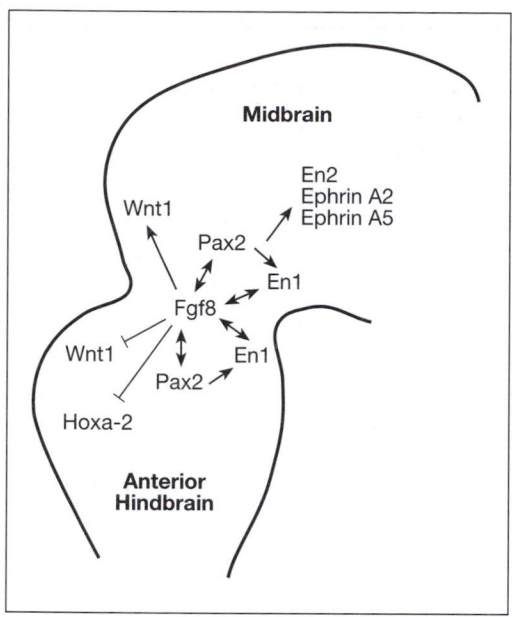

Fig. 2 Regulation of gene expression by FGF8 in posterior midbrain and anterior hindbrain based upon the results of ectopic FGF8 protein expression.

have shown that FGF8 locally respecifies midbrain tissue to an anterior hindbrain phenotype ($Otx2-$, $Gbx2+$ and $Fgf8+$; 68, 71, 72).

The factors that initiate $Fgf8$ expression within this region are as yet uncharacterized, although it has been shown that $En1$, the first of the posterior midbrain genes to be expressed in that region of the chick embryo, is capable of inducing ectopic $Fgf8$ expression (65). Rather more is known about how $Fgf8$ is maintained at the isthmus. Irving and Mason (71) showed that this occurs through a direct interaction between midbrain and rhombomere 1 (r1). An, as yet unidentified, secreted activity from the midbrain is capable of inducing $Fgf8$ in r1 but not other hindbrain rhombomeres. Midbrain and anterior hindbrain are distinguished by expression of the transcription factors, $Otx2$ and $Gbx2$, respectively, whose normal domains of expression abut the isthmus (73, 74). The activities of these two genes seem to be responsible for the normal positioning of the $Fgf8$ domain at the isthmus, as ectopic expression of either shifts the domain anteriorly or posteriorly (75, 76). However, in the absence of either gene, $Fgf8$ is still expressed but fails to refine to a narrow isthmic domain of expression.

However, while Fgf8 provides a very good candidate for the isthmic patterning activity, several questions remain to be answered. The nature of the inductive signal that regulates the onset of expression of genes expressed before $Fgf8$ such as $Pax2$ and $En1$, as well as possibly $Fgf8$ itself, remains unclear. Studies in both $Xenopus$ and chick embryos have suggested that the notochord is required for En expression in the presumptive midbrain territory and that $Fgf4$, transiently expressed by the notochord, is sufficient to rescue $En1$ expression (65). However, whether $Fgf4$ was permissive or instructive in this context is unclear. Moreover, these data are difficult to reconcile with the explant studies of Doniach and colleagues (36) and studies in which the node was ablated in avian embryos (see ref. 129 for discussion) that appear to show that axial mesoderm is not required for *En* expression in the developing brain.

In addition, some workers have suggested that *Fgf8*-responsive genes show graded expression decreasing anteriorly away from the isthmus. This has been reported for *En2* and for proteins that provide instructional cues for the formation of the retinotectal projection: ephrins A2 and A5 (77). While this may be simply a response to a diminishing *Fgf8* signal, it raises the possibility of an antagonistic signal emanating from the midbrain–forebrain junction. Tissue-grafting studies provide some, albeit weak, evidence for the existence of such a patterning influence (78).

5. Patterning the hindbrain

The embryonic hindbrain, or rhombencephalon, produces the structures of the cerebellum and brainstem. This region of the neuraxis has an anterior boundary where it meets the midbrain, and a posterior limit at the occipital/cervical interface, which is at the level of somites 5/6. A central feature of the development of this region of the nervous system is that of segmentation (79). In contrast to other regions of the nervous system, the hindbrain is overtly segmented—being subdivided into eight units, termed rhombomeres, strung out along the anteroposterior axis and numbered 1–8 from anterior to posterior (Fig. 3). The segmental organization of the hindbrain underpins its development, and the rhombomeres are the building blocks with which this region of the neuraxis is constructed. The process of neurogenesis is clearly segmental in the hindbrain, with neurons first being evident in the even-numbered rhombomeres, 2, 4, and 6, and then later in the odd-numbered rhombomeres, 3 and 5 (79). Indeed, in many ways rhombomeres represent modular units within which the same basic components are laid down. For example, each rhombomere contains roughly the same set of interneurons (80). However, rhombomeres are also individual and have specific identities. This is clearly evident in the development of the motor nuclei of the hindbrain. The motor nuclei that innervate the muscles of the branchial arches develop within the confines of pairs of rhombomeres; the trigeminal motor neurons, which innervate first-arch muscles, arise within rhombomeres 2 and 3, while the facial neurons, which innervate the second arch, form within rhombomeres 4 and 5 (79). Similarly, the somatic motor nuclei of the hindbrain also develop within specific rhombomeres, with the trochlear forming in rhombomere 1, and the abducens in rhombomeres 5 and 6 (79). Finally, rhombomere 4 also generates a unique group of motor neurons, the contralateral vestibuloacoustic (CVA) neurons (81), as well as, in fish and amphibia, giving rise to the very prominent Mauthner neurons.

Rhombomeres also seem to represent developmental compartments. Lineage-tracing studies in the chick have shown that there is no mixing of neuroepithelial cells between rhombomeres, and that rhombomeres remain lineage-restricted until late stages (82, 83). However, once a neuron is born, and has acquired its identity, it may cross between rhombomeres (84). Interestingly, the non-mixing of cells between

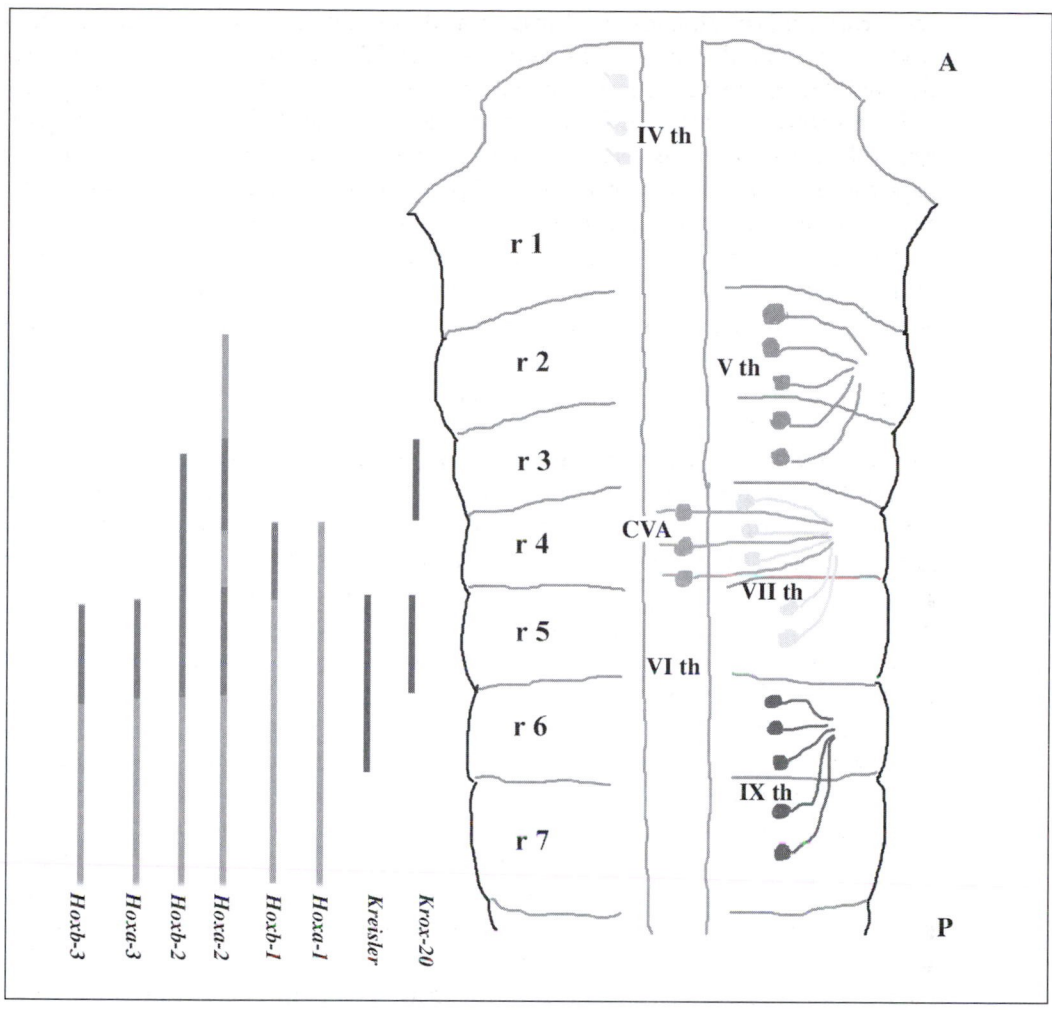

Fig. 3 (See also Plate 3) Schematic of the hindbrain. The rhombomeres are labelled r1 through to r7 from anterior (A) to posterior (P) (r8 not shown). The motor nuclei of the hindbrain are labelled and coloured (in the plate): trochlear (IVth) is light blue; the trigeminal (Vth) red; the facial (VIIth) green; the contral vestibuloacoustic (CVA) pink; the abducens (VIth) yellow; the glossopharyngeal (IXth) dark blue. The expression domains of *Krox-20* and *Kreisler* are shown in blue, and the expression patterns of the *Hox* genes in orange, with regions of elevated expression shown in red.

rhombomeres may be partly due to the differential adhesion of odd-numbered versus even-numbered rhombomeres (85, 86). If two odd-numbered, or two even-numbered, rhombomeres are experimentally juxtaposed then cells will mix between them (85). However, if an odd-numbered and an even-numbered rhombomere are confronted with each other then a new boundary is established and cell mixing is inhibited (85). Correspondingly, it has been shown in cell-mixing experiments that cells from even-numbered rhombomeres sort out from cells of odd-numbered rhombomeres (86).

A molecular explanation of this phenomenon has recently been forthcoming, and it involves an interaction between the Eph receptors, which are receptor tyrosine kinases, and their ligands the ephrins. The Eph receptors and the ephrins fall into two classes (87): EphA receptors bind the ephrin-A ligands which are glycosyl-phosphatidylinositol (GPI) anchored to the membrane, while the EphB receptors bind the transmembrane ephrin-B proteins. However, this subdivision is not absolute, and the EphA4 receptor, besides binding ephrin-A ligands, also binds ephrin-B2 and ephrin-B3. Another notable feature of the Eph receptors and the ephrin-B ligands is that the interaction between these two components results in bidirectional signalling, with pathways being activated in both the receptor-bearing cell and the ligand-bearing cell (88). Importantly, the odd-numbered rhombomeres 3 and 5 express the receptors *EphA4*, *EphB2*, and *EphB3* genes, while the even-numbered rhombomeres express the ligands, *ephrin-B1*, *-B2*, and *–B3* genes (89). Furthermore, experiments in zebrafish have shown that mosaic activation of the ephrins causes cell sorting to the boundaries in even-numbered rhombomeres, while mosaic activation of the Eph receptors results in sorting to the boundaries in odd-numbered rhombomeres (90). Thus it would seem that the restriction in cell mixing between odd-numbered and even-numbered rhombomeres may be due to the bidirectional activation of ephrin and Eph signalling at the interface between rhombomeres.

In contrast to other regions of the developing nervous system, we have a fairly good knowledge of the control genes that underlie the patterning of the hindbrain, and at the forefront of this work has been the analysis of the *Hox* genes (91). It was observed that these homeobox-containing transcription factors showed segmentally restricted expression patterns in the developing hindbrain (Fig. 3). The *Hox* genes are of particular importance, as they seem to play a conserved role in patterning the rostrocaudal axis of all metazoa (91) (see also Chapter 4 for discussion of *Hox* genes and anteroposterior patterning of the vertebral column). These genes are invariably clustered, and within each cluster there is colinearity between the chromosomal order of the genes and their expression domains along the anteroposterior axis, with each successive 3′ gene having a more anterior limit of expression (91). Importantly, those *Hox* genes whose expression extends into the hindbrain display precise rostral limits that coincide with rhombomere boundaries, with each paralogous group of *Hox* genes displaying a similar rostral limit (91). Although, superimposed upon this pattern are differences in expression levels of these genes in particular segments. Thus, the expression domains of the *Hox* genes in the hindbrain form a nested set, and each rhombomere can be defined by its particular *Hox* gene expression profile. For example, rhombomere 2 is marked by its singular expression of *Hoxa-2*, rhombomere 3 by the coexpression of *Hoxa-2* and *Hoxb-2*, and rhombomere 4 by the expression of these two genes plus expression of high levels of *Hoxb-1* and the early expression of *Hoxa-1* (Fig. 3).

Clearly, these expression domains make the *Hox* genes prime candidates for conferring rostrocaudal identity upon the rhombomeres of the hindbrain, and this assertion has been, to some extent, confirmed by mutational analysis of these genes. For example, when the *Hoxa-1* gene, which is normally expressed in r4 and caudal, is ectopically expressed in r2 it causes that segment to develop as if it were r4 (92, 93). In mice, this segment now expresses the *Hoxb-1* at high levels, and develops CVA neurons, while a similar manipulation in zebrafish causes the production of an ectopic Mauthner neuron, a rhombomere 4 specific-cell type, in rhombomere 2. Contrastingly, if *Hoxa-1* is deleted then the regions that normally express this gene are severely affected; r4 is reduced and r5 is deleted (94, 95). Similarly, mutation analysis of the *Hoxb-1* gene, which displays prominent expression in rhombomere 4, has shown that this gene is required for normal rhombomere 4 development (96, 97). In these animals, although the r4 territory is delineated, and it does produce facial and CVA neurons, neither of these neuronal populations mature, they do not undergo their normal migratory behaviour, and r4 derivatives are eventually lost.

The mouse mutants for each of these genes, *Hoxa-1* and *Hoxb-1*, reveal some interesting points. First, although both genes are expressed in rhombomere 4 and caudal of there, the defects are concentrated at the anterior end of the expression domains, and this is generally true for other *Hox* gene mutants. For example, when *Hoxa-2* is knocked out, although this gene is expressed from the rhombomere 1/2 boundary to the posterior tip of the nervous system, the defects are focused upon rhombomeres 2 and 3 (98). These results demonstrate that each *Hox* gene only exerts its effect within its rostral domain of expression, and this has been termed 'posterior prevalence' (99). This concept asserts that posterior *Hox* genes impose their function in a dominant fashion on overlapping anteriorly expressed *Hox* genes. Therefore, in the case of rhombomere 4 that expresses *Hoxa-2*, *Hoxb-2*, *Hoxa-1*, and *Hoxb-1*, it is those genes with the more posterior expression pattern, *Hoxa-1* and *Hoxb-1*, which impose their actions. The second point that emerges is that *Hox* genes have multiple roles in directing hindbrain development. Thus, although *Hoxa-1* and *Hoxb-1* are expressed from very early stages in rhombomere 4, the loss of function of each gene does not result in a complete failure of rhombomere 4 development, but rather in specific defects. It is only in animals that are mutant for both *Hoxa-1* and *Hoxb-1* that rhombomere 4 fails to develop (100). This suggests that these *Hox* genes act synergistically to promote rhombomere 4 development, and this situation seems to be generally true. This also helps to explain the less dramatic nature of the defects observed than would simply be predicted from expression patterns.

Studies on the *Hox* genes have also given us insights into genetic hierarchies that act to control anterior to posterior patterning in the hindbrain. Perhaps surprisingly, this work has shown that the full *Hox* expression domains are, to an extent, built up piecemeal, with specific enhancers directing expression in particular segments. For example, dissection of sequences located upstream of *Hoxb-2* have identified one enhancer that directs expression of this gene in rhombomeres 3 and 5 and another enhancer for rhombomere 4 expression (101, 102). Similarly, it has been shown that the normal expression pattern of *Hoxb-1* is due to the influence of a 3' enhancer that directs early high-level expression in the neural tube, and another enhancer that promotes the later rhombomere 4 specific expression of this gene (103).

Besides identifying the elements of DNA that direct expression to specific segments, these studies have also pinpointed some of the transcription factors that act upon these elements to control gene expression. In-depth analysis of the rhombomere 3/5 enhancer, which is found not only in the *Hoxb-2* gene but also in *Hoxa-2*, demonstrated that this element contained a binding motif for the zinc-finger transcription factor, *Krox-20* (101, 104). Importantly, *Krox-20* is expressed very early in hindbrain development in two stripes, which correspond to rhombomeres 3 and 5 (see Fig. 3), before the establishment of the *Hox* gene expression domains (105). If this motif is mutated in either the *Hoxa-2*, or *Hoxb-2*, 3/5 enhancer, such that *Krox-20* can no longer bind to this element, then expression is no longer directed to these rhombomeres, showing that this transcription factor is necessary for these genes to be expressed in these segments (101, 104). Another transcription factor that has been shown to directly regulate *Hox* gene expression is the b-zip transcription factor, *Kreisler* (106). Again this gene is expressed early in the developing hindbrain, before the *Hox* genes, although *Kreisler* is found in rhombomeres 5 and 6 (Fig. 3) (106). This transcription factor has been found to be necessary for directing the expression of *Hoxa-3* and *Hoxb-3* in these two segments, which constitute their anterior expression domain (107, 108). Dissection of the regulatory regions of these two genes identified an enhancer associated with each that drives expression in rhombomeres 5 and 6. Importantly, each of these elements contained consensus binding sites for *Kreisler*, which were shown to be essential for these enhancers to function.

Further evidence that these two transcription factors are required for directing *Hox* gene expression in the developing hindbrain, and indeed that their function is required for normal hindbrain patterning, comes from mutational studies. In animals lacking the *Krox-20* gene, although the presumptive rhombomere 3 and rhombomere 5 territories are specified, they do fail to develop normally, and with time are lost (109). Moreover, the expression profiles of both *Hoxa-2* and *Hoxb-2*, which have enhancers with *Krox-20* binding sites associated with them, are altered, with neither of these genes showing high-level expression in these segments. Similarly, in *Kreisler* mutants the hindbrain is severely abnormal (110, 111). In these animals rhombomere 5 fails to develop, and a region of neural tissue of indeterminate identity, although with some rhombomere 6 characteristics, forms between rhombomeres 4 and 7. *Hox* gene expression is also perturbed in these mutants and, as predicted from the studies on gene regulation, the anterior domains of *Hoxa-3* and *Hoxb-3* expression are severely affected.

Although these studies have given us great insight into the genetic regulation of rhombomere identity, we still don't know how the hindbrain comes to be segmented. *Krox-20* was considered to be a strong candidate for fulfilling such a role, as it is expressed in the presumptive rhombomere 3 and 5 territories before morphological segmentation. However, these two segments are still delineated in the *Krox-20* mutants (109). In fact the process of segmentation is probably linked with that of anteroposterior regionalization. Support for this assertion comes from studies of animals mutant for *Kreisler*. In both mouse and zebrafish embryos, which lack this gene function, the posterior hindbrain fails to undergo normal segmentation, and, concomitant with this, is a failure in normal anteroposterior patterning (110, 112). Further support also comes from the analysis of retinoid-deficient quail embryos. In this situation the posterior hindbrain is not specified and, instead, the cells, which would have normally contributed to this structure, participate in the formation of an enlarged anterior hindbrain (113). Similarly, mice mutant for *RALDH2*, the retinoic acid synthesis enzyme, also show a lack of posterior hindbrain specification, and the presence of anterior hindbrain markers caudally (114).

The fact that retinoid-deficient quail embryos lack normal anteroposterior patterning of the hindbrain is also interesting because it suggests that retinoic acid plays a role in this process, and indeed there is persuasive evidence for this. It has been known for many years that *Hox* genes are differentially sensitive to retinoic acid, such that those that lie towards the 3′ end of the *Hox* clusters respond to lower levels of this factor than the more 5′ genes (91) (see also Chapter 4). It has also been shown that the ectopic application of retinoic acid in mice can result in the presence of two stripes of *Hoxb-1* expression in the hindbrain (114); one in the normal rhombomere 4 position and another ectopically in what would have been rhombomere 2. In these animals there was also evidence for a general transformation of the identity of rhombomeres 2 and 3 towards that of rhombomeres 4 and 5 (115). The effect of retinoic acid on the *Hox* genes is also likely to be direct as retinoic-acid response elements, RAREs, have been found in the regulatory apparatus of both *Hoxa-1*, *Hoxb-1*, and *Hoxb-4* (103, 116). Mutational analysis of these elements has shown that they are required to establish the early expression patterns of these genes. Moreover, direct evidence for retinoic acid playing a role in determining the anteroposterior limits of expression of individual *Hox* genes comes from experiments in which RAREs were exchanged between two members of the family (116). If the *Hoxb-4* RARE element is exchanged for the *Hoxb-1* RARE then the normal domain of *Hoxb-4* expression was extended to the rhombomere 3/4 level, which is the usual anterior limit of *Hoxb-1* expression (116). Retinoic acid can also affect *Kreisler* function. If retinoic acid-soaked beads are placed opposite the anterior hindbrain, then expression of this gene can be induced in the rhombomere 2/3 territory (117). Interestingly, *Kreisler* expression can also be induced in the anterior hindbrain by grafting somites to this area, and in fact this manipulation has also been shown to induce caudal hindbrain expression (117). Given that these structures express high levels of the retinoid biosynthetic enzyme, *RALDH2*, it is likely that these are the endogenous source of retinoic acid in the hindbrain territory (36).

The above discussion of hindbrain development may give the impression that rhombomeres develop as autonomous units. A number of experiments, however, suggest that this is not completely true. When rhombomeres are transplanted anteriorly within the hindbrain they maintain their identity, so suggesting autonomy. For example, if rhombomere 4 is grafted to the position of rhombomere 2, then the grafted tissue still expresses high levels of *Hoxb-1*, indicative of its rhombomere 4 identity (118). However, if rhombomeres are transplanted posteriorly then they can change their identity; if anterior segments are placed in the postotic environment then they will start to express *Hoxb-4* (119). This latter result is in keeping with the somites playing a major role in patterning the neuraxis, as the grafted anterior hindbrain tissue is now surrounded by somites, and it seem that these are the source of the posteriorizing signal. In addition, inter-rhombomeric signalling seems not only to be involved in reinforcing rhombomeric boundaries but also in determining aspects of their phenotype. A notable feature of rhombomeres 3 and 5 is that they are depleted in their production of neural crest, which arises from the lateral edges of the neural folds during closure of the dorsal neural tube and then emigrates away to form a large number of different cell types (see Chapters 8 and 10) (120). In these two segments the majority of the crest is apoptotically deleted, and interestingly this cell-death programme is induced by an interaction with their flanking even-numbered rhombomeres (120). If either rhombomere 3 or 5 are removed from their neighbouring segments, either by placing them in culture or by transposition within the embryo, then they become unconstrained and produce a vast amount of neural crest. The molecular effector of this interaction is as yet unidentified, but the death programme itself involves *Bmp4* inducing *msx-2* expression and finally neural crest apoptosis (121). Other aspects of rhombomere 3 are also controlled by neighbour interaction (122). Rhombomere 3 normally exhibits high level *Krox-20* expression and the absence of *follistatin* expression. However, this segment is removed from the influence of its neighbours then *Krox-20* becomes downregulated, while *follistatin* becomes strongly expressed. The important point about these results is that they demonstrate that rhombomeres are not isolated entities, but that they talk to each other and regulate aspects of their development.

6. Influence of the isthmus on the anterior hindbrain

Rhombomere 1 develops at the interface between two distinct patterning strategies; graded signalling from the isthmus confronts segmentation of the hindbrain, raising the question as to whether or not the isthmus also influences its development. This most anterior hindbrain segment shares few properties with its posterior neighbours. It is the only hindbrain segment that lacks any *Hox* gene expression. Its motor nucleus (IV or trochlear nucleus) forms only anteriorly and has a unique axonal projection, and it develops unique neuronal populations such as the locus coeruleus. Perhaps most importantly, rhombomere 1 is the source of the entire cerebellum (123) whose granule cell population forms one-third of the total neuronal complement of the adult brain in humans.

Isthmus tissue induces ectopic cerebellar tissue grafted posteriorly within the hindbrain in chick embryos (124). Moreover, embryos lacking either functional FGF8 or GBX2 proteins also lack cerebella (52, 69, 125). Recently, it has been shown that FGF8 induces patterns of gene expression unique to rhombomere 1 when ectopically expressed in posterior rhombomeres. Moreover, these studies have shown that it also determines the anterior limit of *Hox* expression within the developing brain (see Fig. 2), thereby maintaining rhombomere 1 as a *Hox*-free zone and setting aside that region of the neural tube from which the cerebellum will develop. These studies also show that the inducing signal for *Hoxa-2*, possibly a retinoid (see above), normally extends within the prospective rhombomere 1 territory but is antagonized in that region by FGF8 (126).

7. Patterning the spinal cord

The spinal cord, which has an anterior limit at the level of somites 5/6, is by far the simplest and the most uniform region of the central nervous system. However, it does also show anteroposterior regionalization, and this is probably most marked in the distribution of motor neuron subtypes. Motor neurons can be subdivided based upon their segregation into discrete columns (Fig. 4). The somatic motor neurons can be split into four columns: the medial and lateral components of the medial motor column, MMC_M and MMC_L, respectively; and the medial and lateral components of the lateral motor column, LMC_M and LMC_L, respectively. In addition, there is a further group of visceral motor neurons, which form the column of Terni, CT. Of all these motor columns, only the MMC_M, which projects to the axial muscles, is found along the length of the spinal cord. The others are discontinuous. The lateral motor columns are found at the forelimb (brachial) and hindlimb (lumbar) levels, with the LMC_M innervating ventrally derived limb muscles and the LMC_L innervating the dorsally derived muscles. By contrast, the CT visceral motor neurons, which innervate the sympathetic ganglia, are only found at thoracic levels.

Rostrocaudal inversion of the lumbosacral neural tube at relatively early stages of development, once the neural tube has closed but well before motor neuron differentiation has occurred, results in a respecification of motor neuron subtypes (127). The lumbar segments develop as if they were sacral. Interestingly, it has recently been shown that each of these motor neuron subtypes can be identified prior to target

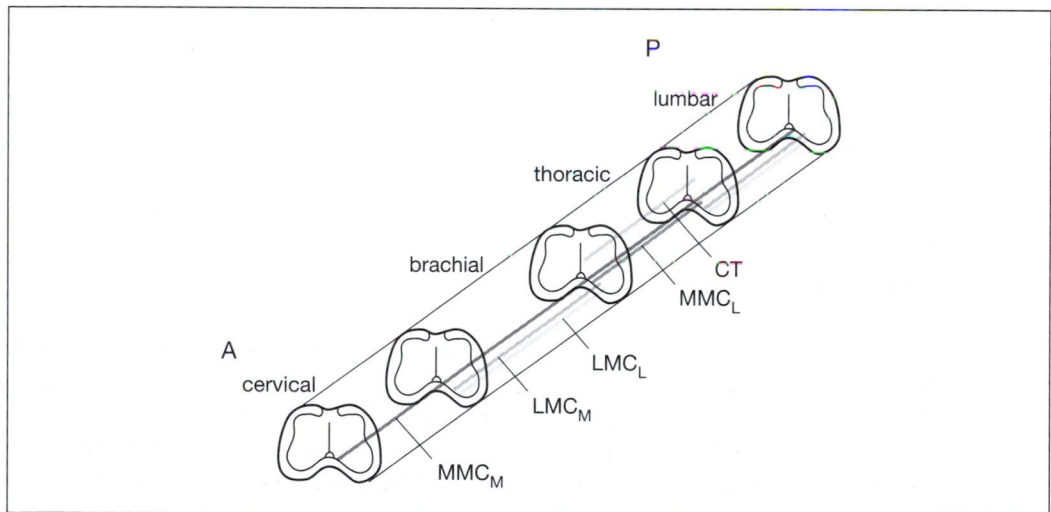

Fig. 4 Schematic of the spinal cord, highlighting the positions of the motor columns along the axis. The medial component of the medial motor column (MMC_M), which runs the length of the spinal cord, is shown in dark gray. The lateral component of the medial motor column (MMC_L), which is found primarily at thoracic levels, is shown in medium gray. The medial and lateral components of the lateral motor column (LMC_M, LMC_L), which are found at limb levels, are shown in shades of light gray. The visceral motor column, the column of Terni, CT, which is found at thoracic levels, is shown dorsally in light gray.

innervation through their differential expression of the LIM-homeodomain proteins (128), and this fact has been exploited to look in detail at how different motor neurons subtypes arise at distinct anteroposterior levels. If neural tube from thoracic levels is transplanted to brachial levels, then the grafted tissue now forms motor neurons of the LMC_M and LMC_L, while brachial neural tube when moved to the thoracic region produces derivatives indicative of this level, the motor neurons of the column of Terni (129). Furthermore, switching neural tube between thoracic and brachial levels also resulted in a change in *Hox* gene expression, such that the transplanted tissue took on the *Hox* gene expression profile of its new location (129). These studies suggest that anteroposterior identity is not fixed early in development, but that it is assigned to the developing spinal cord by environmental cues.

There are a number of tissues flanking the developing spinal cord that could play a role in its patterning: the axial mesoderm, the paraxial mesoderm, and the lateral plate mesoderm. The axial mesoderm is unlikely to be able to confer high-resolution patterning information upon the developing spinal cord. It shows uniform motor neurons inducing activity along its length, and, although it also induces different types of ventral neurons at different rostrocaudal levels, this is due to differential responses by the neural tissue. By contrast, both the paraxial mesoderm and the lateral plate mesoderm show pronounced anteroposterior regionalization, and could conceivably impart this to the nervous system. Of these two, however, experimental studies have shown that it is the paraxial mesoderm that patterns the spinal cord (129). If thoracic somites are transplanted in the place of their brachial counterparts, then the neural tube that lies alongside does not form LMCs but rather produces the motor neurons of the column of Terni, which would normally only be found at thoracic levels. Similarly, if thoracic neural tube is flanked by brachial somites, the neurons of the column of Terni do not arise, and instead the neurons of the LMC are evident. These experiments have pinpointed the paraxial mesoderm as the source of the regionalizing influence on the spinal cord, but the molecular effector of this interaction has not been uncovered. There are, of course, the usual suspects, FGFs and retinoids, but currently data demonstrating a role for either of these activities are lacking.

8. Conclusions

It is clear from the above discussion that several strategies are involved in the anteroposterior regionalization of the vertebrate nervous system, and prominent amongst these are the use of segmentation and the use of graded signalling from local organizers. However, it is important to note that, by and large, we have little knowledge as yet of the identities of the signalling molecules that regionalize the main territories of the neuraxis.

Acknowledgements

Work in the authors' laboratories is supported by the MRC, BBSRC, The Wellcome Trust, and Human Frontier Science Program.

References

1. Mangold, O. (1933) Uber die Induktionsfahigkeit der verschiedenen Bezirke der Neurula von Urodelen. *Naturwissenschaften*, **21**, 761.
2. Nieuwkoop, P. D., Boterenbrood, E. C., Kremer, A., Bloesma, F. F. S. N., Hoessels, E. L. M. J., Meyer, G. and Verheyen, F. J. (1952) Activation and organisation of the central nervous system in amphibians. *J. Exp. Zool.*, **120**, 1.
3. Sive, H. L., Hattori, K. and Weintraub, H. (1989) Progressive determination during formation of the anteroposterior axis in *Xenopus laevis*. *Cell*, **58**, 171.
4. Muhr, J., Graziano, E., Wilson, S., Jessell, T. M., and Edlund, T. (1999) Convergent signals specify midbrain, hindbrain, and spinal cord identity in gastrula stage chick embryos. *Neuron*, **23**, 689.
5. Bally-Cuif, L., Gulisano, M., Broccoli, V. and Boncinelli, E. (1995). c-otx2 is expressed in two different phases of gastrulation and is sensitive to retinoic acid treatment in chick embryo. *Mech. Dev.*, **49**, 49.
6. Niss, K. and Leutz, A. (1998) Expression of the homeobox gene GBX2 during chicken development. *Mech. Dev.*, **76**, 151.
7. Shamim, H. and Mason, I. (1998) Expression of Gbx-2 during early development of the chick embryo. *Mech. Dev.*, **76**, 157.
8. Bouwmeester, T., Kim, S., Sasai, Y., Lu, B. and DeRobertis, E. M. (1996) Cerberus is a head-inducing secreted factor expressed in the anterior endoderm of Speman's organizer. *Nature*, **382**, 595.
9. Foley, A. C., Storey, K. G. and Stern, C. D. (1997) The prechordal region lacks neural inducing ability, but can confer anterior character to more posterior neuroepithelium. *Development*, **124**, 2983.
10. Knoetgen, H., Viebahn, C. and Kessel, M. (1999) Head induction in the chick by primitive endoderm of mammalian, but not avian origin. *Development*, **126**, 815.
11. Thomas, P. and Beddington, R. (1996) Anterior primitive endoderm may be responsible for patterning the anterior neural plate in the mouse embryo. *Curr. Biol.*, **6**, 1487.
12. Jones, C. M., Broadbent, J., Thomas, P. Q., Smith, J. C., Beddington, R. S. (1999) An anterior signalling centre in *Xenopus* revealed by the homeobox gene XHex. *Curr. Biol.*, **9**, 946.
13. Ang, S-L. and Rossant, J. (1994) Hnf-3B is essential for node and notochord formation in mouse development. *Cell*, **78**, 561.
14. Acampora, D., Mazan, D., Lallemand, Y., Avataggiato, V., Murray, M., Simeone, A., and Brulet, P. (1995) Forebrain and midbrain regions are deleted in Otx2 mutant due to a defective anterior neuroectoderm specification during gastrulation. *Development*, **121**, 3279.
15. Belo, J. A., Bouwmeester, T., Leyns, L., Kertesz, L., Gallo, N., Folliette, M., and de Robertis, E. M. (1997) Cerberus-like is a secreted factor with neuralising activity expressed in the anterior primitive endoderm of the mouse gastrula. *Mech. Dev.*, **68**, 45.
16. Thomas, P., Brown, A., and Beddington, R. S. P. (1998) Hex: a homeobox gene revealing pre-implantation asymmetry in the mouse embryo and an early transient marker of endothelial cell precursors. *Development*, **125**, 85.
17. Ding, J., Yang, L., Yam, Y. T., Chen, A., Desai, N., Wynshaw-Boris, A., and Shen, M. M. (1998) Cripto is required for correct orientation of the anterior-posterior axis in the mouse embryo. *Nature*, **395**, 702.
18. Shawlot, W. and Behringer, R. (1995) Requirement for Lim1 in head organizer function. *Nature*, **374**, 425.

19. Ang, S-L., Jin, O., Rhinn, M., Daigle, N., Stevenson, L., and Rossant, J. (1996) A targeted mouse Otx mutation leads to severe defects in gastrulation and formation of axial mesoderm and to deletion of rostral brain. *Development*, **122**, 243.

20. Rhinn, M., Dierich, A., Shawlot, W., Behringer, R. R., Le Meur, M., and Ang, S. (1998) Sequential roles for Otx2 in visceral endoderm and neuroectoderm forebrain and midbrain induction and specification. *Development*, **125**, 845.

21. Glinka, A., Wu, W., Onichtchouk, D., Blumenstock, C., and Niehrs, C. (1997) Head induction by simultaneous repression of Bmp and Wnt signalling in *Xenopus*. *Nature*, **389**, 517.

22. Leyns, L., Bouwmeester, T., Kim, S-H., Piccolo, S. and De Robertis, E. M. (1997) Frz-b is a secreted antagonist of wnt signaling in the Spemann organizer. *Cell*, **88**, 747.

23. Glinka, A. P., Wu, W., Delius, H., Monaghan, A. P., Blumenstock, C., and Miehrs, C. (1998) Dickopf-1 is a member of a family of secreted proteins and functions in head induction. *Nature*, **391**, 357.

24. Itoh, K. and Sokol, S. Y. (1999) Axis determination by inhibition of Wnt signaling in *Xenopus*. *Genes Dev.*, **17**, 2328.

25. Lamb, T. M., Knecht, A. K., Smith, W. C., Stachel, S. E., Economides, A. N., Stahl, N., Yancopolous, G. D., and Harland, R. M. (1993) Neural induction by the secreted polypeptide noggin. *Science*, **262**, 713.

26. Hemmati-Brivanlou, A., Kelly, O. G., and Melton, D. A. (1994) Follistatin, an antagonist of activin, is expressed in the Spemann Organiser and displays direct neuralising activity. *Cell*, **77**, 283.

27. Sasai, Y., Lu, B., Steinbeisser, H., and DeRobertis, E. M. (1995) Regulation of neural induction by Chd and Bmp-4 antagonistic patterning signals in *Xenopus*. *Nature*, **376**, 333.

28. Muhr, J., Graziano, E., Wilson, S., Jessell, T. M., and Edlund, T. (1999) Convergent inductive signals specify midbrain, hindbrain, and spinal cord identity in gastrula stage chick embryos. *Neuron*, **4**, 689.

29. Cox, W. G. and Hemmati-Brinvanlou, A. (1995) Caudalization of neural fate by tissue recombination and bFGF. *Development*, **121**, 4349.

30. Woo, K. and Fraser, S. E. (1997) Specification of the zebrafish nervous system by nonaxial signals. *Science*, **277**, 254.

31. Lamb, T. M. and Harland, R. M. (1995) Fibroblast growth factor is a direct neural inducer, which combined with noggin generates anteroposterior neural pattern. *Development*, **121**, 3627.

32. Alvarez, I. S., Araujo, M., and Nieto, M. A. (1998) Neural induction in whole chick embryo culture by FGF. *Dev. Biol.*, **199**, 42.

33. Storey, K. G., Goriely, A., Sargent, C. M., Brown, J. M., Burns, H. D., Abud, H. M., and Heath, J. K. (1998) Early posterior neural tissue is induced by FGF in the chick embryo. *Development*, **125**, 473.

34. Blumberg, B., Bolado, J., Moreno, T. A., Kintner, C., Evans, R. M., and Papalopulu, N. (1997) An essential role for retinoid signaling in anteroposterior neural patterning. *Development*, **124**, 373.

35. Swindell, E. C., Thaller, C., Sockanathan, S., Petkovich, M., Jessell, T. M., and Eichele, G. (1999) Complementary domains of retinoic acid production and degradation in the early chick embryo. *Dev. Biol.*, **216**, 282.

36. Doniach, T., Phillips, C. R., and Gerhart, J. C. (1992) Planar induction of anteroposterior pattern in the developing central nervous system of *Xenopus laevis*. *Science*, **257**, 542.

37. Herrick, C. J. (1933) Morphogenesis of the brain. *J. Morphol.*, **54**, 233.

38. Figdor, M. C. and Stern, C. D. (1993) Segmental organisation of the embryonic diencephalon. *Nature*, **363**, 630.

39. Puelles, L. and Rubenstein, J. L. R. (1993) Expression patterns of homeobox and their putative regulatory genes in the embryonic mouse forebrain suggest a neuromeric organisation. *Trends Neurosci.*, **16**, 472.

40. Fernandez, A. S., Pieau, C., Reperant, J., Boncinelli, E., and Wassef, M. (1998) Expression of emx1 and dlx1 homeobox genes define 3 molecularly distinct domains in the telencephalon of mouse, chick, turtle and frog embryos: implications for the evolution of telencephalic subdivisions in amniotes. *Development*, **125**, 2099.

41. Puelles, L., Kuwana, E., Puelles, E., and Rubenstein, J. L. R. (1999) Comparison of the mammalian and avian telencephalon from the perspective of gene expression. *Eur. J. Morphol.*, **37**, 139.

42. Wilson, S. W., Brennan, C. B., Macdonald, R., Brand, M., and Holder, N. (1997) Analysis of axon tract formation in zebrafish brain: the role of territories of gene expression and their boundaries. *Cell Tissue Res.*, **290**, 189.

43. Golden, J. A. and Cepko, C. L. (1996) Clones in the chick diencephalon contain multiple cell types and siblings are widely dispersed. *Development*, **122**, 65.

44. Walshe, C. and Cepko, C. L. (1992) Widespread dispersion of neuronal clones across functional regions of the cerebral cortex. *Science*, **255**, 434.

45. Anderson, S. A., Eisenstat, D., Shi, L., and Rubenstein, J. L. R. (1997) Interneuron migration from the basal telencephalon to neocortex: dependence on DLx genes. *Science*, **278**, 474.

46. O'Rourke, N. A., Chenn, A., and McConnel, S. K. (1997) Postmitotic neurons migrate tangentially in the cortical ventricular zone. *Development*, **124**, 997.

47. Fishell, G. (1995) Striatal precursors adopt cortical identities in response to local cues. *Development*, **121**, 803.

48. Na, E., McCarthy, M., Meyt, C., Lai, E., and Fishell, G. (1998) Telencephalic progenitors maintain anteroposterior identity autonomously. *Curr. Biol.*, **8**, 987.

49. Campbell, K., Olsson, M., and Bjorkland, A. (1995) Regional incorporation and site-specific differentiation of striatal precursors transplanted into the embryonic forebrain ventricle. *Neuron*, **15**, 1259.

50. Houart, C., Westerfield, M., and Wilson, S. W. (1998) A small population of anterior cells pattern the forebrain during zebrafish gastrulation. *Nature*, **391**, 788.

51. Shimamura, K. and Rubenstein J. L. R. (1997) Inductive interactions direct early regionalisation of the mouse forebrain. *Development*, **124**, 2709.

52. Meyers, E. N., Lewandoski, M., and Martin, G. R. (1998) An Fgf8 mutant allelic series generated by Cre- and Flp-mediated recombination. *Nat. Genet.*, **18**, 136.

53. Zelster, L. M., Larsen, E. W., and Lumsden, A. (2001) A new developmental compartment in the forebrain regulated by Lunatic fringe. *Nat. Neurosci.*, **4**, 683.

54. Pannese, M., Polo, C., Andreazolli, M., Vignalli, R., Kablar, B., Barsacchi, G., and Boncinelli, E. (1995) The *Xenopus* homolog of Otx-2 is a maternal homeobox gene that demarcates and specifies anterior body regions. *Development*, **121**, 707.

55. Thisse, C., Thisse, B., Halpern, M. E., and Postlethwait, J. H (1994) Goosecoid expression in neuroepithelium in neurectoderm and mesendoderm is disrupted in zebrafish Cyclops gastrulas. *Dev. Biol.*, **164**, 420.

56. Dale, J. K., Vesque, C., Lints, T. J., Sampath, T. K., Furley, A., Dodd, J., and Placzek, M. (1997) Cooperation of BMP7 and SHH in the induction of forebrain ventral midline cells by prechordal mesoderm. *Cell*, **90**, 257.

57. Feldman, B., Gates, M. A., Egan, E. S., Dougan, S. T., Renneback, G., Sirotkin, H. I., Schier, A. F., and Talbot, W. S. (1998) Zebrafish organiser development and germ-layer formation requires nodal-related signals. *Nature*, **395**, 181.

58. Rebagliati, M. R., Toyama, R., Fricke, C., Haffter, P., and Dawid, I. B. (1998) Zebrafish nodal-related genes are implicated in axial patterning and establishing left-right asymmetry. *Dev. Biol.*, **199**, 261.

59. Sampath, K., Rubenstein, A. L., Cheng, A. M. S., Liang, J. O., Fekany, K., Solnica-Krezel, L., Korzh, V., Halpern, M. E., and Wright, C. V. E. (1998) Induction of the zebrafish ventral brain and floorplate requires Cyclops/nodal signalling. *Nature*, **395**, 185.

60. Joyner, A. L. (1996) Engrailed, Wnt and Pax genes regulate midbrain–hindbrain development. *Trends Genet.*, **12**, 15.

61. Bally-Cuif, L. and Wassef, M. (1995) Determination events in the nervous system of the vertebrate embryo. *Curr. Opin. Genet. Dev.*, **5**, 450.

62. Wassef, M. and Joyner, A. L. (1997) Early mesencephalon/metencephalon patterning and development of the cerebellum. *Perspect. Dev. Neurobiol.*, **5**, 3.

63. Crossley, P. H. and Martin, G. R. (1995) The mouse Fgf8 gene encodes a family of polypeptides and is expressed in regions that direct outgrowth and patterning in the developing embryo. *Development*, **121**, 439.

64. Mahmood, R., Bresnick, J., Hornbruch, A., Mahony, K., Morton, N., Colquhoun, K., Martin, P., Lumsden, A., Dickson, C., and Mason, I. (1995) FGF-8 in the mouse embryo: a role in the initiation and maintenance of limb bud outgrowth *Curr. Biol.*, **5**, 797.

65. Shamim, H., Mahmood, R., Logan, C., Doherty, P., Lumsden, A., and Mason, I. (1999) Sequential roles for Fgf4, En1 and Fgf8 in specification and regionalisation of the midbrain. *Development*, **126**, 945.

66. Crossley, P. H., Martinez, S., and Martin, G. R. (1996) Midbrain development induced by FGF8 in the chick embryo. *Nature*, **380**, 66.

67. Sheikh, H. and Mason, I. (1996) Polarising activity of FGF-8 in the avian midbrain. *Int. J. Dev. Biol.* Suppl. 1, 117S.

68. Martinez, S., Crossley, P. H., Cobos, I., Rubenstein, J. L., and Martin, G. R. (1999) FGF8 induces formation of an ectopic isthmic organiser and isthmocerebellar development via a repressive effect on Otx2 expression. *Development*, **126**, 1189.

69. Riefers, F., Bohli, H., Walsh, E., Crossley, P., Stanier, D., and Brand, M. (1998) *Fgf8* is mutated in zebrafish *acerebellar* (*ace*) mutants and is required for maintenance of midbrain-hindbrain boundary development and somitogenesis. *Development*, **125**, 2381.

70. Meyers, E. N., Lewandoski, M., and Martin, G. R. (1998) An Fgf8 mutant allelic series generated by Cre- and Flp-mediated recombination. *Nat. Genet.*, **18**, 136.

71. Irving, C. and Mason, I. (1999) Regeneration of isthmic tissue is the result of a specific and direct interaction between rhombomere 1 and midbrain. *Development*, **126**, 3981.

72. Liu, A., Losos, K., and Joyner, AL (1999) FGF8 can activate Gbx2 and transform regions of the rostral mouse brain into a hindbrain fate. *Development*, **126**, 4827.

73. Bally-Cuif, L., Gulisano, M., Broccoli, V., and Boncinelli, E. (1995) c-otx2 is expressed in two different phases of gastrulation and is sensitive to retinoic acid treatment in chick embryo. *Mech. Dev.*, **49**, 49.

74. Shamim, H. and Mason, I. (1998) Expression of Gbx-2 during early development of the chick embryo. *Mech. Dev.*, **76**, 157.

75. Brocolli, V., Boncinelli, E., and Wurst, W. (1999) The caudal limit of Otx2 expression positions the isthmic organiser. *Nature*, **401**, 164.

76. Millet, S., Campbell, K., Epstein, D. J., Losos, E., Harris, E., and Joyner, A. L. (1999) A role

for Gbx2 in repression of Otx2 and positioning of the mid/hindbrain organizer. *Nature*, **401**, 161.

77. Logan, C., Wizenmann, A., Drescher, U., Monschau, B., Bonhoeffer, F., and Lumsden, A. (1996) Rostral optic tectum acquires caudal characteristics following ectopic engrailed expression. *Curr. Biol.*, **6**, 1006.

78. Marin, F. and Puelles, L. (1994) Patterning of the embryonic avian midbrain after experimental inversions: a polarizing activity from the isthmus. *Dev. Biol.*, **163**, 19.

79. Lumsden, A. and Keynes, R. (1989) Segmental patterns of neuronal development in the chick hindbrain. *Nature*, **337**, 424.

80. Clarke, J. D. W. and Lumsden, A. (1993) Segmental repetition of neuronal phenotype sets in the chick embryo hindbrain. *Development*, **118**, 151.

81. Simon, H. and Lumsden, A. (1993) Rhombomere-specific origin of the contralateral vestibulo-acoustic efferent neurons and their migration across the embryonic midline. *Neuron*, **11**, 209.

82. Fraser, S. E., Keynes, R., and Lumsden, A. (1990) Segmentation in the chick embryo hindbrain is defined by cell lineage restrictions. *Nature*, **344**, 431.

83. Wingate, R. J. T. and Lumsden, A. (1996) Persistence of rhombomeric organisation in the postsegmental hindbrain. *Development*, **122**, 2143.

84. Birgbauer, E. and Fraser, S. E. (1994) Violation of cell lineage restriction compartments in the chick hindbrain. *Development*, **120**, 1347.

85. Guthrie, S. and Lumsden, A. (1991) Formation and regeneration of rhombomere boundaries in the developing chick hindbrain. *Development*, **112**, 221.

86. Wizenmann, A. and Lumsden, A. (1997) Segregation of rhombomeres by differential chemoaffinity. *Mol. Cell. Neurosci.*, **9**, 448.

87. Gale, N. W., Holland, S. J., Valenzuela, D. M., Flenniken, A., Pan L., Ryan, T. E., Henkemeyer, M., Strebhardt, K., Hirai, H., Wilkinson, D. G., Pawson, T., Davis, S., and Yancopoulos, G. (1996) Eph receptors and ligands comprise two major specificity subclasses and are reciprocally compartmentalized during embryogenesis. *Neuron*, **17**, 9.

88. Mellitzer, G., Xu, Q., and Wilkinson, D. G. (1999) Eph receptors and ephrins restrict cell intermingling and communication. *Nature*, **400**, 77.

89. Xu, Q. and Wilkinson, D. G. (1997) Eph-related receptors and their ligands: mediators of contact dependent cell interactions. *J. Mol. Med.*, **75**, 576.

90. Xu, Q., Mellitzer, G., Robinson, V., and Wilkinson, D. G. (1999) *In vivo* cell sorting in complementary segmental domains mediated by Eph receptors and ephrins. *Nature*, **399**, 267.

91. Krumlauf, R. (1994) Hox genes in vertebrate development. *Cell*, **78**, 191.

92. Zhang, M., Kim, H.-J., Marshall, H., Gendron-Maguire, M., Lucas, A. D., Baron, A., Gudas, L., Gridley, T., Krumlauf, R., and Grippo, J. F. (1994) Ectopic Hoxa-1 induces rhombomere transformation in mouse hindbrain. *Development*, **120**, 2431.

93. Alexandre, D., Clarke, J. D., Oxtoby, E., Yan, Y.-L., Jowett, T., and Holder, N. (1996) Ectopic expression of *Hoxa-1* in the zebrafish alters the fate of the mandibular arch neural crest and phenocopies a retinoic acid-induced phenotype. *Development*, **122**, 735.

94. Carpenter, E. M., Goddard, J. M., Chisaka, O., Manley, N. R., and Capecchi, M. R. (1993) Loss of Hoxa-1 (Hox 1.6) function results in the re-organization of the murine hindbrain. *Development*, **118**, 1063.

95. Mark, M., Lufkin, T., Vonesch, J.-L., Ruberte, E., Olivo, J.-C., Dolle, P., Gorry, P., Lumsden, A., and Chambon, P. (1993) Two rhombomeres are altered in *Hoxa-1* mutant mice. *Development*, **119**, 319.

96. Goddard, J., Rossel, M., Manley, N., and Capecchi, M. (1996) Mice with targeted disruption of *Hoxb1* fail to form the motor nucleus of the VIIth nerve. *Development*, **122**, 3217.

97. Studer, M., Lumsden, A., Ariza-McNaughton, L., Bradley, A., and Krumlauf, R. (1996) Altered segmental identity and abnormal migration of motor neurons in mice lacking *Hoxb1*. *Nature*, **384**, 630.

98. Gavalas, A., Davenne, M., Lumsden, A., Chambon, P., and Rijli, F. M. (1997) Role of Hox-a2 in axon pathfinding and rostral hindbrain patterning. *Development*, **124**, 3693.

99. Gonzalez-Reyes, A., Urquia, N., Gehring, W. J., Struhl, G., and Morata, G. (1990) Are cross-regulatory interactions between homeotic genes functionally significant. *Nature*, **344**, 78.

100. Gavalas, A., Studer, M., Lumsden, A., Rijli, F. M., Krumlauf, R., and Chambon, P. (1998) *Hoxa1* and *Hoxb1* synergize in patterning the hindbrain, cranial nerves and second pharyngeal arch. *Development*, **125**, 1123.

101. Sham, M. H., Vesque, C., Nonchev, S., Marshall, H., Frain, M., Das Gupta, R., Whiting, J., Wilkinson, D., Charnay, P., and Krumlauf, R. (1993) The zinc finger gene *Krox-20* regulates *Hoxb-2* (*Hox2.8*) during hindbrain segmentation. *Cell*, **72**, 183.

102. Maconochie, M. K., Nochev, S., Studer, M., Chan, S.-K., Popperl, H., Sham, M.-H., Mann, R., and Krumlauf, R. (1997) Cross-regulation in the mouse *HoxB* complex: the expression of *Hoxb2* in rhombomere 4 is regulated by *Hoxb1*. *Genes Dev.*, **11**, 1885.

103. Studer, M., Gavalas, A., Marshall, H., Ariza-McNaughton, L., Rijli, F. M., Chambon, P., and Krumlauf, R. (1998) Genetic interactions between Hoxa1 and Hoxb1 reveal new roles in regulation of early hindbrain patterning. *Development*, **125**, 1025.

104. Nonchev, S., Vesque, C., Maconochie, M., Seitanidou, T., Ariza-McNaughton, L., Frain, M., Marshall, H., Sham, M. H., Krumlauf, R., and Charnay, P. (1996) Segmental expression of *Hoxa-2* in the hindbrain is directly regulated by *Krox-20*. *Development*, **122**, 543.

105. Wilkinson, D. G., Bhatt, S., Chavrier, P., Bravo, R., and Charnay, P. (1989) Segment-specific expression of a zinc finger gene in the developing nervous system of the mouse. *Nature*, **337**, 461.

106. Cordes, S. P. and Barsh, G. S. (1994) The mouse segmentation gene *kr* encodes a novel basic domain-leucine zipper transcription factor. *Cell*, **79**, 1025.

107. Manzanares, M., Cordes, S., Kwan, C.-T., Sham, M.-H., Barsh, G. S., and Krumlauf, R. (1997) Segmental regulation of *Hoxb3* by *kreisler*. *Nature*, **387**, 191.

108. Manzanares, M., Cordes, S., Ariza-McNaughton, L., Sadl, V., Maruthainar, K., Barsh, G. S., and Krumlauf, R. (1999) Conserved and distinct roles of kreisler in regulation of the paralogous *Hoxa3* and *Hoxb3* genes. *Development*, **126**, 759.

109. Schneider-Manoury, S., Topilko, P., Seitanidou, T., Levi, G., Cohen-Tannoudji, M., Pournin, S., Babinet, C., and Charnay, P. (1993) Disruption of Krox-20 results in alteration of rhombomeres 3 and 5 in the developing hindbrain. *Cell*, **75**, 1199.

110. McKay, I. J., Muchamore, I., Krumlauf, R., Maden, M., Lumsden, A., and Lewis, J. (1994) The kreisler mouse: a hindbrain segmentation mutant that lacks two rhombomeres. *Development*, **120**, 2199.

111. Manzanares, M., Trainor, P. A., Nonchev, S., Ariza-McNaughton, L., Brodie, J., Gould, A., Marshall, H., Morrison, A., Kwan, C.-T., Sham, M.-H., Wilkinson, D. G., and Krumlauf, R. (1999) The role of *kreisler* in segmentation during hindbrain development. *Dev. Biol.*, **211**, 220.

112. Moens, C. B., Cordes, S. P., Giorgianni, M. W., Barsh, G. S., and Kimmel, C. B. (1998) Equivalence in the genetic control of hindbrain segmentation in fish and mouse. *Development*, **125**, 381.

113. Gale, E., Zile, M., and Maden, M. (1999) Hindbrain respecification in the retinoid deficient quail. *Mech. Dev.*, **89**, 43.

114. Niederreither, K., Vermot, J., Schuhbaur, B., Chambon, P., and Dolle, P. (2000) Retinoic acid synthesis and hindbrain patterning in the mouse embryo. *Development*, **127**, 75.

115. Marshall, H., Nonchev, S., Sham, M.-H., Muchamore, I., Lumsden, A., and Krumlauf, R. (1992) Retinoic acid alters hindbrain Hox code and induces transformation of rhombomeres 2/3 into 4/5 identity. *Nature*, **360**, 737.

116. Gould, A., Itasaki, N., and Krumlauf, R. (1998) Initiation of rhombomeric Hoxb4 expression requires induction by somites and a retinoid pathway. *Neuron*, **21**, 39.

117. Grappin-Botton, A., Bonnin, M.-A., Sieweke, M., and LeDouarin, N. M. (1998) Defined concentrations of a posteriorising signal are critical for *MafB/Kreisler* segmental expression in the hindbrain. *Development*, **125**, 1173.

118. Guthrie, S., Muchamore, I., Kuriowa, A., Marshall, H., Krumlauf, R., and Lumsden, A. (1992) Neurectodermal autonomy of Hox-2.9 expression revealed by rhombomere transpositions. *Nature*, **356**, 157.

119. Grappin-Botton, A., Bonnin, M. A., McNaughton, L. A., Krumlauf, R., and LeDouarin, N. M. (1995) Plasticity of transposed rhombomeres: Hox gene induction is correlated with phenotypic modifications. *Development*, **121**, 2707.

120. Graham, A., Heyman, I., and Lumsden, A. (1993) Even-numbered rhombomeres control the apoptotic elimination of neural crest cells from odd-numbered rhombomeres in the chick hindbrain. *Development*, **119**, 233.

121. Graham, A., Francis-West, P., Brickell, P., and Lumsden, A. (1994) The signalling molecule Bmp 4 mediates apoptosis in the rhombencephalic neural crest. *Nature*, **372**, 684.

122. Graham, A. and Lumsden, A. (1996) Interactions between rhombomeres modulate *Krox-20* and *Follistatin* expression in the chick embryo hindbrain. *Development*, **122**, 473.

123. Wingate, R. J. T. and Hatten, M. E. (1999) The role of the rhombic lip in avian cerebellum development. *Development*, **126**, 4395.

124. Martinez, S., Marin, F., Nieto, M. A., and Puelles, L. (1995) Induction of ectopic engrailed expression and fate change in avian rhombomeres: intersegmental boundaries as barriers. *Mech. Dev.*, **51**, 289.

125. Wassarman, K. M., Lewandoski, M., Campbell, K., Joyner, A. L., Rubenstein, J. L., Martinez, S., and Martin, G. R. (1997) Specification of the anterior hindbrain and establishment of a normal mid/hindbrain organizer is dependent on Gbx2 gene function. *Development*, **124**, 2923.

126. Irving, C. and Mason, I. (2000) Signalling by FGF8 from the isthmus patterns anterior hindbrain and establishes the anterior limit of Hox gene expression. *Development*, **127**, 177.

127. Matisse, M. P. and Lance-Jones, C. (1996) A critical period for the specification of motor pools in the chick lumbrosacral spinal cord. *Development*, **121**, 659.

128. Tsuchida, T., Ensini, M., Morton, S. B., Baldassare, M., Edlund, T., Jessell, T. M., and Pfaff, S. L. (1994) Topographic organisation of embryonic motor neurons defined by expression of LIM homeobox genes. *Cell*, **79**, 957.

129. Ensini, M., Tsuchida, T. N., Belting, H.-G., and Jessell, T. M. (1998) The control of rostro-caudal pattern in the developing spinal cord: specification of motor neuron subtype identity is initiated by signals from paraxial mesoderm. *Development*, **125**, 969.

7 | Axon guidance in the developing vertebrate nervous system

SARAH GUTHRIE

1. Differentiation and axonogenesis

During development, neurons originate in the inner, proliferative ventricular zone of the neural tube and then move away to form the outer mantle zone. Both progenitor cells and neurons may migrate radially or tangentially with respect to the developing neuroepithelium. Neuronal migrations occur over a protracted period and take place in a complex landscape containing many cells which have already differentiated and extended axons (1). As a general rule, large projection neurons are generated first, and smaller interneurons and glial cells are produced later. Neuronal migration is an active process, which results in neurons taking up their correct positions within laminar or nuclear structures (2). Axonogenesis often begins during this migration so that the processes of migration, differentiation, and axonogenesis are intimately linked. Thus the processes of migration and axon extension are regulated coordinately and are likely to depend on the same, or similar, receptor–ligand systems.

Following their last cell division, the acquisition of differentiated properties by neurons occurs in several stages. Aspects of differentiation encompass: the expression of repertoires of genes encoding transcription factors, neuronal surface receptors, and neurotransmitters; morphology of the cell body, dendrites, and axons; and patterning of synaptic connections with neurons or other targets. Within the central nervous system, the pathway of differentiation depends on a 'matrix' of anteroposterior and dorsoventral patterning information, which in turn depends on cell lineage and on signalling molecules in the environment (3) (see Chapter 6 for discussion of anteroposterior patterning). Cells that lie at the border between the neural tube and the adjacent ectoderm are specified to form the neural crest, which migrates into the periphery to form many derivatives, including neurons and glia of the autonomic and sensory ganglia (see Chapter 8). Some aspects of cell identity are acquired early, while neurons are still in the ventricular or subventricular zone, but other hallmarks of neuronal differentiation are gradually accrued over a protracted period and may depend on local factors (4). Axon extension can occur before, during, and after neuronal migration, and is concomitant with the process of acquisition of other aspects of neuronal fate.

The link between the assumption of regional identity and axon pathway is nicely demonstrated by the motor neurons of the spinal cord. A subpopulation of cells in the ventral portion of the ventricular zone of the cord are induced to form motor neurons under the influence of sonic hedgehog (SHH) signalling from the ventral floor plate and underlying axial mesoderm, the notochord (5, 6). Exposure of neural plate cells to a gradient of SHH protein leads to the expression of a set of homeodomain-containing proteins in distinct progenitor domains along the dorso-ventral axis of the neural tube, which in turn leads to the differentiation of various neuronal types at distinct positions (7). In the ventral domain of the spinal cord, some of the neuronal types thus generated are subsets of motor neurons, which express different combinations of LIM homeobox genes concomitant with their axonal pathfinding to various peripheral targets (8). All postmitotic motor neurons express *Islet-1*, while the progenitors of spinal motor neurons destined to project their axons ventrally express *Lhx3* and *Lhx4* (9). A large body of evidence from both invertebrate and vertebrate systems implicates LIM genes in various aspects of neuronal identity and axonal pathfinding (10). For example, loss of *Islet-1* function leads to a failure of differentiation, culminating in the loss of all motor neurons (11). By contrast, elimination of the function of *Lhx3* and *Lhx4* does not impair differentiation, but the ventral axonal pathways fail to form. Instead, there is a fate conversion to a dorsal motor neuron phenotype, in which neurons exit the neuroepithelium via dorsal exit routes, a manner more characteristic of cranial (and a subset of spinal) motor neurons (9). Later in development, *Lhx3* is expressed by a subset of motor neurons located in the medial motor column that innervate axial muscles. Forcing expression of *Lhx3* in all motor neurons converts their identity to that of the axial subtype, based on the position of their cell bodies, repertoires of gene expression, and innervation of axial muscle targets (12). Another LIM gene, *Lim1*, has recently been shown to be involved in the pathfinding of motor neurons of the lateral motor column, which project via a dorsal nerve branch to the dorsal limb muscles. Loss of *Lim1* function does not affect the cell body position of these neurons, but abolishes the selectivity of nerve projections to the limb, and mutant neurons project via both the dorsal and ventral nerve branches (13). Thus the axon pathway chosen by a neuron may depend on transcription factors expressed in progenitor cells or soon after the final mitosis. The remainder of the chapter, however, will focus on the extrinsic guidance cues and their axonal receptors that are thought to shape axon pathways during development.

2. Axon guidance

Axonal growth and guidance is known to depend on molecular cues in the developing embryo. Responses to these guidance cues depend on an exploratory structure on the tip of the axon, the growth cone, whose significance was first appreciated by the Spanish neuroscientist Santiago Ramon y Cajal in the 1890s. He realized that growth cones are able to sense environmental cues which specify the routes along which they grow (1, 14). Studies of growth cones *in vitro* and *in vivo* have revealed some of the elements of the cascade that translates external cues into growth-cone steering and axon extension. Growth cones contain a number of finger-like protrusions, the filopodia, and veil-like structures, the lamellipodia. These structures are highly dynamic: their preferential stabilization on one side of the growth cone will eventually lead to turning and axonal growth in that direction. Underlying this behaviour are a multiplicity of cytoskeletal proteins, including microtubules, which are localized within the organelle-rich central domain of the growth cone and extend into the base of the filopodia. The most important feature for growth-cone steering appears to be actin filaments, however, which exist as a meshwork within the lamellipodia and as bundles within the filopodia (14, 15). Actin filaments undergo treadmilling, with polymerization at the leading edge of the growth cone, retrograde transport, and depolymerization in the body of the growth cone. Anisotropic distribution of guidance cues in the environment leads to the asymmetrical activation of receptors, which affects the cytoskeleton and exerts a positive or negative effect on growth core motility, causing deflection towards or away from the stimulus.

3. Axon guidance molecules

Physical features of the landscape in the developing embryo are likely to provide guidance cues for growing axons; for example, enlarged extracellular spaces between neuroepithelial cells might provide routes for growing axons (16). The significance of such cues *in vivo* is difficult to evaluate, however, and such pathways are over-whelmingly likely to collaborate with specific molecular cues to guide axons, for example in the intercellular spaces that are colonized by the transverse axon tracts of the developing hindbrain (17, 18). In the later part of the twentieth and early part of the twenty-first century, attention has therefore focused largely on testing the role of various tissues or candidate molecules for their positive or negative effects on axon growth and navigation. To date, four major modes of action of axon guidance molecules have been distinguished. Guidance cues may affect axons by contact with adjacent cells, or by diffusion of signals from more distant sites, and such contact-mediated or diffusible cues may act in either a positive or a negative manner. While positive effects constitute the promotion of growth and/or chemoattraction, negative effects constitute inhibition of growth and/or chemorepulsion (19).

Studies of the molecular biology of axon guidance originate to some extent from the isolation of nerve growth factor (NGF), which was shown to have a potent effect in promoting the outgrowth of sensory neurons in the chick embryo (20). *In vitro* studies also showed that NGF was chemoattractive when released from a micropipette into the path of a growing dorsal root ganglion (DRG) sensory axon (21). NGF is now well established as a target-derived factor with potent effects on the survival and differentiation of sympathetic neurons and specific subsets of sensory neurons, but as yet there is no compelling evidence that NGF has a role in axon pathfinding *in vivo* (22). Relatives of NGF within the neurotrophin (NT) family include brain-derived neurotrophic factor (BDNF), NT-3, NT-4/5, and NT-6, all of which have distinct effects on various groups of neurons (23). Although the initial characterization of the influence of these factors depended on axon outgrowth assays, more recent data have also supported the notion that some neurotrophins and other growth factors are effective in orienting axons. For example, the chemoattractant effect of the branchial arches on trigeminal sensory axons appears to be mediated by BDNF and NT-3 (24), and both these factors can reorient growth cones *in vitro* (25). Moreover, hepatocyte growth factor (HGF) is expressed in peripheral tissues and can chemoattract spinal and cranial motor axons *in vitro* (26, 27). These observations remain to be investigated further *in vivo*, since mice mutant for these factors show only minor defects in the relevant nerve projections (24, 26, 27).

During the last decade, the field of axon guidance has progressed with amazing rapidity, owing to the characterization of families of axon guidance molecules. This axon guidance revolution has also been fuelled by the discovery of the amazing degree of conservation of molecules and mechanisms across vertebrate and invertebrate phyla (28). In particular, the identification of candidate genes in genetically tractable organisms such as the fruitfly *Drosophila* and the nematode worm *Caenorhabditis elegans* has allowed researchers studying vertebrates to make precise predictions about function in their own model systems. Important families of axon guidance molecules include the netrins, Semaphorins, ephrins, Slit proteins, and diverse extracellular matrix and cell-adhesion molecules (29–31).

Netrins are secreted proteins, which may act as chemoattractants or chemorepellents, and which were first characterized based on their effects in promoting the outgrowth of spinal commissural axons (32). Receptors for netrins in vertebrates include DCC ('deleted in colorectal carcinoma') and neogenin (33), Unc5 homologues (34), and the adenosine A2b receptor (35). The Semaphorins are a large family of molecules with both transmembrane and secreted forms (36), whose structural hallmark is a 500-amino acid 'Sema' domain. Semaphorins signal via a receptor complex that includes the neuropilins (36, 37), thought to be required for determining the specificity of binding, and the plexins (38, 39), thought to transduce signals (40). Like netrins, Semaphorins have been shown to be capable of eliciting chemoattraction or chemorepulsion. The ephrins comprise a group of molecules that are transmembrane or GPI-linked (glycosyl-phosphatidylinositol) to the cell membrane, and which bind the family of transmembrane Eph receptors (41, 42). These are receptor tyrosine kinases, which are classified as EphAs or EphBs depending on whether they bind to GPI-anchored (ephrin-A) or transmembrane (ephrin-B) ligands (see also Chapter 6). Ephrin–Eph interactions are now well known to be involved in axon repulsion, and can act in a graded fashion to allow topographic mapping of axon projections to their targets. Extracellular matrix molecules include laminin, F-spondin, anosmin, and the recently characterized Slit proteins (31, 43). Among cell-adhesion molecules are a large family of immunoglobulin-related molecules that include the neural cell-adhesion molecule (NCAM), L1, NgCAM, SC1, TAG-1, the netrin receptors DCC and Unc5 homologues, and fibroblast growth factor (FGF) receptors (44). These transmembrane molecules are implicated in promoting axon outgrowth, and/or axon fasciculation by binding homophilically or heterophilically to other immunoglobulin family members. In some cases interactions between these molecules in the plane of the membrane have also been indicated, and a wide range of modulatory interactions between axon guidance molecule receptors are likely.

A large number of experimental paradigms have been used to investigate the roles of axon guidance molecules. Examples of *in vitro* assays are collagen gel co-cultures (Fig. 1), in which gradients of diffusible molecules can be established, and neuronal cultures on planar substrata, which may be presented as alternating stripes of candidate guidance molecules—the 'stripe assay' (Fig. 2). *In vivo* approaches include the use of targeted mutations in mice, misexpression of genes in early chick embryos, and an open-brain preparation in early frog embryos. One interesting combination of *in vivo* and *in vitro* techniques involves injecting mRNAs encoding guidance receptors into early frog embryos. Spinal cord neurons are subsequently isolated, and their responses to guidance molecules tested *in vitro*, thus revealing the effects of misexpressing the receptors of interest. This review can only briefly describe the

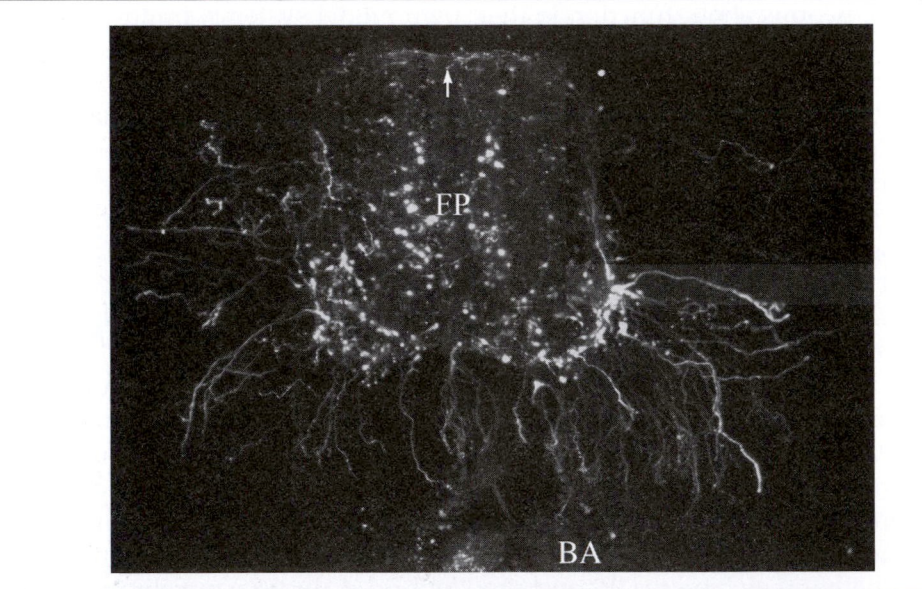

Fig. 1 (See also Plate 4) Collagen gel co-culture, used to demonstrate the influence of branchial arch-derived chemoattractants on hindbrain motor neurons. Cranial motor axons were retrogradely labelled using fluorescent tracers, and hindbrain explants containing these neurons were co-cultured with explants of branchial arch (BA) at a distance. Motor axons normally emerge from the lateral sides of the explant, i.e. from the right and left, since they grow away from the midline tissue of the floor plate (FP) under the influence of chemorepulsion. In the presence of a branchial arch explant, many axons still emerge laterally, but some exit the explant adjacent to the branchial arch explant and turn towards this tissue.

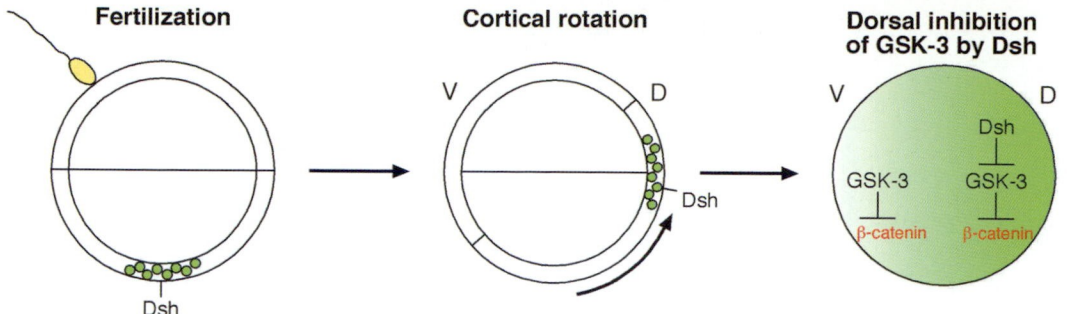

Accumulation of β-catenin in dorsal nuclei leads to transcription of dorsal specific genes

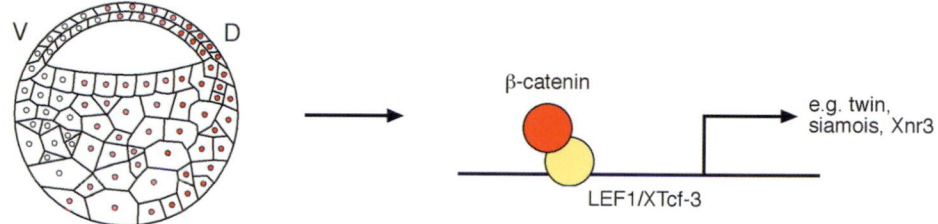

Plate 1 Model of the mechanism of localized Wnt pathway activation during dorsoventral axis specification in Xenopus. Dsh associates with a specific class of vesicles at the vegetal pole and these vesicles are transported dorsally along the subcortical microtubule array during cortical rotation. This translocation contributes to the asymmetrical distribution of Dsh along the dorsoventral axis and the localized activation of a maternal Wnt signally pathway. Activation of Wnt signalling leads to the downregulation of GSK3 activity, thereby promoting the stabilization of ß-catenin, ß-Catenin then accumulates in dorsal nuclei where, in combination with XTcf-3, it activates the transcription of dorsal-specific regulatory genes. (Adapted with permission from ref. 51.) D=dorsal; V=ventral.

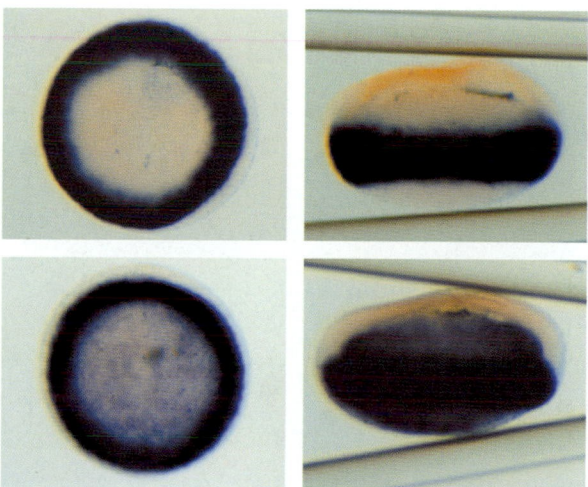

Plate 2 *Brachyury* and its targets. Comparison of expression patterns of *Xbra* (upper panels) and *Bix1* (lower panels) at early gastrula stage 10. *Xbra* is expressed in only the equatorial zone of the embryos, whereas *Bix1* is expressed in both the marginal one and the vegetal hemisphere. (Reproduced with permission from ref. 106).

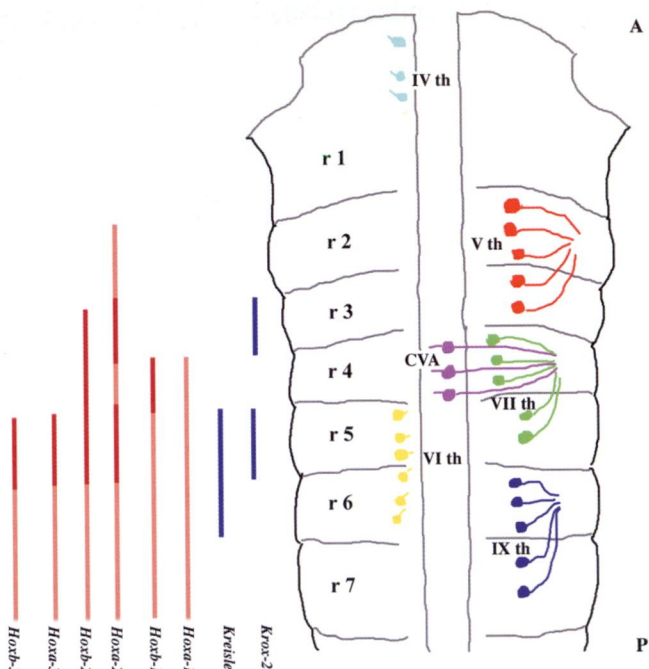

Plate 3 Schematic of the hindbrain. The rhombomeres are labelled r1 through to r7 from anterior (A) to posterior (P). The motor nuclei of the hindbrain are labelled and coloured: the trochlear (IVth) is light blue; the trigeminal (Vth) red; the facial (VIIth) green; the contral vestibuloacoustic (CVA) mauve; the abducens (VIth) yellow; the glossopharyngeal (IXth) dark blue. The expression domains of *Krox-20* and *Kreisler* are shown in blue, and the expression patterns of the *Hox* genes in orange, with regions of elevated expression shown in red.

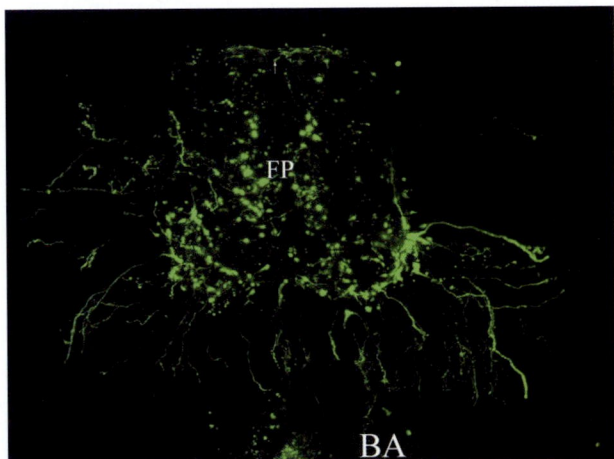

Plate 4 Collagen-gel co-culture, used to demonstrate the influence of branchial arch-derived chemoattractans on hindbrain motor neurons. Cranial motor axons were retrogradely labelled using fluorescent tracers, and hindbrain explants containing these neurons were co-cultured with explants of branchial arch (BA) at a distance. Motor axons normally emerge from the lateral sides of the explant, i.e. from the right and left, since they grow away from the midlne tissue of the floor plate (FP) under the influence of chemorepulsion. In the presence of a branchial arch explant, many axons still emerge laterally, but some exit the explant adjacent to the branchial arch explant and turn towards this tissue.

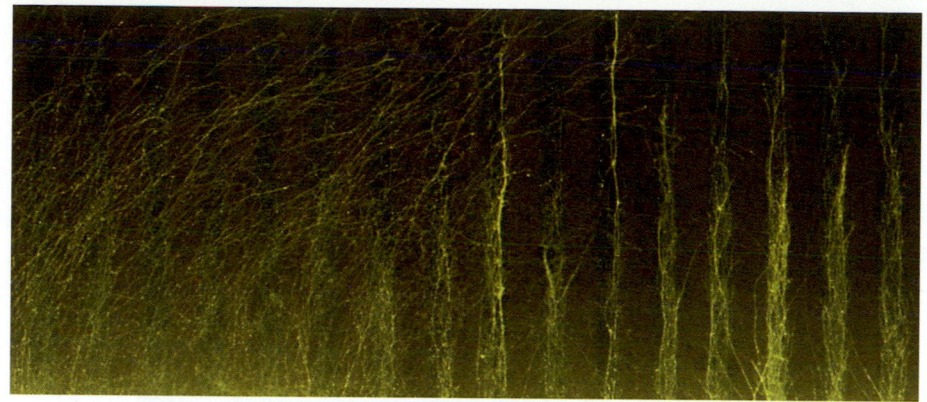

Plate 5 'Stripe assay' used to demonstrate avoidance by temporal axons of posterior tectal membranes. Alternating lanes consisting of membranes from anterior and posterior tectal thirds were prepared on a nucleopore filter. To discriminate between the two membrane types for later analyses, membranes derived from the posterior tectum were marked with fluorescein isothlocyanate (FITC)-labelled flourescent beads. Retinal stripes were stained with N-4-4-(4-didecylaminostryryl) M-methyl-pyridium iodide (DIAsp) before arranging them perpendicular to the membrane stripes. Temporal retinal axons (right) show a clear striped outgrowth and avoid membrane lanes containing posterior membranes, whereas nasal axons (left) do not discriminate between these lanes. (Figure courtesy of Dr Uwe Drescher.)

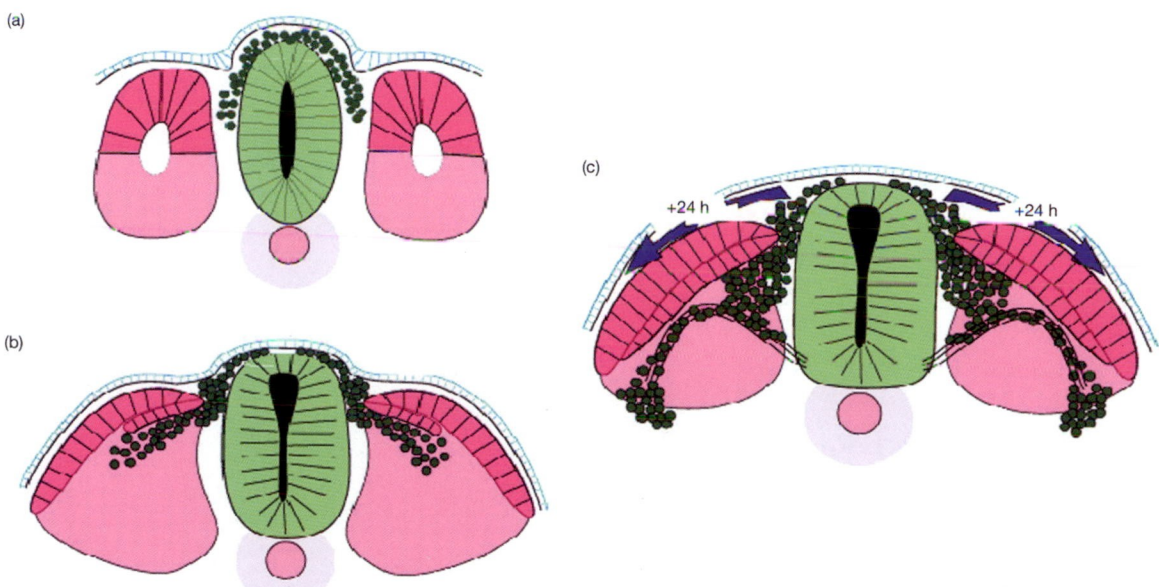

Plate 6 Schematic depiction of the migratory pathways of the trunk neural crest. (a) Neural crest cells (dark green) detach from the dorsal surface of the neural tube (light green) and migrate ventrally between the neural tube and somites (pink). (b) When they reach the somites, they invade along the developing myotome. (c) The ventrally migrating neural crest cells differentiate into the neurons and glial cells of the sensory and sympathetic ganglia, and spread out along the ventral root motor fibres, where they differentiate into glial cells. Then, 24 h after migrating ventrally, another population of neural crest cells migrates dorsolaterally beneath the ectoderm and these differentiate into the pigment cells of the skin. (Figure courtesy of Martha Spence. Modification of Fig. 1 in ref. 1.)

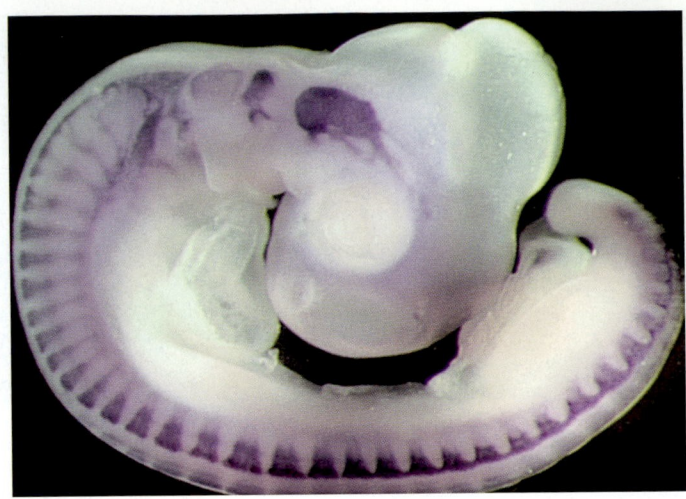

Plate 7 *In situ* hybridization in a stage-20 chick embryo revealing the expression of FoxD3. FoxD3 is a transcription factor that is expressed by all neural crest cells except melanoblasts. This embryo demonstrates the contribution of the neural crest to the head cranial ganglia and the segmental migration of the neural crest through the somites in the trunk. (Micrograph courtesy of Robert Kos.)

Control-Fc

Ephrin-B1-Fc

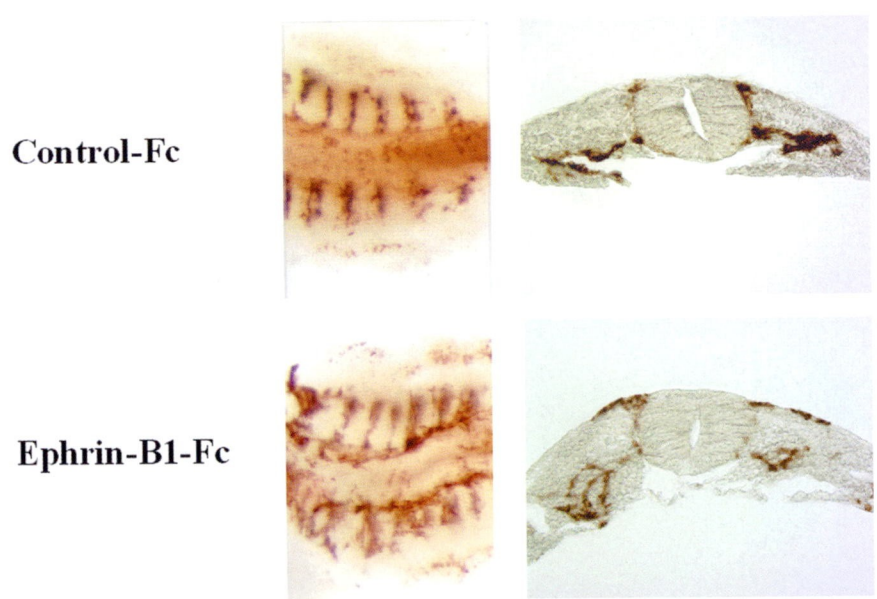

Plate 8 Pertubation of ephrin function in trunk tissue pieces results in the loss of segmental migration through the somites, and the premature migration into the dorsolateral path. (Control) Pieces of stage-13 trunk were placed in culture; after 24 hours neural crest cells (labelled dark brown) had migrated normally through the anterior half of each somite, but had not yet invaded the dorsolateral path. (Ephrin–B1–Fc) When the explants are treated with a fusion protein comprising the Eph-receptor binding portion, the Eph receptors on the neural crest cells are occupied with soluble ligand and so are insensitive to ephrins in their pathways. Consequently, the segmental migration through somites was pertubed and neural crest cells invaded the dorsolateral path precociously. (Micrograph courtesy of Alicia Santiago.)

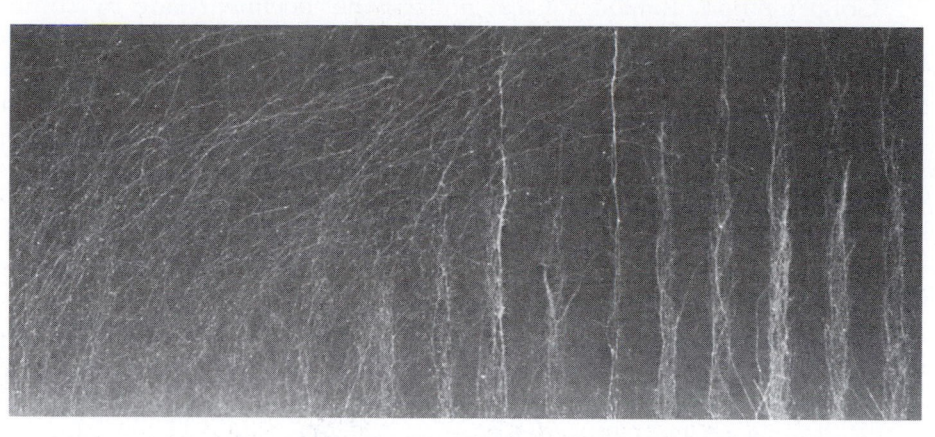

Fig. 2 (See also Plate 5) 'Stripe assay' used to demonstrate avoidance by temporal axons of posterior tectal membranes. Alternating lanes consisting of membranes from anterior and posterior tectal thirds were prepared on a nucleopore filter. To discriminate between the two membrane types for later analyses, membranes derived from the posterior tectum were marked with fluorescein isothiocyanate (FITC)-labelled fluorescent beads. Retinal stripes were stained with *N*-4-4-(4-didecylaminostryryl) *M*-methyl-pyridium iodide (DiAsp) before arranging them perpendicular to the membrane stripes. Temporal retinal axons show a clear striped outgrowth and avoid membrane lanes containing posterior membranes, whereas nasal axons do not discriminate between these lanes. (Figure courtesy of Dr Uwe Drescher.)

roles and contexts of a few axon guidance molecules. To do this, it will focus on three major systems in which the molecular basis of axon guidance has started to emerge; guidance at the midline, formation of sensory and motor pathways to and from the spinal cord, and the visual system. Finally, some selected aspects of downstream signalling from axon guidance molecules to the cytoskeleton within growth cones will be discussed.

4. Axon guidance in the CNS

4.1 Guidance at the midline of the central nervous system

Commissures are the axonal connections that join the two sides of the nervous system. In the spinal cord, commissural axons extend from dorsal cell bodies along a ventral trajectory, crossing the developing motor columns to reach the floor plate, a region of non-neuronal cells lying at the ventral midline. Once commissural axons have reached the floor plate, they cross it and then make a sharp right-angled turn to project rostrally in a longitudinal axon tract that extends along the border of the floor plate. Abundant evidence shows that the midline floor plate plays an important role in orchestrating the patterns of projections on either side, including those of axons that cross the midline, as well as those that remain ipsilateral, and there is extensive conservation of midline guidance mechanisms between vertebrates and invertebrates (19, 28). Based on sections of embryonic chick spinal cord stained using the

Golgi method, Ramon y Cajal noticed the 'beeline' made by commissural axons towards the floor plate, and proposed that this pathway might depend on chemo-attractive substances (45). Nearly 100 years later, this idea was investigated directly by co-culturing explants of dorsal spinal cord with floor plate explants in collagen gels. Commissural axons grew out of these dorsal explants and in separate experiments reoriented in proximity to the ventral spinal cord explant, suggesting that the floor plate produces molecules that chemoattract and promote the outgrowth of commissural axons (46).

Biochemical purification from chick brains was later used to isolate an activity that promoted the outgrowth of commissural axons, which yielded two related molecules, named netrin-1 and netrin-2 (32). Netrins were found to be expressed in the ventral spinal cord, and netrin-secreting COS cells presented as a focal source in collagen gels were found to chemoattract commissural axons (47). Sequence comparisons indicated high homology of netrin with laminin and also with unc-6, a gene product previously characterized in the nematode, *C. elegans* (48, 49). In *unc-6* mutants there were defects in the pathways of ventrally-projecting axons, analogous to those of commissural neurons, but some dorsal projections were also affected, implicating netrins more widely in axon guidance (48). To test the idea that netrin might influence dorsal axon growth in vertebrates, hindbrain trochlear motor neurons were chosen, since their cell bodies occupy a ventral position, but their axons extend dorsally within the neural tube. When cultured alone, explants containing trochlear motor neurons extended a swathe of axons, whilst in the presence of floor plate explant or netrin-1-secreting cells, axon outgrowth into the collagen was inhibited (50). Other populations of hindbrain neurons also showed responses to netrin-1, with commissural neurons showing chemoattraction and subsets of hindbrain motor neurons showing chemorepulsion (51, 52). These striking findings show that netrin-1 produced at the ventral midline is bifunctional, chemoattracting or chemorepelling subpopulations of circumferentially-growing axons.

Insights from the nematode also pointed to two candidate genes, *unc-40* and *unc-5*, which were known to encode transmembrane molecules which could represent netrin receptors (33). Whilst *unc-40* mutants exhibited defects in the guidance of ventrally-projecting neurons, *unc-5* mutants showed defects in the projection patterns of dorsally-projecting neurons, implicating the gene products in the chemoattractive and chemorepulsive effects of UNC-6 (netrin), respectively (48). Vertebrate homologues of *unc-40* (*DCC* and *neogenin*) and of *unc-5* (*unc5h1*, *unc5h2*, and *rcm*) have now been identified (34, 53, 54). *DCC* is expressed within dorsal commissural neurons, and the protein is present on commissural axons as they extend towards the floor plate. More importantly, anti-DCC antibodies block the netrin-dependent outgrowth of commissural axons *in vitro*, indicating that DCC mediates the responses of these axons to netrin (53). Recent studies, however, suggest that a component of a putative DCC receptor complex is the adenosine A2b receptor, which may be primarily responsible for binding netrin (35). By contrast with *DCC*, *Unc5h1* is expressed in ventral regions of the spinal cord (34), but since spinal motor neurons are not repelled by netrin-1 in culture (51), the significance of this expression pattern remains uncertain, and a requirement for *Unc5* homologues in trochlear motor axon repulsion remains to be shown. Nevertheless, experiments on cultured *Xenopus* spinal neurons have shown that *Unc5h1* can mediate the repulsive effects of netrins *in vitro* ((55); and see later).

A dramatic demonstration of the role of netrins in axon pathfinding has been provided by studies on transgenic mice. In *netrin-1* mutants, commissural axons were found to stall *en route* to the floor plate, with some axons projecting aberrantly towards the central canal of the spinal cord and few extending far enough to form a ventral commissure (56). Although these defects might have been anticipated based on netrin's known function at the midline, there were more extensive disruptions to axon pathfinding, reflected by the absence of the corpus callosum, hippocampal commissure, and the anterior commissure. These defects were echoed in *DCC* mutants, which showed complete loss of the corpus callosum and hippocampal commissure and only partial formation of the anterior commissure (57). In *netrin-1* mutants, trochlear motor axon pathfinding appeared normal, although the positioning of trochlear motor neuron cell bodies was abnormal and the migration of the rhombic lip progenitors of the pontine nuclei and the inferior olive was also affected (56, 58). Recent studies have also highlighted a possible role of netrin-1 in the ipsilateral growth of axons, since both the formation of hippocampal projections and thalamocortical projections were impaired in *netrin-1*-deficient mice (59, 60).

Other diffusible molecules that are candidates for influencing midline guidance in vertebrates are the Slit proteins, which are expressed at high levels by the floor plate (61) and play a number of roles in axon guidance and cell migration (43). Studies on guidance at the *Drosophila* midline had previously shown that Slit acts as a gatekeeper, acting via the Robo receptor to segregate axons along ipsilateral or commissural pathways (62). In vertebrates, commissural axons become sensitive to Slit proteins only after crossing the floor plate (63). Slits (and Semaphorins) are present in midline and ventral tissues; thus they confine commissural axons to tracts in a narrow repellent-free corridor on either side of the floor plate (63). The acquisition of sensitivity to Slits and Semaphorins appears to be tightly linked to a loss of sensitivity to netrin-mediated chemoattraction, which occurs when axons cross the floor plate (64). Recent identification of further Robo receptors in the fruitfly, with roles in the lateral positioning of axon tracts relative to the midline, will no doubt add impetus to studies of the role played by Slit in vertebrates (65).

Netrins and Slits are not the sole determinants of midline guidance: contact-dependent mechanisms have also been implicated. In the chick embryo, *axonin-1* (the chick homologue of *TAG-1*) and *NgCAM* are expressed by spinal commissural neurons and are involved in the process of crossing the midline. These molecules can bind homophilically, mediating axon fasciculation, whilst axonin-1 can also bind heterophilically to NrCAM, which is present on the floor plate. Antibody perturbation of axonin-1 or NrCAM function in chick embryos *in vivo* led to pathfinding errors, such that some axons failed to cross and instead grew along the ipsilateral border of the floor plate (66). Thus, interactions between axonin-1 and NrCAM are required to ensure that axons cross the midline. By contrast, perturbation of NgCAM or of F-spondin, a secreted protein present on the floor plate, led to defasciculation throughout the pathway but not to failure to cross the midline (66, 67), implicating these molecules in axon–axon interactions rather than axon guidance *per se*.

Ephrin-Bs are expressed at the midline (68, 69) and have recently emerged as candidates in the pathfinding of a number of axonal subpopulations at this region. For example, EphB receptors are expressed on postcrossing commissural axons within the longitudinal tracts, and the ephrin-Bs can induce the growth-cone collapse of commissural axons *in vitro* (69). This could provide a mechanism to constrain commissural axons in positions close to the midline. A slightly different scenario applies to corticospinal axons, which also express EphB receptors, but utilize ephrin signalling to prevent inappropriate recrossing of the spinal cord, since mice mutant for *ephrin-B3* exhibited bilateral innervation of motor neurons by the corticospinal tract (70). This general theme is reinforced in other studies showing that formation of the anterior commissure depends on signalling from EphB-receptor pathway cells to ephrin-B-bearing axons (71), in a repulsive manner, with loss of signalling resulting in axons aberrantly invading the surrounding territory. Guidance of the contralateral vestibuloacoustic afferents across the midline of the hindbrain also requires an ephrin-B-mediated interaction, and in mice mutant for *EphB2* these projections do not form correctly (72).

4.2 Formation of sensory and motor pathways to and from the spinal cord and brainstem

From their origin in the dorsal root ganglia (DRGs), sensory afferent axons send one branch into the spinal cord and one out into the periphery. The early polarization of these neurons may depend on a kind of 'surround repulsion', in which diffusible chemorepellent guidance cues emanate from the dermomyotome on one side and the notochord on the other (73). Central processes of DRG neurons first grow longitudinally along the spinal cord, before sending branches into the dorsal horns; these discrete stages in the axon outgrowth might be reliant on different mechanisms. Early projections from DRG neurons are those of 1a afferents, which are dependent for their survival on NT-3, whilst later projections are formed by the nociceptive and thermoceptive afferents which depend on NGF. In order to identify factors involved in branch formation, dissociated cultures of DRG neurons in the presence of NGF were used as a system in which to test the effects of various factors. Extracts of spinal cord were found to induce branching, but this effect could not be mimicked by any of the known molecules tested, pointing to a novel activity (74). A biochemical purification from calf brain was therefore undertaken, and ultimately led to the identification of the Slit protein (74), whilst almost simultaneously three vertebrate *Slit* genes were identified in human and rat models by cloning based on homology with the *Drosophila* gene (61). Evidence that vertebrate Slit proteins are involved in DRG axonal branch formation came from experiments in which Slit proteins caused axon elongation and branching when applied to cultures of DRG neurons (74). Slit is expressed within the dorsal spinal cord at stages of DRG collateral formation, and binds vertebrate homologues of Robo, which are expressed by DRG neurons (61). Interestingly, Slit proteins also have a repulsive action on the migration of inter-neuron precursors from the subventricular zone to the olfactory bulb (75) and GABAergic interneuron precursors from the ganglionic eminence to the cortex (76). Functions in both axon guidance and neuronal migration have thus been revealed for both netrins and Slits, implying fundamental similarities in the mechanisms underlying these two processes.

Once DRG axons enter the spinal cord, the NGF-dependent axons remain dorsally, whilst the NT-3-dependent population grow ventrally into the locale of the motor column. Collagen-gel co-culture experiments showed that the ventral neural tube is repellent for NGF-dependent DRG afferents, acting to confine these projections dorsally (77). Since Semaphorin-3A (Sema3A) is expressed by cells in the ventral part of the cord (see, for example, refs 78, 79), this implies that Sema3A could mediate the repulsion of NGF-responsive axons, confining them to the dorsal part of the cord. Clusters of cells secreting Sema3A repelled NGF-responsive axons, but not the NT-3-responsive population (78), and the repulsive activity of the ventral cord could be neutralized by antibodies against Sema3A (80). However, examination of mice lacking Sema3A function has so far failed to reveal striking defects in the central projection of DRG afferents (81). Thus different molecules act to initiate branching and to ensure the correct termination of DRG afferents. Other factors, hitherto unidentified, may be involved in the initial longitudinal extension of axons along the spinal cord.

By contrast with sensory axons, motor axons project away from the spinal cord to form ventral roots. In the brainstem, cranial motor neurons may project via either dorsal or ventral pathways, but a common pathway for all motor axons is to project initially away from the midline floor plate. Collagen-gel co-culture studies have shown that the floor plate produces a diffusible chemorepellent that excludes both cranial and spinal motor axons from the midline (82). At spinal levels, part of this effect may be mediated by Slit proteins, since a repulsive effect of this molecule was shown in collagen gel co-cultures (61); the effect of Slits on cranial motor axons has not yet been tested. Similar experiments showed that while netrin-1 repels only dorsally projecting cranial motor axons, Sema3A repels both dorsally projecting cranial motor axons and ventrally-projecting cranial and spinal motor axons (50, 51). This provides evidence that the signals controlling the trajectory of ventrally pro-jecting cranial motor neurons are distinct from those which trigger dorsal projection. Despite being repelled by the floor plate, oculomotor neurons (which lie within the midbrain) respond neither to netrin-1 nor to Sema3A (51), implying that additional chemorepellents may operate at midbrain levels. Indeed, recent analysis of mice lacking neuropilin-2 function showed defects in the organization and fasciculation of the oculomotor and the trochlear nerves, which may reflect a loss of response to Semaphorins, including Sema3F (83, 84).

Along the peripheral pathway of spinal motor axons, several tissues are non-permissive for growth, including the posterior somitic sclerotome, the perinoto-chordal mesenchyme (PNM), and, at limb levels, the pelvic girdle (85). These tissues are characterized by enhanced levels of peanut agglutinin (PNA)-binding glyco-conjugates and chondroitin sulfate proteoglycans (86, 87). The inhibitory properties of the posterior sclerotome are of particular note, since this phenomenon gives rise to the segmentation of the spinal nerves by causing motor axons to extend preferentially via the anterior sclerotome (see Chapter 4). Transplantation experiments of somite halves to create double-anterior or double-posterior sclerotomes in chick embryos initially showed the instructive role of the posterior sclerotome, refuting the idea of intrinsic segmentation of the spinal nerves (88). The collapsing activity of somite extracts on sensory axons is abolished by the addition of PNA (86) indicating that PNA-binding proteins have some role in barrier function. Additional players in the barrier repellent functions of the somite have recently come to light in the form of the ephrins. Ephrins-B1 and -B4 in the chick and ephrin-B2 in the mouse are expressed in the posterior sclerotome, and can repel motor axons when coated in stripes alternating with a permissive substratum such as laminin (89). These results implicate the ephrins in motor axon repulsion, although whether the PNA-binding repulsive activity is related to these molecules remains unclear. Corresponding receptors for these ephrins, EphB2 and EphB3, are present on spinal motor axons in both mouse and chick (89–92). Thus spinal motor axons are likely to be repelled by the posterior sclerotome by multiple mechanisms. The anterior sclerotome has also been implicated in the production of chemoattractant cues of unknown identity (26), and epaxial motor neurons may be attracted by their dermomyotome target, since in its absence they fail to project (93).

Further along their pathway, motor axons destined to innervate the hindlimb in the chick embryo collect to form the two lumbosacral plexi, distal to which they segregate into either a ventral or a dorsal nerve trunk, before following stereotyped paths towards muscle targets. Many studies have used experimental manipulations to force spinal motor axons into foreign territory and so characterize the axon guidance process. The results of these experiments broadly support a general scheme, in which motor axons are able to project along common 'highways' yet can respond to specific cues at 'decision' regions (94). Chemoattraction from limb-bud mesenchyme may be required to target spinal motor axons to the correct region, since such an effect can be demonstrated in explant co-cultures (26). Cell clusters secreting HGF mimic this effect, and antibodies against HGF block the chemo-attractant effect of limb-bud mesenchyme. Moreover, at times when motor axons are invading the limb, they express the HGF receptor, Met, and HGF is present in the mesenchymal environment (26).

Several additional aspects of spinal motor axon innervation of the target region may depend on the ephrins and their Eph receptors. *EphA7* is expressed at the limb base, surrounding the site of motor axon convergence, and motor axons in the lateral motor column at lumbar and brachial levels express *ephrin-A5*. When *EphA7* expression is eliminated, the convergence of axons at the plexus is blocked, and instead axons fan out prematurely and fail to enter the limb (95). The EphA4 receptor is also expressed differentially on spinal motor axons that navigate to different regions, with higher expression levels on axons that project dorsally (96). In mice lacking *EphA4* function, axons that would normally grow dorsally navigate instead via the ventral nerve to innervate an inappropriate territory. A subtler role in motor axon targeting to muscles is also suggested by the observation that ephrin-A sub-family members are differentially expressed on muscles along the anteroposterior axis in the cervical region; disruption of ephrin expression patterns perturbs the topography of muscle innervation at the synaptic level (97).

Dorsoventral axon targeting in the limb depends on interactions mediated by L1 and the polysialic acid-bearing form of NCAM (PSA-NCAM). Rather than being mediated by the differential expression of cell-adhesion molecules, the modulation of appropriate axon fasciculation occurs as a result of the modification of NCAM by polysialic acid, which decreases the homophilic interactions of NCAM. Disparate levels of PSA expression on dorsally projecting lumbosacral axons compared to their ventral counterparts, persists from the time of entry into the plexus region to the period of intramuscular branching and synaptogenesis (98). Removal of PSA *in ovo*, by injection of a PSA-specific endoneuraminidase (endo N) during motor axon sorting within the plexus region, introduced a high degree of projection errors in motor pools supplying both dorsal and ventral muscles (99). Removal of PSA in this fashion is thought to lead to increased fasciculation in the plexus, so that growth cones are unable to respond to guidance cues (100). Application of anti-L1 antibodies has been shown to reverse the effects of PSA removal on growth-cone trajectories and projection errors via an effect on decreased fasciculation (99).

In contrast to spinal motor pathways, the peripheral pathways of cranial motor axons do not appear to be influenced by chemorepulsion from the cranial paraxial mesoderm, which shows no overt segmentation. Rather the periodicity of the nerve roots depends on the intrinsic segmentation of the neural tube (101). Transplantation experiments showed a role for the exit point in the navigation of dorsally projecting axons out of the neural tube (102), but no direct evidence for chemoattraction by exit-point cells has been obtained. Once outside the neuroepithelium, the branchiomotor subpopulation of axons grow into the branchial arches, to innervate the muscle plate that lies in the core of the arch. Collagen gel co-cultures have shown that the branchial arches are the origin of chemoattractant cues for branchiomotor axons using *in vitro* co-cultures, with HGF partly responsible for this effect (27). However, other chemoattractant molecules may also be produced by the branchial arches, since *HGF* and *Met* mutant mice show only minor defects in the cranial nerves, and the branchial-arch chemoattraction of motor axons is only partly blocked by anti-HGF antibodies (27). Studies on the projection of trigeminal sensory axons towards the branchial arches proposed that BDNF and NT-3 are produced in the arch and are chemoattractant for sensory axons, but the possible action of these factors on cranial motor neurons has not been tested (24). The detailed patterning of cranial motor axon projections into the branchial arch also depends on Semaphorins, since targeted mutations of Sema3A or its receptor neuropilin-1 produce a defasciculation phenotype (103, 104). This suggests that Semaphorins channel motor axons to the central portion of the branchial-arch muscle plate, and that in the absence of this repulsion, axons ramify into the surrounding territory, which consists of neural crest-derived mesenchyme. Other contingents of cranial motor axons are the visceral motor neurons that form part of the parasympathetic nervous system. These axons diverge from the branchiomotor axon population, and instead grow out to innervate parasympathetic ganglion targets. Recent findings show a requirement for their ganglionic target in the navigation of these axons, since in mice mutant for the homeobox-containing gene *Phox2a*, the ganglia are missing and visceral motor neurons fail to project correctly (105). Although visceral motor neurons themselves express *Phox2a*, raising the possibility that the effect is cell-autonomous, the balance of evidence favours the loss of ganglion-derived factors to explain the phenotype. In addition, transplants of visceral motor neurons to ectopic locations in the chick embryo hindbrain have shown that visceral motor axons can successfully pathfind back to their ganglion targets, strengthening the evidence for a chemoattractant factor or factors (106).

4.3   Axon guidance within the visual system

During the pathfinding of retinal ganglion axons, they initially grow over the neuroepithelial surface of the diencephalon, and then decussate either completely (frog) or partially (other vertebrate classes) within the region of the optic chiasm. In fish, frogs, and chicks, the principal target of the retinal axons is a relay station in the midbrain, the optic tectum, whose homologue in mammals is the superior colliculus. In mammals, retinal ganglion cells project to the superior colliculus, and also via a thalamic relay, the lateral geniculate nucleus (LGN), from which second-order neurons project to the visual cortex. Retinal axons map topographically to particular regions on the tectum or LGN, with nasal and temporal axons projecting predomin-antly to posterior and anterior regions, respectively. The initial projection of retinal axons towards the tectum appears to depend on the presence of laminin (107) and FGF within the pathway (108). At the optic chiasm midline, growth cones show increased complexity, apparently presaging their pathfinding decisions, and a specialized population of glial cells together with growth-retardant properties in this region facilitates turning and decision-making (109). Part of this effect may be medi-ated by Slits, which are expressed at the chiasm and which can inhibit the growth of retinal ganglion axons *in vitro* (110). Recent studies have also shown that the ephrin-Bs act in a selective way on ipsilaterally projecting neurons. The ipsilateral projection of retinotectal axons in *Xenopus* occurs at a developmental stage when ephrin-Bs are starting to be expressed at the chiasm. Premature expression of these *ephrins* hastens the formation of ipsilateral projections by retinal ganglion cells, which express *EphB* receptors (111).

A series of elegant experiments, using sophisticated tissue-culture approaches, first explored the nature of the cues that organize axons along the anteroposterior axis of the tectum. Membranes were prepared from anterior or posterior tectum and layered on to a filter to present alternating stripes. Retinal explants were then placed at one edge of this array, so that the nasotemporal axis lay orthogonal to the membrane stripes. Under these conditions, axons grew out on the membranes in a patterned manner. Whereas nasal axons appeared to grow equally well on the anterior or posterior substrates, temporal axons were channelled so that they only grew on the anterior tectal membranes and avoided the posterior ones (112). Enzymatic removal of GPI-anchored proteins from the membrane carpets enabled temporal axons to grow on posterior membranes, pointing to an inhibitory activity within the posterior tectum. Part of the inhibitory activity was subsequently identified as two of the GPI-anchored ephrins, *ephrin-A2* and *ephrin-A5*, which were shown to be localized in the tectum in an increasing rostral to caudal gradient (113, 114). Correspondingly, the gene encoding the EphA3 receptor, which can bind these ligands, is expressed in a decreasing temporal to nasal gradient among retinal ganglion cells in the chick embryo, implying that retinal axons which express a high level of receptor map to tectal regions with a low level of ligand, and vice versa (113). In the mouse embryo, *EphA3* is not expressed by retinal ganglion cells, but *EphA5* and *EphA6* are expressed in a similar gradient across the nasotemporal axis of the retina (115).

In vitro assays have shown that substrate-bound ephrin-A2 is repellent for temporal axons, whilst ephrin-A5 repels both categories of axons (114, 116, 117). *Ephrin-A5* expression is restricted to a more caudal region of the tectum and also displays a steeper gradient. Temporal axons might be excluded from the caudal tectum due to *ephrin-A5* expression, whilst the termination of nasal axons in specific regions of the caudal tectum depends on the gradient of ephrin-A5 and its interaction with Eph receptors on the growth cone (reviewed in ref. 118). In support of this hypothesis, mice mutant for either or both of these ephrins show defects in retinotectal mapping. In the absence of *ephrin-A5* function, some temporal axons aberrantly project to the caudal tectum, and even project to an inappropriate target, the inferior colliculus, which normally expresses high levels of *ephrin-A5* (119). Mice doubly mutant for *ephrin-A2* and *ephrin-A5* also show defects in retinal axon topographic mapping on to the tectum (120). The relationship between levels of EphA expression and retinotectal mapping was explored in a study in which *EphA3* expression was driven in a subpopulation of retinal ganglion cells. Both *EphA3*-expressing and non-expressing-axons mapped on to the tectal surface along a smooth curve, with the position of an individual termination field determined by the relative, not the absolute, level of EphA expression (121). Axon targeting appeared to depend on the relative level of Eph receptors along with competitive interactions with neighbouring axons, rather than an axon with a particular level of Eph expression mapping to a particular position on the tectal surface.

Another complication to the picture is the observation that nasal axons themselves express *ephrin-A5*. Eliminating this molecule from the axons converts nasal axons to a temporal phenotype, in the sense that they now obey the repulsion promoted by stripes of posterior tectal membranes (122). This raises the possibility that co-expression of ligand on receptor-expressing cells leads to desensitization to repellent guidance cues, allowing for the correct pattern of topographic mapping. Models abound to show how ephrins govern the wiring of retinal axons, but it may be that none as yet are complete. It seems likely that axons growing on to the optic tectum via its rostral end terminate at positions in which they sense a critical level of a repulsive ligand. The region within which growth cones terminate depends on the levels of receptors, and possibly the combinations of receptors expressed. In addition, either some permissive quality of the tectum or a chemoattractant may be required to explain the continued growth of axons, to bring them in proximity to cues determining their termination site.

5. Signal transduction systems for axon guidance

Since the era when families of axon guidance molecules were identified, the major focus of attention has shifted and diversified into understanding how these molecular cues are interpreted by axons to bring about changes in growth and direction. Substantial progress has been made in unravelling the cellular machinery responsible for translating events at the cell surface into cytoskeletal changes that allow axonal growth and growth-cone steering. It is now emerging that growth cone responses depend on a complex web of factors. For example, signals can be modulated in a number of ways, including proteolytic processing of guidance ligands or receptors, and interactions of receptors in the plane of the membrane to form a variety of complexes. In addition, responses depend on which molecular components of downstream signalling pathways are activated, and the internal state of the growth cone at the time when signalling is initiated. Progress in the understanding of these factors is now contributing to knowledge of the way in which axons respond to multiple axon guidance molecules to generate their pathways. Only a few examples of these mechanisms can be given here.

5.1 Modulation of ligand and receptor function

Proteolytic processing is one way of modulating the activity of guidance molecules and their receptors. For some members of the Semaphorin 3 subclass, including Sema3A, the protein is initially synthesized as an inactive precursor, and is then cleaved proteolytically to give rise to a fragment active in axonal repulsion (123). Recent characterization of vertebrate Slit proteins showed proteolytic cleavage *in vitro*, leading to the generation of an amino-terminal fragment that remained associated with cell surfaces, and a carboxy-terminal fragment that was more readily diffusible (61). This processing also appears to occur *in vivo*, since the processed form can be detected in *Drosophila* embryo extracts (61), but it has not yet been resolved whether one or both fragments are active *in vivo*. Recent data also shows that proteases can cleave membrane-bound molecules involved in axon guidance to modulate signalling. In one case this may solve a paradox raised by the fact that the binding between ephrins and their Eph receptors on two adjacent cells can lead to repulsion. Ephrin-A2 was found to form a complex with the metalloprotease Kuzbanian, and contact of the ephrin-bearing cells with axons expressing the appropriate Eph receptor led to ephrin cleavage and retraction of the axon (124). Cleavage of the ephrin ligand may thus represent a way of converting a potentially attractive interaction into a repellent one. Metalloproteases were also found to be capable of cleaving the netrin receptor DCC at the cell surface. For commissural axons, which express DCC and are chemoattracted by netrin, metalloprotease inhibitors potentiated the chemoattractive effect of netrin, providing a potential route to modulate netrin function (125).

Recent findings also point to the idea that individual receptors do not mediate simple positive or negative responses to axon guidance cues, but that the composition of receptor complexes influences responses to guidance cues. Some of these studies have been based on the culture of *Xenopus* spinal neurons, some of which express *DCC*. Single neurons growing on a planar substrate can be chemoattracted by a pipette containing netrin-1, and their turning angle quantitated. This system was used to test the effects of Unc5h1, by injecting mRNA encoding *Unc5h1* into every cell of early frog embryos, and subsequently isolating spinal neurons (55). Coexpression of *Unc5h1* and *DCC* was found to confer a chemorepellent response, causing growth cones to turn away from, rather than towards, a source of netrin-1. Injection of constructs encoding the cytoplasmic domain of Unc5h1, containing a myristoylation sequence, targets this construct to the inner leaflet of the lipid bilayer, and is sufficient to transform the response. This conversion appears to be accomplished by association between the cytoplasmic domains of DCC and Unc5h1 proteins, since a complex of these proteins can be immunoprecipitated following their coexpression in *Xenopus* spinal neurons.

A similar phenomenon of association between cytoplasmic domains seems to apply to DCC and the Slit receptor Robo. In this case, activation of Robo by Slit leads to association with the cytoplasmic domain of DCC, and the cessation of the attractive effect of netrin (126). The relevance of this mechanism may be to convert attraction to the midline by netrin into repulsion from the midline by Slit, ensuring that axons do not cross more than once. A direct interaction between the signalling pathways via Sema3A and L1 was also recently shown, based on the observation that corticospinal axons are repelled by the ventral spinal cord during their decussation at the boundary between the hindbrain and the spinal cord (127). In *L1* mutant mice this repulsive response was absent, and L1 and the Semaphorin receptor neuropilin-1 were subsequently found to form a stable complex that is required for the response to Sema3A (127).

5.2 Signal transduction pathways

For some guidance receptors, tyrosine phosphorylation is likely to be the key event in downstream signalling. For example, the Eph receptors are phosphorylated following ligand binding, although there is also now good evidence that ephrins themselves can transduce intracellular signals despite lacking intrinsic kinase activity (41, 42). Cell-adhesion molecules (CAMs) stimulate axon outgrowth by activating the FGF tyrosine kinase receptor (44). NCAM is thought to interact with the FGF receptor in the plane of the membrane, recruiting phospholipase C-gamma which binds to phosphorylated tyrosine residues, and ultimately leads (via several intermediaries) to calcium influx. This influx acts on calmodulin kinase to trigger neurite outgrowth. Ligand binding may also cause recruitment to the membrane of adapter proteins, which can initiate a range of downstream effects, with the ultimate result of causing increased polymerization or depolymerization of actin filaments.

Prime candidates for influencing cytoskeletal changes are the family of Rho-related small GTPases, which function as molecular switches existing in GTP (active) or GDP-bound (inactive) forms. Replacement of GDP by GTP causes activation, whilst inactivation depends on an intrinsic GTPase activity (128). Of these proteins, three—Rho, Rac, and Cdc42—have been studied most extensively in fibroblasts, in which they are thought to modulate the organization of the cytoskeleton in different ways. Injection of active Cdc42 into quiescent fibroblasts causes filopodia to form, whilst active Rac induces lamellipodial formation, and active Rho causes the formation of stress fibres (129). A number of studies now indicate that the Rho-like GTPases are likely to fulfil similar functions in neural cells, since dominant active or inactive forms of these molecules can influence neuronal morphology (130). *In vitro* studies, based largely on PC12 cells, have led to the idea that activation of Cdc42 and Rac1 lead to filopodia and lamellipodia formation, perhaps in response to positive guidance cues, whilst activation of RhoA leads to growth-cone collapse in response to negative guidance cues. However, recent data have shown additional levels of complexity in the regulation of the Rho family, raising doubts that a simple model for the operation of Rho family members will suffice (131). For example, the model would predict RhoA's involvement in growth-cone collapse, whereas, in DRG neurons, injection of activated Rac1 increases the number of collapsed growth cones in response to added collapsin-1/Sema3A (132). One way of reconciling these data is to propose that different growth-cone collapsing factors might act via different pathways, depending on a specific level of activity of the various Rho GTPases. Indeed, in chick motor neurons, the expression of constitutively active forms of Rac1 or RhoA abolished myelin-induced growth-cone collapse, whereas dominant-negative Rac1 or Cdc42 negated collapsin-induced collapse (133). Ephrin-A5-mediated growth-cone collapse has been shown to occur via Rho and its downstream effector Rho kinase, since treatment of retinal ganglion cells with ephrin-A5 led to an activation of Rho and downregulation of Rac. Application of inhibitors of Rho or Rho kinase led to a strong reduction in growth-cone collapse (134).

How are signals that impinge on the growth cone interpreted in order to govern the activity of the Rho-like GTPases and ultimately to change the organization of the cytoskeleton? Intermediate between receptors and the Rho-like GTPases are a group of regulatory proteins that influence the ratio of the GDP and GTP-bound forms. Guanine exchange factors (GEFs) and GTPase-activating proteins (GAPs) increase and decrease the ratio of GTP-bound protein, respectively, whilst guanine nucleotide dissociation inhibitors (GDIs) stabilize them in the inactive, GDP-bound form (135). An example of the action of such a factor is that the GEF Tiam-1, which is specific for Rac, and can induce cell spreading when overexpressed in PC12 cells. In conjunction with laminin, but not with fibronectin, Tiam-1 induces the development of neurite-like processes with wide lamellipodia (136), implying that the extracellular matrix context is important for the cellular response. Interestingly, plexin receptors for Semaphorins can bind directly to Rho GTPases and may have an intrinsic GAP activity necessary to induce growth-cone collapse, providing a direct route to the actin cytoskeleton (137).

The effects of the Rho-like GTPases are transduced via 20 or more target proteins that directly regulate cytoskeletal conformation, including N-Wasp, Pak, Rho kinase, and myosin light-chain phosphatase (MLCP). These proteins may differentially influence the formation of filopodia, lamellipodia, and growth-cone collapse. For example, coexpression of *N-WASP* with *Cdc42* in COS cells induces the formation of long filopodia (138), while Pak is implicated as an effector of Rac1 in the growth-cone collapse of DRG neurons mediated by Sema3A (139). Another route for regulating actin dynamics is via the Enabled (Ena)/VASP family of actin regulatory proteins These proteins, including mammalian Ena—i.e. 'Mena'—can bind profilin, an actin-binding protein, as well as being directly involved in actin polymerization (140). In mice mutant for *Mena* there are defects in the formation of the corpus callosum and the hippocampal commissure, suggesting a role for *Mena* in axon guidance or growth-cone motility (141). Indeed, Mena protein is strikingly localized in the tips of growth cones, distal to the actin filaments. The link between Ena/VASP proteins and axon guidance signals is best understood in *Drosophila*, in which Ena is regulated by phosphorylation and dephosphorylation by the Abl tyrosine kinase and receptor protein tyrosine phosphatases (RPTPs) respectively (142, 143). The RPTPs are required for growth-cone motility, since mutations in RPTP genes cause a phenotype in which axons stop short of their targets (144). The role of RPTPs in vertebrates is likely to provide future research directions, since homologues have been isolated and their expression patterns mapped, but ligands are as yet unidentified (145). Whilst *in vitro* experiments point to a role for vertebrate RPTPs in axon guidance, firm evidence of an *in vivo* role in this process is presently lacking, though it is currently under investigation using knockout animals (see, for example, ref. 146). In addition to their interactions with the Ena/VASP family, the RPTP, Dlar and the tyrosine kinase, Abl may interact with the trio family of GEFs in order to regulate the Rho GTPases (147). In *Drosophila*, Trio has been shown to bind Rac and in turn to activate Pak to control actin dynamics (148). A version of Trio has been identified in humans that binds to LAR (149), but a possible role in axon guidance has not been elucidated.

5.3   Calcium and cAMP in growth cones

Recent experiments have emphasized that responses to attractant or repellent guidance cues depend on the internal state of the growth cone, and, in particular, on calcium and cAMP levels (150). Lowering extracellular calcium abolished growth-cone turning in *Xenopus* spinal neurons in response to a netrin gradient, implying that calcium influx is a prerequisite of growth-cone steering (151). The requirement for appropriate calcium levels for growth-cone responses was further investigated by manipulating cytoplasmic calcium, either by blocking calcium channels or the release of calcium from internal stores. Various degrees of reduction of calcium levels showed that a dramatic reduction abolished chemoattraction in a netrin gradient, while a more modest reduction converted chemoattraction to chemorepulsion (152). It thus appears that high calcium signalling can cause chemoattraction, whereas low signalling causes chemorepulsion. A reduction in the intracellular levels of cAMP had the same effect—converting chemoattraction in response to netrin, BDNF, or acetylcholine into chemorepulsion (25, 151). Interestingly, the extracellular matrix molecule laminin can also convert netrin-mediated chemoattraction into chemorepulsion, and netrin and laminin, respectively, increase and decrease levels of cAMP in the growth cone (153). One possible model to explain these observations is that a guidance cue, such as netrin impinging on the growth cone, leads to both calcium influx through the membrane and mobilization of calcium from internal stores, which activate calcium-sensitive adenylate cyclase, thereby elevating cAMP levels (150). Cyclic AMP would then cause chemoattraction by inhibiting Rho and preventing growth-cone collapse pathways (30). Nevertheless, an even more complex picture is now emerging. For example, different cyclic nucleotides might play distinct roles in pathfinding in response to different guidance cues; the turning responses in a gradient of NT-3 were not abrogated by a reduction in cAMP, and repulsive responses to Sema3A and myelin-associated glycoprotein gradients were abolished by elevating cGMP rather than cAMP levels (25, 154).

6. Conclusions and future directions

This is a period of rapid and exciting progress within the axon guidance field of research. Improved molecular biological techniques, greater availability of reagents, development of many different animal models, and extensive international collaborations have all contributed to huge advances in the elucidation of molecular mechanisms. Large-scale genetic screens in, for example, the zebrafish and mouse have yielded many mutants with defective development of axon pathways that will provide material for study in many years to come. Although molecular studies have prominence, it is worth pointing out that neuroanatomy still has a long way to go in providing us with a precise description of axon pathways whose underlying mechanisms of formation we wish to investigate. Luckily, neuroanatomical studies have gained impetus from a variety of elegant techniques. For example, the use of transgenic mouse lines harbouring *tau–lacZ* transgenes in particular loci results in the labelling of axon pathways derived from specific subsets of neurons. A similar purpose can be served in chick embryos by generating gene constructs encoding *tau–GFP* (Green fluorescent protein) under the control of general or specific promoters, and incorporating them into neuroepithelial cells prior to axon outgrowth. Such techniques may now be used in conjunction with other approaches, for example tissue culture, to explore the basis of the formation of particular axon tracts. In the case of transgenic mice, a particularly powerful approach is to cross mice expressing a reporter gene (as described above) with lines mutant for a particular axon guidance cue. Analysis of downstream signalling pathways and their modulation will continue to be a huge focus of interest, and understanding signalling components in more detail also holds out the hope of more convenient therapies to potentiate axon regrowth and neural repair.

References

1. Jacobson, M. (1991) *Developmental neurobiology*. Plenum Press, New York.
2. Hatten, M. E. (1999) Central nervous system neuronal migration. *Annu. Rev. Neurosci.*, **22**, 511.
3. Lumsden, A. and Krumlauf, R. (1996) Patterning the vertebrate neuraxis. *Science*, **274**, 1109.
4. Edlund, T. and Jessell, T. M. (1999) Progression from extrinsic to intrinsic signaling in cell fate specification: a view from the nervous system. *Cell*, **96**, 211.
5. Tanabe, Y. and Jessell, T. M. (1996) Diversity and pattern in the developing spinal cord. *Science*, **274**, 1115.
6. Jessell, T. M. and Sanes, J. R. (2000) The decade of the developing brain. *Curr. Opin. Neurobiol. Dev.*, **10**, 599.
7. Briscoe, J., Pierani, A., Jessell, T. M., and Ericson, J. A. (2000) A homeodomain protein code specifies progenitor cell identity and neuronal fate in the ventral neural tube. *Cell*, **101**, 435.
8. Tsuchida, T., Ensini, M., Morton, S. B., Baldassare, M., Edlund, T., Jessell, T. M. and Pfaff, S. L. (1994) Topographic organisation of embryonic motor neurons defined by expression of LIM homeobox genes. *Cell*, **79**, 957.
9. Sharma, K., Sheng, H. Z., Lettiere, K., Li, H., Karavanov, A., Potter, S., Westphal, H. and Pfaff, S. L. (1998) Lim homeodomain factors Lhx3 and Lhx4 assign subtype identities for motor neurons. *Cell*, **95**, 817.
10. Pfaff, S. and Kinter, C. (1998) Neuronal diversification: development of motor neuron subtypes. *Curr. Opin. Neurobiol.*, **8**, 27.
11. Pfaff, S. L., Mendelsohn, M., Stewart, C. L., Edlund, T., and Jessell, T. M. (1996) Requirement for LIM homeobox gene I sl-1 in motor neuron generation reveals a motor neuron-dependent step in interneuron differentiation. *Cell*, **84**, 309.
12. Sharma, K., Leonard, A. E., Lettieri, K., and Pfaff, S. (2000) Genetic and epigenetic mechanisms contribute to motor neuron pathfinding. *Nature*, **406**, 515.
13. Kania, A., Johnson, R. L., and Jessell, T. M. (2000) Coordinate roles for Lim homeobox genes in directing the dorsoventral trajectory of motor axons in the vertebrate limb. *Cell*, **102**, 161.
14. Gordon-Weeks, P. R. (2000) *Neuronal growth cones*. Cambridge University Press, Cambridge.
15. Mackay, D. J., Nobes, C. D., and Hall, A. (1995) The Rho's progress: a potential role during neuritogenesis for the Rho family of GTPases. *Trends Neurosci.*, **18**, 496.
16. Silver, J. and Sidman, R. S. (1980) A mechanism for the guidance and topographic patterning of retinal ganglion cell axons. J. *Comp. Neurol.*, **189**, 101.
17. Heyman, I., Kent, A., and Lumsden, A. (1993) Cellular morphology and extracellular space at rhombomere boundaries in the chick embryo hindbrain. *Dev. Dyn.*, **198**, 241.
18. Heyman, I., Faissner, A., and Lumsden, A. (1995) Cell and matrix specialisation of rhombomere boundaries. *Dev. Dyn.*, **204**, 301.
19. Tessier-Lavigne, M. and Goddman, C. S. (1996) The molecular biology of axon guidance. *Science*, **274**, 1123.

20. Levi-Montalcini, R. and Angeletti, P. U. (1963) Essential role of the nerve growth factor in the survival and maintenance of dissociated sensory and sympathetic nerve cells in vitro. *Dev. Biol.*, **7**, 653.

21. Gundersen, R. W. and Barrett, J. N. (1979) Neuronal chemotaxis: chick dorsal-root axons turn toward high concentrations of nerve growth factor. *Science*, **206**, 1079.

22. Tessier-Lavigne, M. (1994) Axon guidance by diffusable repellents and attractants. *Curr. Opin. Genet. Dev.*, **4**, 596.

23. Henderson, C. E. (1996) Role of neurotrophic factors in neuronal development. *Curr. Opin. Neurobiol.*, **6**, 64.

24. O'Connor, R. and Tessier-Lavigne, M. (1999) Identification of maxillary factor, a maxillary process-derived chemoattractant for developing trigeminal sensory axons. *Neuron*, **24**, 165.

25. Song, H., Ming, G., and Poo, M. (1997) cAMP-induced switching in turning direction of nerve growth cones. *Nature*, **388**, 275.

26. Ebens, A., Brose, K., Leonardo, E. D., Gartz, H. J. M., Bladt, F., Birchmeier, C., Barres, B. A. and Tessier-Lavigne, M. (1996) Hepatocyte growth factor/scatter factor is an axonal chemoattractant and neurotrophic factor for spinal motor neurons. *Neuron*, **17**, 1157.

27. Caton, A., Hacker, A., Naeem, A., Livet, J., Maina, F., Bladt, F., Klein, R., Birchmeier, C. and Guthrie, S. (2000) The branchial arches and HGF are growth-promoting and chemo-attractant for cranial motor axons. *Development*, **127**, 1751.

28. Chisholm, A. and Tessier-Lavigne, M. (1999) Conservation and divergence of axon guidance mechanisms. *Curr. Opin. Neurobiol.*, **9**, 603.

29. Goodman, C. S. and Shatz, C. J. (1993) Developmental mechanisms that generate precise patterns of neuronal connectivity. *Cell*, **72**, 77.

30. Mueller, B. K. (1999) Growth cone guidance: first steps towards a deeper understanding. *Annu. Rev. Neurosci.*, **22**, 351.

31. Varela-Echavarrja, A. and Guthrie, S. (1997) Molecules making waves in axon guidance. *Genes Dev.*, **11**, 545.

32. Serafini, T., Kennedy, T. E., Galko, M. J., Mirzayan, C., Jessell, T. M., Tessier-Lavigne, M. (1994) The netrins define a family of axon outgrowth-promoting proteins homologous to *C. elegans* UNC-6. *Cell*, **78**, 409.

33. Guthrie, S. (1997) Axon guidance: netrin receptors are revealed. *Curr. Biol.*, **7**, R6.

34. Leonardo, E. D., Hinck, L., Masu, M., Keino-Masu, K., Ackerman, S. L., and Tessier-Lavigne, M. (1997) Vertebrate homologues of C. elegans UNC-5 are candidate netrin receptors. *Nature*, **386**, 833.

35. Corset, V., Nguyen-Ba-Charvet, K. T., Forcet, C., Moyse, E., Chedotal, A., and Mehlem, P. (2000) Netrin-1-mediated axon outgrowth and cAMP production requires interaction with adenosine A2b receptor. *Nature*, **407**, 747.

36. Kolodkin, A. L., Matthes, D. J., and Goodman, C. S. (1993) The semaphorin genes encode a family of transmembrane and secreted growth cone guidance molecules. *Cell*, 75:1389.

37. He, Z. and Tessier-Lavigne, M. (1997) Neuropilin is a receptor for the axonal chemo-repellent semaphorin III. *Cell*, **90**, 739.

38. Winberg, M. L., Noordermeer, J. N., Tamagnone, L., Comoglio, P. M., Spriggs, M. K., Tessier-Lavigne, M., *et al.* (1998) Plexin A is a neuronal semaphorin receptor that controls axon guidance. *Cell*, **95**, 903.

39. Comeau, M. R., Johnson, R., DuBose, R. F., Petersen, M., Gearing, P., VandenBos, T., Park, L., Farrah, T., Buller, R. M., Cohen, J. I., Strockbine, L. D., Rauch, C. and Spriggs, M. K. (1998) A poxvirus-encoded semaphorin induces cytokine production from monocytes and binds to a novel cellular semaphorin receptor, VESPR. *Immunity*, **8**, 473.

40. Yu, H. and Kolodkin, A. L. (1999) Semaphorin signaling: a little less per-plexin. *Neuron*, **22**, 11.

41. O'Leary, D. D. M. and Wilkinson, D. G. (1999) Eph receptors and ephrins in neural development. *Curr. Opin. Neurobiol.*, **9**, 65.

42. Frisen, J., Holmberg, J., and Baracid, M. (1999) Ephrins and there Eph receptors: multitalented directors of embryonic development. *EMBO J.*, **18**, 5159.

43. Brose, K. and Tessier-Lavigne, M. (2000) Slit proteins: key regulators of axon guidance, axonal branching, and cell migration. *Curr. Opin. Neurobiol.*, **10**, 95.

44. Walsh, F. S. and Doherty P. (1997) Neural cell adhesion molecules of the immunoglobulin superfamily: role, in axon growth and guidance. *Annu. Rev. Cell Dev. Biol.*, **13**, 425.

45. Ramon, Y. and Cajal, S. (1892) La retine des vertebres. *La Cellule*, **9**, 119.

46. Tessier-Lavigne, M., Placzek, M., Lumsden, A. G. S., Dodd, J., and Jessell, T. M. (1988) Chemotropic guidance of developing axons in the mammalian central nervous system. *Nature*, **336**, 775.

47. Kennedy, T. E., Serafini, T., de la Torre, J. R. and Tessier-Lavigne, M. (1994) Netrins are diffusible chemotropic factors for commissural axons in the embryonic spinal cord. *Cell*, **78**, 425.

48. Hedgecock, E. M., Culotti, J. G. and Halls, D. H. (1990) The unc-5, unc-6 and unc-40 genes guide circumferential migrations of pioneer axons and mesodermal cells on the epidermis in *C. elegans*. *Neuron*, **2**, 61.

49. Ishii, N., Wadsworth, W. G., Stern, B. D., Culotti, J. G., and Hedgecock, E. M. (1992) UNC-6 a laminin-related protein guides cell and pioneer axon migrations in *C. elegans*. *Neuron*, **9**, 873.

50. Colamarino, S. A. and Tessier-Lavigne, M. (1995) The axonal chemoattractant netrin-1 is also a chemorepellent for trochlea motor axons. *Cell*, **81**, 621.

51. Varela-Echavarria, A., Tucker, A., Paschel, A. W., and Guthrie, S. (1997) Motor axon subpopulations respond differentially to the chemorepellents netrin-1 and semaphorin D. *Neuron*, **18**, 193.

52. Shirasaki, R., Mirzayan, C., Tessier-Lavigne, M., and Murakami, F. (1996) Guidance of circumferentially-growing axons by netrin-dependent and independent floor plate chemotropism in the vertebrate brain. *Neuron*, **17**, 1079.

53. Keino-Masu, K., Masu, M., Hinck, L., Leonardo, D., Chan, S. S. Y., Culotti, J. G. and Tessier-Lavigne, M. (1996) Deleted in colorectal cancer (DCC) encodes a netrin receptor. *Cell*, **87**, 175.

54. Ackerman, S. L., Kozak, L. P., Przyborski, S. A., Rund, L. A., Boyer, B. B., and Knowles, B. B. (1997) The mouse rostral cerebellar malformation gene encodes an UNC-5-like protein. *Nature*, **386**, 838.

55. Hong, K., Hinck, L., Nishiyama, M., Poo, M., Tessier-Lavigne, M., and Stein, E. (1999) A ligand-gated association between cytoplasmic domains of UNC5 and DCC family receptors converts netrin-induced growth cone attraction to repulsion. *Cell*, **97**, 927.

56. Serafini, T., Colamarino, S. A., Leonardo, E. D., Wang, H., Beddington, R., Skarnes, W. C., and Tessier-Lavigne, M. (1996) Netrin-1 is required for commissural axon guidance in the developing vertebrate nervous system. *Cell*, **87**, 1001.

57. Fazeli, A., Dickinson, S. L., Hermiston, M. L., Tighe, R. V., Steen, R. G., Small, C. G., Stoeckli, E. T., Keino-Masu, K., Masu, M., Rayburn, H., Simons, J., Bronson, R. T., Gordon, J. I., Tessier-Lavigne, M. and Winberg, R. A. (1997) Phenotype of mice lacking functional deleted in colorectal cancer (DCC) gene. *Nature*, **386**, 796.

58. Bloch-Gallego, E., Ezan, F., Tessier-Lavigne, M. and Sotelo, C. (1999) Floor plate and

netrin-1 are involved in the migration and survival of inferior olivary neurons. *J. Neurosci.*, **19**, 4407.

59. Barallobre, M. J., Del Rio, J. A., Alcantara, S., Borrell, V., Aguado, F., Ruiz, M., Carmona, M. A., Martin, M., Fabre, M., Yuste, R., Tessier-Lavigne, M. and Sonano, E. (2000) Aberrant development of hippocampal circuits and altered neural activity in netrin 1-deficient mice. *Development*, **127**, 4797.

60. Braisted, J. E., Catalano, S. M., Stimac, R., Kennedy, T. E., Tessier-Lavigne, M., Shatz, C. J., and O'Leary, D. D. (2000) Netrin-1 promotes thalamic axon growth and is required for proper development of the thalamocortical projection. *J. Neurosci.*, **20**, 5792.

61. Brose, K., Bland, K. S., Wang, K. H., Arnott, D., Henzel, W., Goodman, C. S., Tessier-Lavigne M. and Kidd, T. (1999) Slit proteins bind Robo receptors and have an evolutionarily conserved role in repulsive axon guidance. *Cell*, **96**, 795.

62. Kidd, T., Bland, K. S., and Goodman, C. S. (1999) Slit is the midline repellent for the robo receptor in Drosophila. *Cell*, **96**, 785.

63. Zou, Y., Stoeckli, E., Chen, H., and Tessier-Lavigne, M. (2000) Squeezing axons out of the grey matter: a role for slit and semaphorin proteins from midline and ventral spinal cord. *Cell*, **102**, 363.

64. Shirasaki, R., Katsumata, R., and Murakami, F. (1998) Change in chemoattractant responsiveness of developing axons at an intermediate target. *Science*, **279**, 105.

65. Tear, G. (2001) A new code for axons. *Nature*, **409**, 472.

66. Stoeckli, E. T. and Landmesser, L. T. (1995) Axonin-1, Nr-CAM, and Ng-CAM play different roles in the *in vivo* guidance of chick commissural neurons. *Neuron*, **14**, 1165.

67. Burstyn-Cohen, T., Tzarfaty, V., Frumkin, A., Feinstein, Y., Stoeckli, E., and Klar, A. (1999) F-Spondin, is required for accurate pathfinding of commissural axons at the floor plate. *Neuron*, **23**, 233.

68. Gale, N. W., Holland, S. J., Valenzuela, D. M., Flenniken, A., Pan, L., Ryan, T., Henkemeyer, M., Strebhardt, K., Hirai, H., Wilkinson, D. G., Pawson, T., Davis, S. and Yancopoulos, G. D. (1996) Eph receptors and ligands comprise two major specificity subclasses and are reciprocally compartmentalized during embryogenesis. *Neuron*, **17**, 9.

69. Imondi, R., Wideman, C., and Kaprielian, Z. (2000) Complementary expression of transmembrane ephrins and their receptors in the mouse spinal cord: a possible role in constraining the orientation of longitudinally projecting axons. *Development*, **127**, 1397.

70. Yokoyama, N., Romero, M. I., Cowan, C. A., Galvan, P., Helmbacher, F., Charnay, P., Parada, L. F. and Henkemeyer, M. (2001) Forward signaling mediated by Ephrin-B3 prevents contralateral corticospinal axons from recrossing the spinal cord midline. *Neuron*, **29**, 85.

71. Henkemeyer, M., Orioli, D., Henderson, J., Saxton, T., Roder, J., Pawson, T. and Klein, R. (1996) Nuk controls pathfinding of commissural axons in the mammalian central nervous system. *Cell*, **86**, 35.

72. Cowan, C. A., Yokoyama, N., Bianchi, L. M., Henkemeyer, M., and Fritzsch, B. (2000) EphB2 guides axons at the midline and is necessary for normal vestibular function. *Neuron*, **26**, 417.

73. Keynes, R., Tannahill, D., Morgenstern, D. A., Johnson, A. R., Cook, G. M. W., and Pini, A. (1997) Surround repulsion of spinal sensory axons in higher vertebrate embryos. *Neuron*, **18**, 889.

74. Wang, K. H., Brose K., Arnott, D., Kidd, T., Goodman, C. S., Henzel, W. and Tessier-Lavigne, M. (1999) Biochemical purification of a mammalian slit protein as a positive regulator of sensory axon elongation and branching. *Cell*, **96**, 771.

75. Wu, W., Wong, K., Chen, J., Jiang, Z., Dupuis, S., Wu, J. Y. and Rao, Y. (1999) Directional

guidance of neuronal migration in the olfactory system by the protein Slit. *Nature*, **400**, 331.

76. Zhu, Y., Li, H-S., Zhou, L., Wu, J. Y., and Rao, Y. (1999) Cellular and molecular guidance of GABAergic neuronal migration from an extracortical origin to the neocortex. *Neuron*, **23**, 473.

77. Fitzgerald, M., Kwait, G. C., Middleton, J., and Pini, A. (1997) Ventral spinal cord inhibition of neurite outgrowth from embryonic rat dorsal root ganglia. *Development*, **117**, 1377.

78. Messersmith, E. K., Leonardo, E. D., Shatz, C. J., Tessier-Lavigne, M., Goodman, C. S., and Kolodkin, A. L. (1995) Semaphorin III can function as a selective chemorepellent to pattern sensory projections in the spinal cord. *Neuron*, **14**, 949.

79. Puschel, A. W., Adams, R. H., and Betz, H. (1995) Murine semaphorin D/collapsin is a member of a diverse gene family and creates domains inhibitory for axonal extension. *Neuron*, **14**, 941.

80. Shepherd, I., Luo, Y., Lefcort, F., Reichardt, L. F., and Raper, J. A. (1997) A sensory axon repellent secreted from ventral spinal cord explants is neutralised by antibodies raised against collapsin-1. *Development*, **124**, 1377.

81. Behar, O., Golden, J. A., Mashimo, H., Schoen, F. J., and Fishman, M. C. (1996) Semaphorin III is needed for growth of nerves, bones and heart. *Nature*, **383**, 525.

82. Guthrie, S. and Pini, A. (1995) Chemorepulsion of developing motor axons by the floor plate. *Neuron*, **14**, 1117.

83. Giger, R. J., Cloutier, J. F., Sahay, A., Prinjha, R. K., Levengood, D. V., Moor, S. E., Pickering, S., Simmons, D., Rastan, S., Walsh, F. S., Kolodkin, A. L., Ginty, D. D. and Geppert, M. (2000) Neuropilin-2 is required *in vivo* for selective axon guidance responses to secreted semaphorins. *Neuron*, **25**, 29.

84. Chen, H., Bagri, A., Zupicich, J. A., Zou, Y., Stoeckli, E., Pleasure, S. J., Lowenstein, D. H., Skarnes, W. C., Chédotal, A. and Tessier-Lavigne, M. (2000) Neuropilin-2 regulates the development of selective cranial and sensory nerves and hippocampal mossy fibre projection. *Neuron*, **25**, 43.

85. Tosney, K. W. (1991) Cells and cell-interactions that guide motor axons in the developing chick embryo. *BioEssays*, **13**, 17.

86. Davies, J. A., Cook, G. M., Stern, C. D., and Keynes, R. J. (1990) Isolation from chick somites of a glycoprotein fraction that causes collapse of dorsal root ganglion growth cones. *Neuron*, **4**, 11.

87. Oakley, R. A. and Tosney, K. W. (1993) Contact-mediated mechanisms of motor axon segmentation. *J. Neurosci.*, **13**, 3773.

88. Keynes, R. J. and Stern, C. D. (1984) Segmentation in the vertebrate nervous system. *Nature*, **310**, 786.

89. Wang, H. U. and Anderson, D. J. (1997) Eph family transmembrane ligands can mediate repulsive guidance of trunk neural crest migration and motor axon outgrowth. *Neuron*, **18**, 383.

90. Henkemeyer, M., Marengere, L. E. M., McGlade, J., Olivier, J. P., Conlon, R., Holmyard, D. P., Letwin, K. and Pawson, T. (1994) Immunolocalization of the Nuk receptor tyrosine kinase suggests roles in segmental patterning of the brain and axonogenesis. *Oncogene*, **9**, 1001.

91. Kilpatrick, T. J., Brown, A., Lai, C., Gassmann, M., Goulding, M., and Lemke, G. (1996) Expression of the Tyro4/Mek4/Cek4 gene specifically marks a subset of embryonic motor neurons and their muscle targets. *Mol. Cell. Neurosci.*, **7**, 62.

92. Ohta, K., Nakamura, M., Hirokawa, K., Tanaka, S., Iwama, A., Suda, T., Ando, M. and Tanaka, H. (1996) The receptor tyrosine kinase, Cek8 is transiently expressed on subtypes of motoneurons in the spinal cord during development. *Mech. Dev.*, **54**, 59.

93. Tosney, K. W. (1987) Proximal tissues and patterned neurite outgrowth at the lumbosacral level of the chick embryo: deletion of the dermamyotome. *Dev. Biol.*, **122**, 540.

94. Landmesser, L. (1984) The development of specific motor pathways in the chick embryo. *Trends Neurosci.*, **7**, 336.

95. Araujo, M., Piedra, M. E., Herrera, M. T., Ros, M. A., and Nieto, M. A. (1998) The expression and regulation of chick EphA7 suggests roles in limb patterning and innervation. *Development*, **125**, 4195.

96. Helmbacher, F., Schneider-Maunoury, S., Topilko, P., Tiret, L., and Charnay, P. (2000) Targeting of the EphA4 tyrosine kinase receptor affects dorsal/ventral pathfinding of limb motor axons. *Development*, **127**, 3313.

97. Feng, G., Laskowski, M. B., Feldheim, D. A., Wang, H., Lewis, R., Frisen, J., Flanagan, J. G. and Sanes, J. R. (2000) Roles for Ephrins in positionally selective synaptogenesis between motor neurons and muscle fibres. *Neuron*, **25**, 295.

98. Tang, J., Landmesser, L., and Rutishauser, U. (1992) Polysialic acid influences specific pathfinding by avian motoneurons. *Neuron*, **8**, 1031.

99. Landmesser, L., Dahm, L., Tang, J., and Rutishauser, U. (1990) Polysialic acid as a regulator of intramuscular nerve branching during embryonic development. *Neuron*, **4**, 655.

100. Landmesser, L., Dahm, L., Schultz, K., and Rutishauser, U. (1988) Distinct roles for adhesion molecules during innervation of embryonic chick muscle. *Dev. Biol.*, **130**, 645.

101. Lumsden, A. G. S. and Keynes, R. (1989) Segmental patterns of neuronal development in the chick hindbrain. *Nature*, **337**, 424.

102. Guthrie, S. and Lumsden, A. (1992) Motor neuron pathfinding following rhombomere reversals in the chick embryo hindbrain. *Development*, **114**, 663.

103. Taniguchi, M., Yuasa, S., Fujisawa, H., Naruse, I., Saga, S., Mishina, M. and Yagi, T. (1997) Disruption of semaphorin III/D gene causes severe abnormality in peripheral nerve projection. *Neuron*, **19**, 519.

104. Kitsukawa, T., Shimizu, M., Sanbo, M., Hirata, T., Taniguchi, M., Bekku, Y., Yagi, T. and Fujisawa, H. (1997) Neuropilin-semaphorin III/D-mediated chemorepulsive signals play a crucial role in peripheral nerve projection in mice. *Neuron*, **19**, 995.

105. Jacob, J., Tiveron, M. C., Brunet, J. F., and Guthrie, S. (2000) Role of the target in the pathfinding of facial visceral motor axons. *Mol. Cell. Neurosci.*, **16**, 14.

106. Jacob, J. and Guthrie, S. (2000) Facial visceral motor neurons display specific rhombomere origin and axon pathfinding behavior in the chick. *J. Neurosci.*, **20**, 7664.

107. Cohen, J., Burne, J. F., Winter, J., and Bartlett, P. (1986) Retinal ganglion cells lose response to laminin with maturation. *Nature*, **322**, 465.

108. McFarlane, S., McNeill, L., and Holt, C. E. (1995) FGF signaling and target recognition in the developing xenopus visual system. *Neuron*, **15**, 1017.

109. Mason, C. A. and Stretavan, D. W. (1997) Glia, neurons, and axon pathfinding during optic chiasm development. *Curr. Opin. Neurobiol.*, **7**, 647.

110. Erskine, L., Williams, S. E., Brose, K., Kidd, T., Rachel, R. A., Goodman, C. S., Tessier-Lavigne, M. and Mason, C. A. (2000) Retinal ganglion cell axon guidance in the mouse optic chiasm: expression and function of Robos and Slits. *J. Neurosci.*, **20**, 4975.

111. Nakagawa, S., Brennan, C., Johnson, K. G., Shewan, D., Harris, W. A., and Holt, C. E. (2000) Ephrin-B regulates the ipsilateral routing of retinal axons at the optic chiasm. *Neuron*, **25**, 599.

112. Walter, J., Henke-Fahle, S., and Bonhoeffer, F. (1987) Avoidance of posterior tectal membranes by temporal retinal axons. *Development*, **101**, 685.

113. Cheng, H. J., Nakamoto, M., Bergemann, A. D., and Flanagan, J. G. (1995) Complementary gradients in expression and binding of ELF-1 and Mek4 in development of the topographic retinotectal projection map. *Cell*, **82**, 371.

114. Drescher, U., Kremoser, C., Handwerker, C., L'schinger, J., Noda, M., and Bonhoeffer, F. (1995) *In vitro* guidance of retinal ganglion cell axons by RAGS, a 25 kDa tactal protein related to ligands for Eph receptor tyrosine kinases. *Cell*, **82**, 359.

115. Feldheim, D. A., Vanderhaeghen, P., Hansen, M. J., Frisen, J., Lu, Q., Barbacid, M. and Flanagan, J. G. (1998) Topographic guidance labels in a sensory projection to the forebrain. *Neuron*, **21**, 1303.

116. Monschau, B., Kremoser, C., Ohta, K., Tanaka, H., and Keneko, T. (1997) Shared and unique functions of RAGS and ELF-1 in guiding retinal axons. *EMBO J.*, **16**, 1258.

117. Nakamoto, M., Cheng, H. J., Friedman, G. C., McLaughlin, T., Hansen, M. J., Yoon, C. H., O' Leary, D. D. and Flanagan, J. G. (1996) Topographically specific effects of ELF-1 on retinal axon guidance *in vitro* and retinal axon mapping in vivo. *Cell*, **86**, 755.

118. Drescher, U., Bonhoeffer, F., and Muller, B. K. (1997) The Eph family in retinal axon guidance. *Curr. Opin. Neurobiol.*, **7**, 75.

119. Frisen, J., Yates, P. A., McLaughlin, T., Friedman, G. C., O'Leary, D. D., and Barbacid, M. (1998) Ephrin-A5 (AL-1/RAGS) is essential for proper retinal axon guidance and topographic mapping in the mammalian visual system. *Neuron*, **20**, 235.

120. Feldheim, D. A., Kim, Y. I., Bergemann, A. D., Frisen, J., Barbacid, M., and Flanagan, J. G. (2000) Genetic analysis of ephrin-A2 and ephrin-A5 shows their requirement in multiple aspects of retinocollicular mapping. *Neuron*, **25**, 563.

121. Brown, A., Yates, P. A., Burrola, P., Ortuno, D., Vaidya, A., Jessell, T. M., Pfaff, S. L., O'Leary, D. D. and Lemke, G. (2000) Topographic mapping from the retina to the midbrain is controlled by relative but not absolute levels of EphA receptor signaling. *Cell*, **102**, 77.

122. Hornberger, M. R., Dutting, D., Ciossek, T., Yamada, T., Handwerker, C., Lang, S., Logan, C., Tanaka, H. and Drescher, U. (1999) Modulation of EphA receptor function by coexpressed ephrinA ligands on retinal ganglion cell axons. *Neuron*, **22**, 731.

123. Adams, R. H., Lohrum, M., Klostermann, A., Betz, H., and Puschel, A. W. (1997) The chemorepulsive activity of semaphorins is regulated by furin-dependent proteolytic processing. *EMBO J.*, **16**, 6077.

124. Hattori, M., Osterfield, M., and Flanagan, J. G. (2000) Regulated cleavage of a contact-mediated axon repellent. *Science*, **289**, 1360.

125. Galko, M. J. and Tessier-Lavigne, M. (2000) Function of an axonal chemoattractant modulated by metalloprotease activity. *Science*, **289**, 1365.

126. Stein, E. and Tessier-Lavigne, M. (2001) Hierarchical organization of guidance receptors: silencing of netrin attraction by Slit through a Robo/DCC receptor complex. *Science*, **291**, 1928.

127. Castellani, V., Chedotal, A., Schachner, M., Faivre-Sarrailh, C., and Rougon, G. (2000) Analysis of the L1-deficient mouse phenotype reveals cross-talk between Sema3A and L1 signaling pathways in axonal guidance. *Neuron*, **27**, 237.

128. Hall, A. (????) Small GTP-binding proteins and the regulation of the actin cytoskeleton. *Annu. Rev. Cell Biol.*, **10**, 31.

129. Nobes, C. D. and Hall, A. (1995) Rho, Rac, and Cdc42 GTPases regulate the assembly of multi-molecular focal complexes associated with actin stress fibres, lamellipodia and filipodia. *Cell*, **81**, 53.

130. Hall, A. (1998) G proteins and small GTPases: distant relatives keep in touch. *Science*, **280**, 2074.

131. Dickson, B. J. (2001) Rho GTPases in growth cone guidance. *Curr. Opin. Neurobiol.*, **11**, 103.

132. Jin, Z. and Stritmatter, S. M. (1997) Rac-1 mediates collapsin-1 induced growth cone collapse. *J. Neurosci.*, **17**, 6256.

133. Kuhn, T. B., Brown, M. D., Wilcox, C. L., Raper, J. A., and Bamburg, J. R. (1999) Myelin and collapsin-1 induce motor neuron growth cone collapse through different pathways: inhibition of collapse by opposing mutants of rac1. *J. Neurosci.*, **19**, 1965.

134. Wahl, S., Barth, H., Ciossek, T., Aktories, K., and Mueller, B. K. (2000) Ephrin-A5 induces collapse of growth cones by activating Rho and Rho kinase. *J. Cell Biol.*, **149**, 263.

135. Van Aelst, L. and D'Souza-schorey, C. (1997) Rho, GTPases and signaling networks. *Genes Dev.*, **11**, 2295.

136. Leeuwen, F. N., Kain, H. E., Kammen, R. A., Michiels, F., Kranenburg, O. W. and Collard, J. G. (1997) The guanine nucleotide exchange factor Tiam 1 affects neuronal morphology; opposing roles for the small GTPases Rac and Rho. *J. Cell Biol.* **139**, 797.

137. Rohm, B., Ottemeyer, A., Lohrum, M., and Puschel, A. W. (2000) Plexin/neuropilin complexes mediate repulsion by the axonal guidance signal semaphorin 3A. *Mech. Dev.*, **93**, 95.

138. Miki, H., Sasaki, T., Takai, Y., and Takenawa, T. (1998) Induction of filopodium formation by a WASP-related actin-depolymerizing protein N-WASP. *Nature*, **391**, 93.

139. Vastrik, I., Eickholt, B. J., Walsh, F. S., Ridley, A., and Doherty, P. (1999) Sema3A-induced growth-cone collapse is mediated by Rac1 amino acids 17–32. *Curr. Biol.*, **9**, 991.

140. Lanier, L. M. and Gertler, F. B. (2000) Actin cytoskeleton: thinking globally, actin' locally. *Curr. Biol.*, **10**, R655.

141. Lanier, L. M., Gates, M. A., Witke, W., Menzies, A. S., Wehman, A. M., Macklis, J. D., Kwiatkowski, D., Soriano, P. and Gertler, F. B. (1999) Mena is required for neurulation and commissure formation. *Neuron*, **22**, 313.

142. Wills, Z., Bateman, J., Korey, C. A., Comer, A., and Van Vactor, D. (1999) The tyrosine kinase Abl and its substrate enabled collaborate with the receptor phosphatase Dlar to control motor axon guidance. *Neuron*, **22**, 301.

143. Gertler, F. B., Comer, A. R., Juang, J. L., Ahern, S. M., Clark M. J., Liebl, E. C. and Hoffmann, F. M. (1995) Enabled, a dosage-sensitive suppressor of mutations in the Drosophila Abl tyrosine kinase, encodes an Abl substrate with SH3 domain-binding properties. *Genes Dev.*, **9**, 521.

144. Krueger, N. X., Van Vactor, D., Wan, H. I., Gelbart, W. M., Goodman, C. S., and Saito, H. (1996) The transmembrane tyrosine phosphatase DLAR controls motor axon guidance in Drosophila. *Cell*, **84**, 611.

145. Stoker, A. and Dutta, R. (1998) Protein tyrosine phosphatases and neural development. *BioEssays*, **20**, 463.

146. Yeo, T. T., Chua-Couzens, J., Butcher, L. L., Bredesen, D. E., Cooper, J. D., Valletta, J. S., Mobley, W. C. and Longo, F. M. (1997) Absence of p75NTR causes increased basal forebrain cholinergic neuron size, choline acetyltransferase activity, and target innervation. *J. Neurosci.*, **17**, 7594.

147. Lin, M. Z. and Greenberg, M. E. (2000) Orchestral maneuvers in the axon: Trio and the control of axon guidance. *Cell*, **101**, 239.

148. Newsome, T. P., Schmidt, S., Dietzl, G., Keleman, K., Asling, B., Debant, A. and Dickson, B. J. (2000) Trio combines with dock to regulate Pak activity during photoreceptor axon pathfinding in Drosophila. *Cell*, **101**, 283.

149. Debant, A., Serra-Pages, C., Seipel, K., O'Brien, S., Tang, M., Park, S. H. and Streuli, M. (1996) The multidomain protein Trio binds the LAR transmembrane tyrosine phosphatase, contains a protein kinase domain, and has separate rac-specific and rho-specific guanine nucleotide exchange factor domains. *Proc. Natl Acad. Sci. USA*, **93**, 5466.

150. Petersen, O. H. and Cancela, J. M. (2000) Nerve guidance: attraction or repulsion by local Ca^{2+} signals. *Curr. Biol.*, **10**, R311.

151. Ming, G., Song, H., Berninger, B., Holt, C. E., Tessier-Lavigne, M., and Poo, M. (1997) cAMP-dependent growth cone guidance by netrin-1. *Neuron*, **19**, 1225.

152. Hong, K., Nishiyama, M., Henley, J., Tessier-Lavigne, M., and Poo, M. (2000) Calcium signalling in the guidance of nerve growth by netrin-1. *Nature*, **403**, 93.

153. Hopker, V. H., Shewan, D., Tessier-Lavigne, M., Poo, M., and Holt, C. (1999) Growth-cone attraction to netrin-1 is converted to repulsion by laminin-1. *Nature*, **401**, 69.

154. Song, H., Ming, G., He, Z., Lehmann, M., McKerracher, L., Tessier-Lavigne, M. and Poo, M. (1998) Conversion of neuronal growth cone responses from repulsion to attraction by cyclic nucleotides. *Science*, **281**, 1515.

8 | Patterning of the neural crest

C. A. ERICKSON

1. Introduction

Neural crest cells are mesenchymal cells that emigrate from the dorsal portion of the neural tube, and are remarkable both for their extensive migration and the multiple derivatives to which they give rise. For these reasons, the neural crest has been a popular model system with which to investigate problems of patterning. At the trunk level, neural crest cells take two major pathways of migration. The first is ventral between the neural tube and somites, and the cells that follow this path differentiate primarily into the neurons and glial cells of the peripheral nervous system, the secretory cells of the adrenal medulla, and the enteric nervous system of the gut. About a day later, neural crest cells migrate dorsolaterally between the ectoderm and the somites where they differentiate into the pigment cells of the skin. At the head level, the situation is even more complicated, because many more derivatives arise from the cranial neural folds and the patterns of migration are much more complex.

The question that will be addressed in this review is what controls the positioning of these different phenotypes of the neural crest in the appropriate places during embryogenesis? Because morphogenesis is somewhat simpler at the trunk level, this review will largely focus on patterning of the trunk neural crest. Moreover, since most of the experimental studies have used the chicken as a model system, this chapter will primarily consider the avian neural crest, although recent genetic studies in mouse and zebrafish will be referenced when appropriate.

The generally accepted model for patterning neural crest derivatives proposes that neural crest cells are multipotent and that they migrate haphazardly into the ventral or dorsolateral paths, where they differentiate according to the cues encountered in these paths (reviewed in ref. 1). This model is supported by experimental evidence from heterotopic grafting experiments, as well as back-transplantation of crest-derived structures into the early migratory crest pathways, which show that, as a population, neural crest cells migrate and differentiate according to their new environment and not their origin.

Recent studies have revealed that an alternative mechanism directs the migration of chicken melanoblasts (precursors to the epidermal melanocytes) into the dorsolateral path: melanoblasts are fate-restricted when they depart from the neural tube (2, 3); primarily neural crest cells that are specified as melanocytes enter the dorsolateral path (3–5); and melanoblasts are the only neural crest cells that can exploit the dorsolateral path under experimental conditions (6). Thus, another model for positioning of neural crest derivatives, based on the study of melanocytes, is that their final distribution is directed by cell-autonomous migratory properties unique to each subpopulation. This means that some neural crest subpopulations are specified prior to or shortly after they separate from the neural tube, and owing to molecular changes accompanying specification, they are able to select the appropriate pathway.

This review will explore the extent to which each of these models accounts for the patterning of the subpopulations of the trunk neural crest that migrate dorsolaterally versus those that migrate ventrally.

2. Migration of melanoblasts: is the exception really the rule?

2.1 Evidence for cell-autonomous pathfinding of melanoblasts

At the trunk level, neural crest cells first migrate ventrally between the somites and neural tube, beginning at Hamburger and Hamilton stage 13 (7); these are the crest cells that will differentiate into the neurons and glial cells of the peripheral nervous system (Fig. 1). However, the dorsolateral path at the trunk level is initially refractory to neural crest-cell migration (6, 8, 9), possibly because the dermamyotome produces molecules that act as barriers to the immediate dispersal of neurogenic or gliogenic precursors (10–13).

By stages 19–20, beginning at the wing-bud level, dorsolateral migration is initiated as ventral migration gradually ceases (8, 11, 14). The reason for the onset of neural crest migration into the dorsolateral path was originally suggested to be a gradual loss of inhibitory molecules, namely chondroitin sulfate proteoglycan and peanut agglutinin (PNA)-binding substances, which appeared to retreat from the dorsolateral path in advance of neural crest invasion (11), thus rendering the dorsolateral path permissive for migration. Earlier experimental evidence seemed to support this view. If the dermamyotome, which presumably produced the inhibitory cues, was ablated prior to the onset of neural crest migration, precocious invasion of the dorsolateral path was observed (9). However, the latter results were perplexing because rather than immediately invading the dorsolateral path, the neural crest cells still migrated ventrally first, and only migrated dorsolaterally after dermis from adjacent segments filled in the ablated portions of the somite. These studies suggested the surprising possibility that timing of invasion of the dorsolateral path was not dependent solely upon a molecular change in the pathway itself, but also required a change in the neural crest cells themselves.

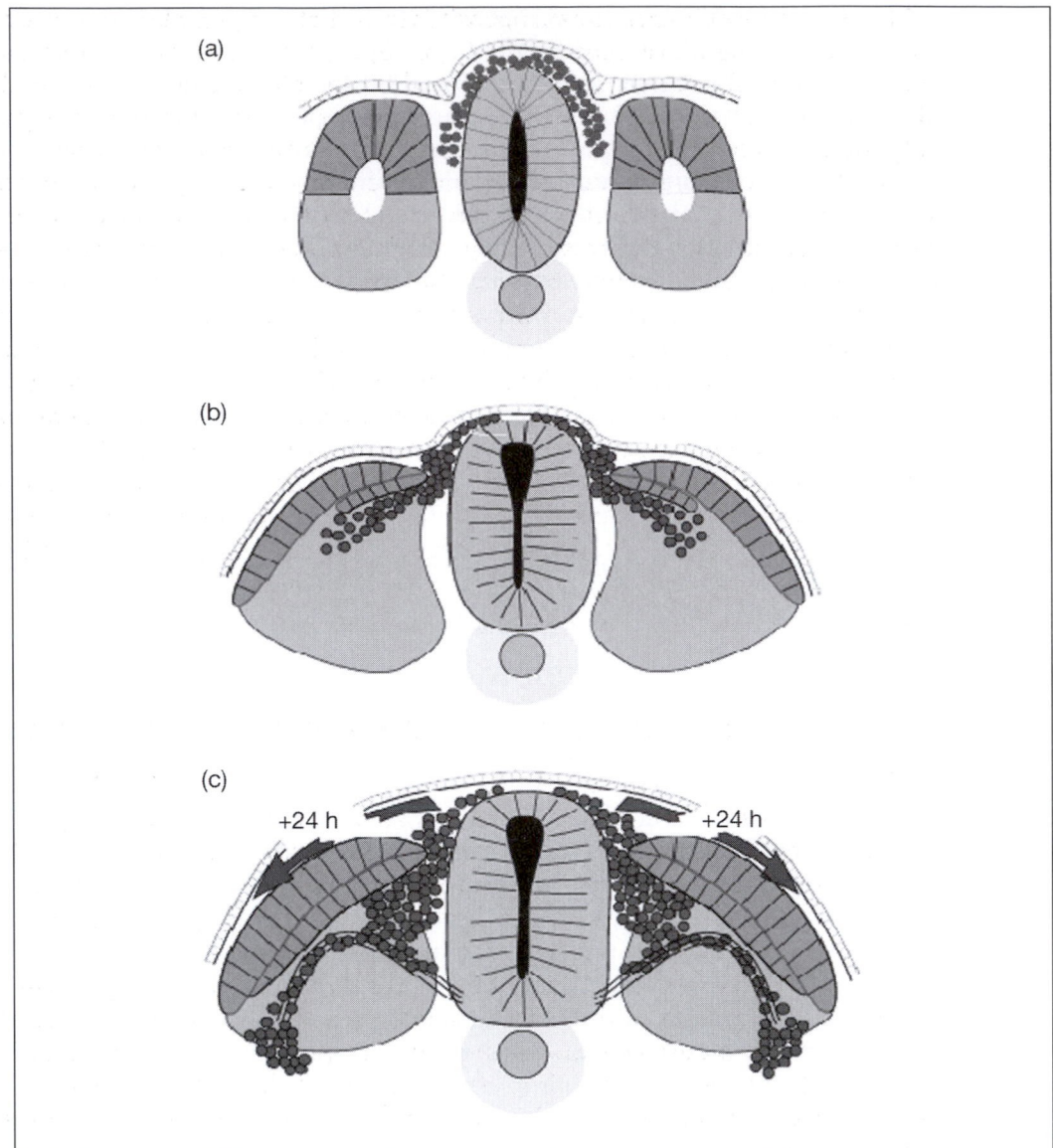

Fig. 1 (See also Plate 6) Schematic depiction of the migratory pathways of the trunk neural crest. (a) Neural crest cells (dark green in the plate) detach from the dorsal surface of the neural tube (light green) and migrate ventrally between the neural tube and somites (pink). (b) When they reach the somites, they invade along the developing myotome. (c) The ventrally migrating neural crest cells differentiate into the neurons and glial cells of the sensory and sympathetic ganglia, and spread out along the ventral root motor fibres, where they differentiate into glial cells. Then, 24 h after migrating ventrally, another population of neural crest cells migrates dorsolaterally beneath the ectoderm and these differentiate into the pigment cells of the skin. (Figure courtesy of Martha Spence. Modification of Fig. 1 in ref. 1.)

The first clear evidence for this possibility is that when melanoblasts are grafted into the early migratory pathway at stage 13/14, they immediately migrate dorso-laterally (6), well in advance of the time that dorsolateral migration usually takes place. In contrast, non-melanogenic neural crest cells grafted into the early pathway only migrate ventrally. In fact, melanoblasts are the only neural crest subpopulation tested that is capable of migrating into the dorsolateral path. Moreover, there appears to be no change in the environment to facilitate dorsolateral migration, because non-melanogenic crest cells grafted into the embryo at stage 19, which is when dorsolateral migration begins, still only migrate ventrally (6). This study suggested that melanoblasts acquire distinctive properties as they are specified that permit them to migrate dorsolaterally. Other studies, which exploited markers for melanoblasts (serum from Smyth line chickens, MEBL-1, MITF), revealed that neural crest cells are specified as pigment cells prior to entering the dorsolateral path, and the vast majority of neural crest cells in the dorsolateral path at all stages express melanoblast-specific markers (3, 4, 14). Taken together, these data show that in order to migrate into the dorsolateral path, a neural crest cell must already be differen-tiating into a pigment cell. This is the first and, to date, only unequivocal evidence that the cell-autonomous migratory properties of a neural crest subpopulation allow it to select the appropriate migratory pathway.

2.2 What molecular properties allow melanoblasts to access the dorsolateral path?

At present, the mechanisms that direct or position melanoblasts in the dorsolateral path are largely unknown. However, some reasonable speculations will be con-sidered here. All these hypotheses are based on the assumption that whatever molecular changes occur prior to the onset of dorsolateral migration, they are in the neural crest cells themselves rather than in the environment.

One possible change in melanoblasts is that they develop unique sensitivity to positive guidance cues from the dermamyotome, allowing them to migrate dorso-laterally. In the mouse, melanoblasts depend upon steel factor produced by the dermamyotome for their initial dispersal on to the dorsolateral path (15–17), and they express the receptor for steel factor, c-*kit*, prior to embarking on the dorsolateral path. In the avian embryo, however, c-*kit* is first expressed by melanoblasts long after they have migrated dorsolaterally (stage 25); steel factor is not produced by the dermamyotome, but rather by the ectoderm and only after stage 25 (18). Thus, acquiring responsiveness to steel factor owing to the expression of c-*kit* is unlikely to control the timing of dorsolateral migration in the chicken.

A second possible molecular change is that the motility of melanoblasts becomes sufficiently enhanced for them to overcome inhibitory or relatively less appealing substrata in the dorsolateral path. Melanoblasts or fully differentiated melanocytes disperse further and faster than early migrating neural crest cells, and are also highly invasive (6, 19–21). Proteases are known to be upregulated in highly invasive and metastatic tumour cells (22, 23), and are localized in focal contacts in cells in culture (24, 25), suggesting that proteolytic activity can enhance cell motility. Neural crest cells are known to produce several proteases, including the neutral serine protease uPA (urokinase-type plasminogen activator) (26–28) and the matrix metalloprotease MMP-2 (our unpublished results). However, these are only expressed in the early dispersing neural crest and appear to be important in the epithelial–mesenchymal transformation and early migration (Duong and Erickson, unpublished observations). Neither is expressed in melanoblasts. Recently, we have found that an integral membrane metalloprotease, ADAM10, is expressed on melanoblasts, but not on the ventrally migrating neural crest (Hall and Erickson, unpublished observations). ADAM10 may contribute importantly to melanoblast motility because bovine ADAM10 possesses type IV collagenase activity (29), which may permit invasion of the epidermal basal lamina; the *Drosophila* orthologue of ADAM10 (*kuzbanian*) is required for axon extension (30).

Another mechanism by which melanoblast migration could be enhanced is by changing adhesive interactions with the extracellular matrix in the dorsolateral path, or with the cells of the dermamyotome that border the path. This might be accomplished in several ways. If adhesion to the extracellular matrix is increased, this may allow the melanoblasts to gain additional traction to overcome what is a less satisfactory, or even inhibitory, migratory substratum. In support of this notion is the observation that S180 cells, when grafted into the early migratory pathway of the chicken embryo, only migrate ventrally (19). However, if these cells are stably transfected with either α_5 or α_4 integrins, both subunits that comprise fibronectin receptors, and then grafted into the early migratory path, they immediately disperse on the dorsolateral path as well (31). This latter study demonstrates that a change in the composition of functional integrins can alter cellular behaviour in the embryonic environment. Alternatively, melanoblasts may develop greater adhesiveness to the cells of the dermamyotome. Cell–cell adhesions in epithelial tissues are largely mediated by cadherins (32, 33), but recently mesenchymal cells, including neural crest cells, have been shown to express cadherins (34, 35). Moreover, at least in culture, some mesenchymal cells can migrate on a derivatized substratum of cadherins. Thus an interesting possibility is that unique cadherins may be expressed on melanoblast precursors, which allow these cells to migrate using cadherins expressed on the dermamyotome. In support of the role of mesenchymal cadherins in positioning neural crest cells, it has recently been shown that different subpopulations of murine melanoblasts occupying the dermis, the epidermis, or the hair follicles each express the same cadherin as the tissue in which they are localized (36).

Finally, because melanoblasts can migrate into a region that is refractory to the migration of early-dispersing neural crest cells, we speculate that melanoblasts are missing a receptor that normally mediates the inhibition of migration. In the last few years a number of guidance molecules have been identified that cause the collapse or repulsion of growth cones, filopodia, or lamellipodia (37) (see Chapter 7 for discussion). These include the netrins, the collapsin/semaphorins, and the Eph-receptor family of ligands, coined 'ephrins'. To date, three laboratories have shown that ventrally migrating neural crest cells are constrained to migrate in the anterior half of somites because of inhibitory ephrins expressed in the posterior somite (13, 38, 39). Similarly, collapsin-1 is expressed in the posterior half of the somite and causes the retraction of lamellipodia of early migrating neural crest cells (40). Ephrins and collapsins are also found in the dorsolateral path, which suggests that in order to invade the dorsolateral space, melanoblasts must not be susceptible to the inhibitory effects of these ligands. Alternatively, because collapsin/semaphorin receptors can mediate attractive behaviour in some circumstances (41), it is also possible that the signalling pathway is altered in melanoblasts so that their migration is now enhanced by binding ephrins and collapsins (see, for example, ref. 42). In support of this latter possibility, we have discovered, using a Boyden chamber chemotaxis assay, that melanoblasts are attracted to soluble ephrin-B2, whereas early migratory neural crest cells are repulsed. Additionally, when ephrin-B2 is adsorbed to a fibronectin substratum, melanoblast adhesion is enhanced, while early migratory crest adhesion to fibronectin is reduced (A. Santiago and C. A. Erickson, *Development*, in press)). These are intriguing results because they show that two different neural crest lineages respond to ephrins in opposite ways. Further, they can account for the unique ability of the melanoblasts to invade the dorsolateral path.

2.3   When and how is the melanoblast lineage specified?

From the above discussion it is clear that the timing of melanoblast migration into the dorsolateral space is dependent upon the timing of melanoblast specification. Because melanoblast invasion of the dorsolateral space occurs late compared to ventral neural crest migration, we hypothesized that perhaps there were no neural crest cells with melanogenic potential amongst the earliest migrating crest. This is, in fact, the case. When we cultured quail neural tubes and assessed the fate of those neural crest cells that emigrated during the first 6 hours, we discovered that no melanocytes differentiated, even after 6 days in culture conditions permissive for melanogenesis (3). Our studies revealed that there is heterogeneity in the developmental potential of the premigratory neural crest, and that early migrating neural crest cells are not melanogenic and therefore cannot invade the dorsolateral space. Henion and Weston (2) further showed, by labelling individual neural crest cells emigrating from the neural tube during the first 6 hours with rhodamine dextran, that the resulting clones do not contain pigment cells. Only later emigrating neural crest cells give rise to pigment cells, and the clones derived from these late migrators are generally 100% pigmented, showing that these cells are fate-restricted (that is, the progeny from a single neural crest cell all have the same phenotype). Together, these two studies reveal that many neural crest cells are developmentally restricted when they leave the neural tube, and that the melanogenic crest cells leave the neural tube relatively late, thus accounting for their delayed dorsolateral migration. Similar observations have been made in zebrafish (43) and in axolotls (44).

The distribution of melanocytes ultimately depends upon the mechanism that controls the specification of the melanogenic lineage. To date, the best evidence for what this mechanism may be comes from studies in zebrafish. The zebrafish neural crest differs from other vertebrates in several respects, which have facilitated an analysis of when and how neural crest-cell specification occurs. First, different neural crest subpopulations are spatially segregated, at least in the cranial region. At the neural plate stage, when the neural epithelium and epidermal ectoderm are a contiguous sheet, melanogenic crest cells reside closest to the neural plate and neurogenic crest cells are adjacent to the epidermal ectoderm (45). Thus, different subpopulations of the crest can be identified prior to migration because of their position in the ectoderm. Second, zebrafish neural crest cells are quite large, enabling relatively easy injections of single cells in order to alter their fate. In an elegant study, Dorsky and colleagues (46) showed that Wnt-1 and Wnt-3a are expressed in the neural plate closest to the melanogenic precursors and that bone morphogenetic proteins (BMPs) are expressed in the ectoderm adjacent to the neurogenic precursors, suggesting that the melanogenic and neurogenic crest were specified by the Wnt and BMP signalling pathways, respectively. If they activated the Wnt signalling pathway in the neurogenic crest by injecting a single cell with β-catenin mRNA, that cell differentiated into a pigment cell at the expense of the neuronal cell. Conversely, if the mRNAs encoding a dominant-negative Wnt or a truncated form of Tcf-3 were injected into melanogenic precursors to inhibit the Wnt signalling pathway, these cells differentiated into neuronal cells.

Although there is no evidence of a physical segregation of neurogenic and melanogenic precursors in the dorsal neural tube of mouse and chick embryos, there is nevertheless evidence that the Wnt signalling pathway triggers melanogenesis and BMP signalling inhibits it. In the absence of Wnt-1 and Wnt-3a expression in the mouse, not only is there a general reduction in the total number of neural crest cells, but these embryos fail to generate melanocytes (47). Similarly in the chick, neural crest cells cultured in Wnt-3a-conditioned medium generate a greater number of melanocytes (48). Moreover, the expression patterns of Wnts and BMPs are compatible with a role in specifying these crest derivatives. BMP4 is expressed in the dorsal neural tube when neuroblasts are migrating (49), but is then downregulated by the time that the melanoblasts leave (48). Wnt-3a and Wnt-1 are expressed in the dorsal neural tube during the entire time when neural crest cells escape the neural tube (50). However, at the time that neurogenic precursors are leaving the neural tube, an antagonist to the Wnt signalling pathway, *frzb1*, is also expressed in the dorsal neural tube and ventrally migrating neural crest (51, 52). Later, when melano-genic crest cells are emigrating, several Wnt family members are still expressed, but *frzb1* is now extinguished (48). It remains to be determined if the Wnt and BMP signalling pathways in the chick are involved in the process of specification, work further downstream in the differentiation process, or act as trophic factors.

2.4 Mistakes will be made!

Although the vast majority of cells that invade the dorsolateral path are melano-blasts, occasional non-melanogenic neural crest cells apparently go astray. For example, Richardson and Sieber-Blum (5) isolated neural crest cells from quail skin and found that a small number of these cells did not give rise to pigment cells. Wakamatsu and colleagues (4) similarly showed that a few neurogenic crest cells invade the dorsolateral pathway at the time that the melanoblasts migrate (averaging from a total of 14 to 34 neuroblasts along the entire trunk), and that all these cells undergo apoptosis, possibly because the neuroblasts on the dorsal path are not exposed to the growth factors that would normally maintain them on the ventral path. Thus, the control of melanocyte patterning is dependent upon positive cues that direct melanoblasts to the correct place, and negative controls that prune inappropriate lineages.

3. Patterning along the ventral pathway

This review first considered the migration of the neural crest along the dorsolateral path in order to present the most convincing evidence for cell-autonomous guidance cues. It will now consider the behaviour of early migrating neural crest cells as they leave the neural tube and enter the ventral pathway.

The separation of the crest from the neural tube occurs in an anterior-to-posterior wave, so that the earliest stages of crest morphogenesis occur in the most posterior end of the embryo (53–55). At the thoracic level, neural crest cells emerge on the dorsal surface of the neural tube at the axial level of the last-formed somite (also referred to as somite -I (56) or somite -1 (54)), and from there spread along the dorsal and lateral surface of the neural tube. The first cells to migrate ventrally invade the intersomitic space between somites -V and -VI and rapidly reach the dorsal aorta (54, 57) (Fig. 2A–C). More anteriorly (at the somite-VII level), the vanguard of neural crest cells invade the somite at the interface between the myotome and sclerotome (54), and then migrate in a medial-to-lateral trajectory (58), initially using the myotome as a substratum for migration (59) (Fig. 2D–G).

It is during the invasion through the somites that the first important patterning event occurs. The neural crest cells do not invade along the entire expanse of the somite, but rather only penetrate the anterior half while scrupulously avoiding the posterior (54, 58, 60, 61). This restriction establishes, at a very early stage, the segmental layout of the adult peripheral nervous system (Fig. 3). As neural crest cells migrate through the somite, they begin to coalesce to form the ganglia of the peripheral nervous system: some cells aggregate adjacent to the dorsal neural tube and form the sensory or dorsal root ganglia. Other neural crest cells migrate ventrally and stop at the dorsal aorta to form the primary sympathetic chain. At a slightly later stage in development these latter crest cells migrate dorsally to form the secondary sympathetic chain, and at somite level 18–24 some of these cells migrate ventrally and laterally and coalesce adjacent to the kidney to give rise to the adrenal medulla (62, 63). Other ventrally migrating cells spread out along the ventral root motor fibres and differentiate into glial cells. Finally, at the sacral level of the trunk (posterior to somite 28), neural crest cells migrate ventral to the dorsal aorta and into the mesenchyme dorsal to the gut, from which point they will eventually invade the

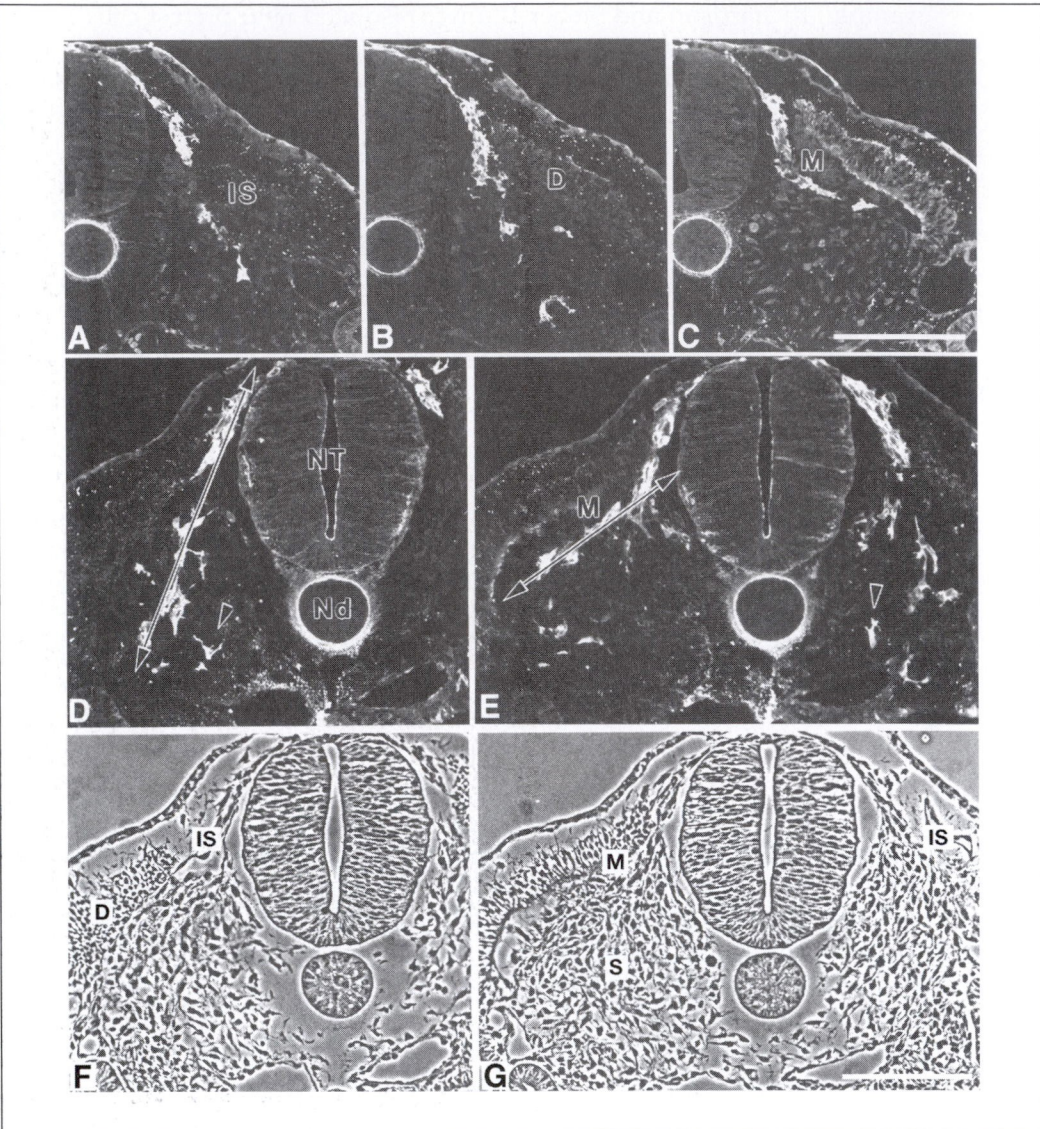

Fig. 2 Early stages of trunk neural crest migration. (A–C) Sections labelled with the HNK-1 antibody reveal the distribution of neural crest cells as they first invade the somites. The section in (A) is through the intersomitic space ('IS'). (B) The most anterior section through the somite just grazing the edge of dermomyotome ('D'), and (C) is through the anterior half of the same somite. Neural crest cells in the intersomitic space disperse ventrally, whereas the neural crest cells that invade the somite migrate on the myotome ('M'), which directs them mediolaterally across the somite. Panels (D–G): sections through somites where neural crest migration is more extensive. The same ventral trajectory in the intersomitic space (panels D, F) and the mediolateral dispersion through the somite (panels E, G), avoiding the sclerotome, is even more pronounced. Scale bar, 100 μm. (Modified with permission from ref. 59.)

postumbilical gut and differentiate into the neurons and glial cells of the enteric nervous system (64–67).

Correct patterning of the ventrally migrating neural crest cells requires: molecular mechanisms to control the pathway of migration through the somite; and cues to control the localization and differentiation of the many different lineages (sensory, sympathoadrenal, glial, enteric) that develop along the ventral path. These two patterning events will be considered separately, although mechanisms that control them are not likely to be mutually exclusive.

3.1 Migration through the somite

As neural crest cells migrate through the somites their distribution becomes asymmetrically patterned along the anteroposterior and dorsoventral axes. First, they only invade the anterior half of the somite and not the posterior half; and second, they migrate along the basal surface of the myotome and do not penetrate the sclerotome deeply. The molecular basis for each of these events will be considered.

3.1.1 Anteroposterior segmental migration

Experimental studies have revealed that the failure of neural crest cells and ventral root motor fibres to invade the posterior somite is controlled by environmental cues in the somite, rather than cell-autonomous migratory properties of the neural crest. If a strip of segmental plate or a cluster of the last-formed somites is excised and then rotated 180 degrees, neural crest cells and the ventral root fibres still invade what had been previously specified as posterior somite (68). Direct observations of migrating neural crest cells labelled with DiI (1,1'-dioctadecyl-3,3,3',3'-tetramethyl carbocyanine) reveal that they retract their lamellipodia and reorient their migration when they contact the posterior somite (67). Together, these data suggest that the posterior somite contains barrier molecules that are inhibitory to neural crest-cell migration.

A number of candidate molecules have been localized to the posterior sclerotome and could possibly inhibit neural crest migration. Of these, five are extracellular matrix (ECM) molecules: chondroitin-6-sulfate proteoglycan (CSPG) (10, 70), peanut agglutinin (PNA)-binding molecules (10, 71), collagen type IX (72), versican (73), and F-spondin (12). Another molecule expressed on posterior sclerotome cells is T-cadherin, a cell–cell adhesion molecule (74). Recently, two ligands that inhibit neuronal migration have also been identified in the posterior sclerotome: ephrin-B1, a ligand for the Eph family of receptor tyrosine kinases (13, 39), and collapsin-1, one of the many members of the collapsin/semaphorin family (40). Although most of these molecules can inhibit neural crest-cell migration *in vitro* (reviewed in ref. 75), several of them are probably not involved in controlling segmental migration *in vivo*. Collagen IX inhibits neural crest-cell migration, but it is not expressed until after they begin to invade the somite. T-cadherin mediates homophilic cell–cell adhesion and is also expressed on motor neurons (76), but is not expressed by the neural crest, so it is unlikely to affect neural crest-cell migration by a homophilic mechanism. Both these molecules may control ventral root motor fibre morphogenesis, however.

To confirm if any of these posteriorly expressed molecules inhibits neural crest migration, their function needs to be perturbed in the embryo. Such an analysis has been greatly facilitated by the development of an *in vitro* explant assay in which a segment of a chick trunk is excised, the neural crest cells are labelled with DiI, and their behaviour directly observed as development proceeds *in vitro* (39, 77, 78). Additionally, these explants can be exposed to perturbing agents in the bathing medium, obviating the need for injecting them, and assuring that the agents will permeate the tissue thoroughly. Using this assay, two different molecules appear to have a role in the segmental migration of the crest.

PNA-binding molecules, which have not been identified, were the first for which a barrier function was suggested (10, 79). When trunk pieces are treated with PNA, neural crest cells enter the posterior sclerotome, presumably because the inhibitory epitopes to which PNA binds are camouflaged (69). However, migration into the posterior sclerotome does not exactly mimic anterior migration, suggesting that PNA-binding molecules alone are not responsible for this pattern. For example, neural crest cells are still delayed compared to their entry in the anterior half, and their direction and rate of migration are different from the cohorts in the anterior half.

Other ligands that may play a role in inhibiting neural crest migration are those that bind to the Eph receptors. This family of receptors and their ligands have been implicated in many axonal guidance events (80) (see Chapter 7). One member of the Eph family of receptor tyrosine kinases, EphB3, is expressed on migrating chick neural crest cells, whereas one of its complementary ligands, ephrin-B1, is expressed in the posterior sclerotome (39). A similar distribution of Eph-family receptors and ligands has been discovered in the mouse (13). Furthermore, when neural crest cells are confronted *in vitro* with a substratum derivatized with stripes of ephrin-B1, neural crest cells avoid the ephrin-B1 stripes and migrate preferentially on the control lanes. Using the explant assay, Krull and colleagues treated trunk segments with monomeric, soluble ephrin-B1 (39), which binds to the receptor on the neural crest and presumably competes with extracellular ephrin-B1. In treated explants, neural crest cells invade both the anterior and posterior somite halves (Fig. 4). These data persuasively argue that ephrins inhibit neural crest-cell invasion of the posterior somite.

At least one ECM molecule appears to influence anteroposterior segmentation. F-spondin is expressed in the posterior sclerotome, the perinotochordal matrix in the anterior somite and the dermamyotome—all regions that are refractory to early neural crest migration (12). When function-blocking antibodies are injected into the space between the somite and neural tube, the neural crest cells invade the perinotochordal mesenchyme, the posterior sclerotome, and even the derma-myotome from the sclerotome, suggesting that F-spondin is yet another inhibitory signal that patterns neural crest migration through the somite. It is worth noting that this is one of the few studies where injection of antifunctional antibodies into the trunk had a discernible affect on morphogenesis.

Recently, another possible inhibitor of migration has been added to the mix. The collapsins/semaphorins are secreted molecules that were first identified by their ability to collapse the growth cones of migrating neurons (41, 81, 82) (see Chapter 7).

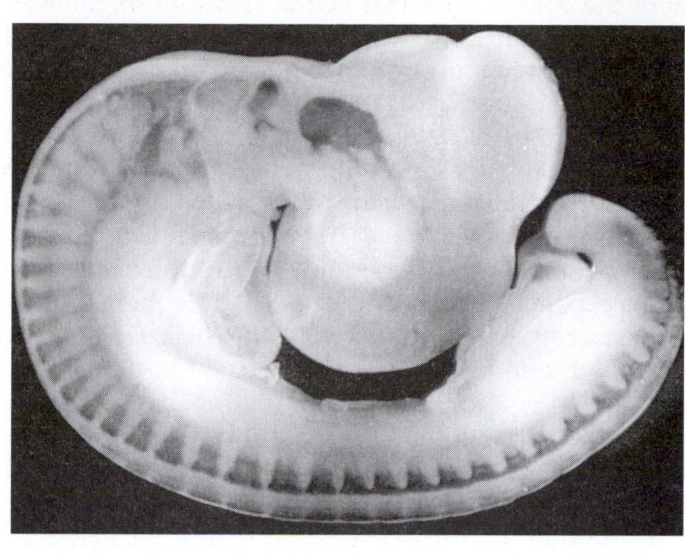

Fig. 3 (See also Plate 7) *In situ* hybridization in a stage-20 chick embryo revealing the expression of FoxD3. FoxD3 is a transcription factor that is expressed by all neural crest cells except melanoblasts. This embryo demonstrates the contribution of the neural crest to the head cranial ganglia and the segmental migration of the neural crest through the somites in the trunk. (Micrograph courtesy of Robert Kos.)

Eickholt and co-workers (40) report that collapsin-1 is expressed in the posterior half of the somite (although this distribution was not seen in another study (83)), and that migrating neural crest cells express neuropilin-1, which is the receptor for collapsin-1 (84, 85). When collapsin-1 is immobilized in stripes alternating with fibronectin, explanted neural crest cells avoid the collapsin stripe. No attempts to perturb function *in vivo* were described in this report.

In summary, a number of molecules can prevent neural crest migration into the posterior somite, suggesting there is considerable redundancy in establishing segmental migration. This redundancy is also suggested by loss-of-function mouse mutants for ephrin-B2 (86), collapsin-1 (87), and neuropilin-1 (88), none of which by themselves appear to perturb the ventral migration of the neural crest, at least when analysed at late stages in development. It might be informative to knock out several of these genes together, or perturb several of these molecules at the same time in the explant system, to observe a more dramatic phenotype. Given this redundancy, it is perplexing that, in the explant assay, there is a robust effect when either the EphB3 receptor or PNA-binding molecules are perturbed individually. This observation suggests that perhaps each of these signalling systems has a unique role to play in segmental migration through the somite, or that different subpopulations of the crest are differentially sensitive to the many cues. Alternatively, it raises the disturbing and confounding possibility that antibody perturbation experiments yield fundamentally different results from genetic loss-of-function studies.

Most of the data suggest that neural crest-cell migration is inhibited through the posterior somite, and so they migrate by default through the anterior somite. It is important to ask what substratum in the anterior somite is permissive for migration. The anterior somite (as well as the posterior) contains a number of ECM molecules on which neural crest cells can migrate, including fibronectin, laminin, collagen, and vitronectin (reviewed in ref. 75). Which of these is used as the predominant migratory substratum is still not known. Numerous studies have attempted to perturb cell–matrix interactions by injecting function-blocking antibodies into the lumen of the somite, but these have had little effect. Whether this is because the antibody is expelled from the somite or whether there is functional redundancy is not known. However, recently, again using the chick explant system, α_4 integrin function was perturbed with function-blocking antibodies or a peptide that competes with fibronectin binding to the α_4 receptor, and neural crest-cell migration was substantially inhibited (89). Because the $\alpha_4\beta_1$ integrin primarily mediates adhesion to fibronectin, it is reasonable to conclude that the neural crest cells at the trunk level use fibronectin as their primary migratory substratum (although thrombospondin, whose distribution is strictly correlated with neural crest migratory pathways, also binds to α_4). There is still much to clarify, however, because neural crest cells are known to express other integrins, including those that mediate attachment to laminin (90) and vitronectin (91). Moreover, not all neural crest cells express α_4 (89, 92, 93). This analysis is bound to be complicated, because integrins can functionally compensate for each other (94).

Although migration into the anterior somite is generally considered to be a default pathway, there is evidence that some molecules may actually promote or enhance neural crest migration. In order for a specific ECM molecule to positively stimulate migration, it most likely should only be present in the anterior half of the somite. Thrombospondin is the only extracellular matrix molecule that fits this criterion. It is distributed in the basal lamina of the neural tube along which neural crest cells first migrate, and is associated with the basal surface of the myotome, on which the neural crest cells spread once they enter the somite (95). Moreover, thrombospondin promotes adhesion and migration *in vitro* to approximately the same extent as fibronectin. Thus, thrombospondin-1 is the only ECM molecule whose distribution and functional properties are consistent with a role in promoting neural crest migration in the anterior somite.

3.2 Dorsal migration through the somite

Besides the anteroposterior patterning discussed above, the neural crest cells have a characteristic dorsoventral distribution in the somite, which subsequently defines the dorsoventral position of the peripheral ganglia. Initially, neural crest cells adhere to and follow the basal surface of the myotome, which directs them on a mediolateral path through the somite to the dorsal aorta and the gut. If the myotome is ablated, neural crest cells will now migrate ventrally and penetrate the sclerotome more deeply (59). Once the basal surface of the myotome is occupied by neural crest cells,

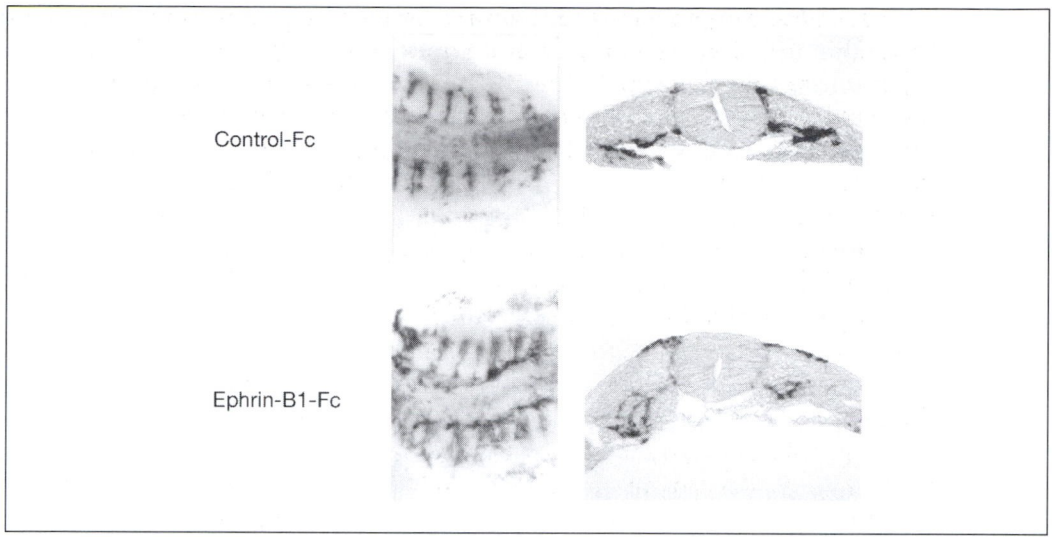

Fig. 4 (See also Plate 8) Pertubation of ephrin function in trunk tissue pieces results in the loss of segmental migration through the somites, and the premature migration into the dorsolateral path. (Control) Pieces of stage-13 trunk were placed in culture; after 24 hours neural crest cells (labelled dark brown) had migrated normally through the anterior half of each somite, but had not yet invaded the dorsolateral path. (Ephrin–Fc) When the explants are treated with a fusion protein comprising the Eph-receptor binding portion, the Eph receptors on the neural crest cells are occupied with soluble ligand and so are insensitive to ephrins in their pathways. Consequently, the segmental migration through somites was pertubed and neural crest cells invaded the dorsolateral path precociously. (Micrograph courtesy of Alicia Santiago.)

they will continue to penetrate the somite by infiltrating the sclerotome, but they still remain dorsal and fail to invade the perinotochordal mesenchyme (54, 58, 60).

The reasons for this persistent dorsal migration are unclear, but may be due to the sum of several mechanisms. First, the ventral sclerotome contains molecules known to be relatively inhibitory to neural crest migration, such as CSPG and PNA-binding molecules (10, 71). Moreover, the notochord produces a specialized extracellular matrix, which creates a cellular exclusion zone around itself *in vivo* and in culture (71, 96). However, experimental studies show that if the neural crest cells are grafted beneath the somite, they readily penetrate the ventral sclerotome (our unpublished results), so this may be only a relatively undesirable substratum and not an absolute barrier. Another possibility is that neural crest cells maintain a mediolateral directionality as they lose touch with the myotome, because they persist in the direction of migration established by their initial trajectory as they invade the somite. Consistent with this view, neural crest cells line up in a head-to-tail formation as they cross the somite with very little dorsoventral deviation (54, 69). Finally, soon after the crest cells invade the somite, the dorsal root motor fibres exit from the ventral neural tube and extend through the sclerotome (58). Eventually, neural crest cells are found spread along the ventral root, but are never found ventral to it (54). This distribution suggests that the ventral root fibres may block ventral migration.

A combination of cues—stimulatory, inhibitory, and permissive—determine the paths that neural crest cells follow ventrally. But the proper pathway of migration is not sufficient to generate the peripheral nervous system. In addition, other factors must control when neural crest cells should stop migrating and how they must differentiate, which will now be considered.

3.3 Control of lineage segregation and migratory behaviour on the ventral path

It has been argued above that there are two possible ways in which the correct neural crest phenotype will become localized and differentiate at the appropriate site: either the neural crest cells are pluripotent and differentiate according to the environmental cues along their path of migration; or they are already specified before they disperse on a particular path, and arrive at the appropriate destination because of cell-autonomous migratory properties that direct them. The melanoblasts are an example of a neural crest lineage that is specified prior to entering its path of migration. The question that will now be addressed is whether there is any evidence that the ventrally migrating neural crest cells are also specified at the time they segregate from the neural tube (97), and if so, do these lineages have unique migratory and pathfinding capabilities (see also ref. 98).

3.3.1 When are neural crest lineages specified?

The preponderance of evidence has supported the notion that, as a population, the neural crest is pluripotent. The classic heterotopic transplant studies of Le Douarin and Teillet (99) show clearly that neural crest cells from one axial level will migrate and differentiate according to another axial level if they are transplanted there. Subsequent studies, in which neural crest-derived structures, such as ganglia, were back-transplanted to the early migratory pathway, also reveal that neural crest cells that had already migrated to one spot are capable of remigrating and giving rise to a variety of phenotypes (100–102). These studies all focused on populations of neural crest cells and not individual cell capabilities, but at the very least they demonstrated that the neural crest population, as a whole, is extremely plastic.

The developmental potential of individual neural crest cells has been tested by employing the limit-dilution cloning strategies first developed by Sieber-Blum and Cohen (103). By taking neural crest cells that had emigrated from neural tubes in culture and subcloning them, they and later investigators have established that some neural crest cells are capable of differentiating into a range of phenotypes. For example, neural crest cells derived from the thoracic axial level can give rise to clones that contain neurons, glial cells, and pigment cells (103). Clones derived from cranial neural folds are even more diverse in their developmental repertoire (104–106). Similarly, *in vivo* labelling of single dorsal neural epithelial cells with rhodamine dextran as a lineage marker, has revealed that resulting clones give rise to neural crest cells that are found in a variety of structures, although how they are actually differentiating has not been rigorously tested because of the eventual dilution of the injected marker and the lack of availability of cell-type specific markers (107, 108). These studies give the clear impression that at least some neural crest cells are multipotent. Moreover, even differentiated neural crest cells have been observed to switch their phenotype when placed in a different environment (109). Consequently, the general view has prevailed that the patterning of the ventrally distributed neural crest is dependent upon environmental cues that control neural crest-cell differentiation.

There is an emerging notion, however, that the neural crest is a much more heterogeneous population of cells than the above studies predicted (97, 110, 111). First, numerous markers for many of the crest lineages have been developed, and these have revealed a level of heterogeneity in the early migratory crest that was not initially suspected. However, these markers cannot be used to infer developmental potential or specification (110). Second, when data from the cloning studies cited above is examined, a large percentage of clones are discovered to comprise only one cell type, which suggests that at the time a particular neural crest cell was isolated, it was already fate-restricted. But it is a recent study by Henion and Weston (2) that unequivocally demonstrates that a large number of neural crest cells are already fate-restricted at the time they leave the neural tube and, together with some other studies (43), suggests that we should re-evaluate the mechanisms that control patterning of the ventral migrating crest.

Henion and Weston (2) addressed the state of specification of trunk neural crest cells by culturing quail neural tubes, labelling a single neural crest cell with rhodamine dextran as it emerged from the neural tube, and then assessing the phenotypes of the cells in the resulting clone, using lineage-specific markers for neurons, glial cells, and melanocytes. There are several advantages of this approach over traditional cloning techniques: limit-dilution cloning studies generally use 24-hour outgrowths, which have already changed a great deal in their developmental potential; and the neural crest cells in the Henion and Weston study remain as an interacting population and so could influence each other's development, as they might in the embryo. Further, they assessed the phenotypes that arise from clones whose founder cells were labelled at different times after their emergence from the neural tube: at 6 hours, between 13 and 16 hours, and at 30 to 36 hours. Their results were surprising. When the phenotypic capabilities were assessed after only 6 hours of migration, 44.5% of the clones gave rise to only one cell type (i.e. neuron, glial, or melanocyte), showing that at the initiation of migration almost half the neural crest cells were already lineage-restricted. By 13–16 hours after the initiation of migration, 72.2% gave rise to only one cell type. The rest of the clones at 6 hours were partially restricted, containing either neurons and glial cells or glial cells and melanocytes, showing that the neuronal and melanocyte lineages are already segregated at the time they initiate migration. Only one clone produced neurons, glial cells, and pigment cells, which is in contrast to the results of the limit-dilution studies (103), and suggests that the earlier cloning procedures select for multipotent neural crest cells. A similar clonal analysis performed *in vivo* in the zebrafish also revealed that the majority of the premigratory crest cells behave as fate-restricted precursors (43), although the sampling procedure was not as extensive.

The second important observation they made is that the developmental potential of neural crest cells changes with the time when they detach from the neural tube. Almost no pigment cells differentiate from cells migrating during 1–6 hours. In a complementary study, Reedy *et al.* (3) showed that when quail neural tubes are serially replated, no pigment cells emerge during the first 6 hours, and only during later stages of migration do pigment cells progressively appear.

The unsuspected level of heterogeneity in the developmental potential of the early migratory neural crest, revealed by Henion and Weston (2), implies that the patterns of neural crest-cell migration may be, in part, regulated by cell-autonomous migratory properties that accompany lineage-restriction. In the following discussion, the evidence will be reviewed that this model accounts for the migration, localization, proliferation, and maintenance of sensory neurons, sympathetic neurons, and enteric neurons.

3.3.2 Sensory neuron lineage

Sensory neurons differentiate from the neural crest cells that coalesce adjacent to the dorsal neural tube and form the segmental dorsal root (DRG) or sensory ganglion (61). The dorsal positioning of the DRG is almost certainly due to some cue from the dorsal neural tube. If the neural tube is transplanted into non-specific mesoderm, such as the limb bud (112) or lateral plate mesoderm (113), or the neural tube is rotated around its dorsoventral axis (68, 114), the DRG always develops adjacent to the dorsal neural tube and independent of any mesodermal influence. Conversely, the segmentation of the DRG is controlled by the somite. The DRG develops in the anterior half of the somite, due primarily to the posterior somite inhibiting neural crest migration. If somites are constructed that comprise two anterior halves, a large unsegmented ganglion develops (115, 116). If a somite comprises two posterior halves, a very small unsegmented ganglion forms dorsal to the somite itself (116). Clearly, environmental cues are critical for the proper position of the DRG. However, an important questions remains: are sensory neuron precursors determined early, and are they uniquely able to respond to these cues to form a ganglion?

When does the sensory lineage arise?
In vitro clonal analysis reveals that at least some neural crest cells are able to give rise to both sensory neurons and catecholaminergic neurons (117). However, this same study suggests that there are also cells that become fate-restricted early in the culture period. Clonal analysis of migratory neural crest cells *in vivo* also shows that fully 25% of all clones analysed contribute only to the DRG (118). Back-transplantations of sympathetic ganglia or enteric neurons fail to give rise to any sensory neurons, again pointing to an early segregation of this lineage (119, 120). These conflicting results suggest there may be multiple lineages that give rise to sensory neurons.

A number of studies support the idea that the DRG comprises a variety of lineages that are spatially, temporally, and molecularly distinct. First, the ganglion consists of neurons in the dorsomedial and ventrolateral quadrants that are morphologically dissimilar and undergo apoptosis at different times. Second, using the stage-specific embryonic antigen (SSEA) marker for sensory neurons, two sources of sensory neurons have been identified in outgrowth cultures: cells that differentiate early in the centre of the outgrowth in association with remnants of the neural tube; and cells that differentiate later that are on the periphery of the outgrowth (117, 121). Third, analysis of cells that emigrate from the neural tube show that a late-migrating population of cells from the neural tube (long after the neural crest cells stop migrating) contribute to the DRG and also to pigment cells (122). Finally, as discussed in the next paragraph, two recent molecular studies reveal the presence of markers that identify at least a portion of the sensory lineage, and that these markers are already expressed in premigratory neural crest cells.

One marker expressed by chick premigratory crest cells and later in some cells that settle in the DRG is the tyrosine kinase-C (trkC) receptor for the neurotrophin, NT-3 (123). This subpopulation requires NT-3 for its survival, at least in culture. Other markers of a subpopulation of sensory neurons are two members of the basic helix–loop–helix transcription factor, neurogenin (124). These are expressed in some cells in the neural tube and in a subpopulation of early migratory crest cells that localize to the DRG. The neurogenins are important in DRG development, because the sensory lineage does not develop in neurogenin-null mice, and the ectopic expression of chick neurogenins in other tissues, including non-neuronal tissue such as the dermamyotome, results in the expression of sensory neuron-specific markers (124). It is not known if trkC and neurogenin are in different sensory lineages or if their expression overlaps.

It seems fair to conclude that, although some neural crest cells retain multipotency and can give rise to a variety of neurons, one or several subpopulations of the crest are already restricted to the sensory lineage at the time they initiate migration.

Do these lineages selectively migrate to the site of the DRG?

The expression of these sensory lineage markers early in migration suggests that one or several distinct sensory lineages may already be specified. If this is so, is there any evidence that they selectively migrate to the site of the developing DRG? In a provocative experiment, Perez *et al.* (124) overexpressed the neurogenins in the premigratory crest using RCAS (replication-competent variant of the avian leukaemia virus) viruses, and found that most of the crest cells expressing neurogenin were selectively localized in the DRG. Possibly, neurogenin expression promotes the migration of these cells to the DRG. Similarly, those cells expressing the trk-C receptor might be attracted haptotactically or chemotactically to the DRG, in a manner analogous to the other receptor tyrosine kinases having this function, since the neural tube is a source of NT-3 (125). On the other hand, expression of these markers may cause the localization of these lineages to the DRG because they selectively result in their proliferation or survival at that site (98, 126). For example, if a silastic barrier is placed between the neural tube and somite the DRG fails to develop (127). If the membrane is coated with brain-derived growth factor (BDGF), another neurotrophic factor, the sensory neurons now survive and a DRG develops (128).

Careful analysis will be required to distinguish amongst these possibilities. But it is interesting that both these markers are expressed prior to migration, both are expressed at about the time when neural crest cells that will form the DRG begin their migration (112), and neither of these markers is seen in neural crest cells that get by the DRG and localize to the ventral roots or sympathetic ganglia. Together, these observations suggest that these lineages 'home' to the developing DRG.

3.3.3   Sympathoadrenal lineage

Neural crest cells give rise to at least three catecholamine-containing cells types in the peripheral nervous system: the adrenergic neurons of the sympathetic ganglia, the chromaffin cells (or phaeochromocytes) of the adrenal gland, and a third cell type called 'small intensively fluorescent' (SIF) cells, which are found in the sympathetic ganglia, the adrenal medulla, and also in the carotid body and in small paraganglia in the gut (129). All these catecholaminergic cells are thought to arise from a common progenitor cell, which initially localizes in the primary sympathetic chain by the dorsal aorta (62). From this site, the cells then disperse dorsally to form the definitive sympathetic ganglia and also ventrally to the kidney, where they contribute to the phaeochromocytes of the adrenal gland. Experimental evidence is consistent with the notion that this lineage first differentiates into neurons (62), and that the neuronal traits are subsequently lost when they migrate to the site of the adrenal. The loss of neuronal traits is probably due to exposure to glucocorticoids in the developing adrenal gland (63), but the cause of the initial neuronal differentiation is unclear. Experimental studies have identified environmental signals from the ventral neural tube, notochord, and somite that are required for sympathoblast differentiation (12, 130, 131). Numerous studies suggest that a series of inductive interactions, including exposure to noradrenaline produced by the notochord (132) and BMPs produced at the dorsal aorta (133–136), provide instructive signals to control neuronal differentiation.

There is no direct evidence for a fate-restricted sympathoadrenal precursor cell in the early migratory neural crest (110). Rather, evidence from studies using chick and rat model systems suggest that the sympathoadrenal lineage is derived from multipotent, partially restricted, progenitors that can also produce some sensory neurons and glial cells. For example, limit-dilution clonal studies by Sieber-Blum (117) produced single clones that contained both adrenergic and sensory neurons. Moreover, markers for adrenergic neurons do not appear until well after the crest cells have stopped migrating. The earliest-known marker for these neurons is the uptake system for noradrenaline, which, in the chick, appears by day 3 in those cells that have already coalesced near the dorsal aorta (133). Other definitive markers, such as presence of catecholamines and *MASH1* and *Phox2* genes appear even later (138–140). Some evidence has been used to support an early segregation of the sensory and autonomic lineages in the early migratory neural crest. For example, when 6-day sympathetic ganglia were back-grafted into the early migratory pathway, these crest cells remigrated and gave rise to glial cells and sympathetic neurons, but never to sensory neurons (101). However, such studies could just as easily be interpreted to mean that lineage restrictions occur relatively late and after migration is complete. The most parsimonious conclusion to be drawn from the current data is that cells of the sympathoadrenal lineage are derived from precursors that are multipotent as they are migrating, and that their specification and differentiation is controlled largely by environmental signals. This suggests that, unlike melanoblasts and some sensory neuroblasts, there are probably no cell-autonomous migratory cues that direct these cells ventrally to the dorsal aorta.

However, a recent study suggests that even though the sympathoadrenal lineage may not be fate-restricted at the time the neural crest cells initiate migration, there still may be cell-autonomous migratory cues that direct some neural crest cells ventrally. Britsch *et al.* (141) have shown that in knockout mice for the EGF-like growth factor neuregulin-1, or for its receptors ErbB2 or ErbB3, neural crest cells fail to arrive at the site of sympathetic ganglion formation adjacent to the dorsal aorta, and that these mice never develop an adrenal medulla or secondary ganglia. The absence of ventral neural crest cells is not believed to be due to apoptosis, as revealed by TUNEL (TdT (terminal deoxynucleotidyl transferase) X-dUTP nick end-labelling) assay, although apoptosis may have been missed since it occurs so quickly. The accumulation of excess neural crest cells in the region of the sensory ganglion, suggests that the phenotype is due to the failure of neural crest cells to move ventrally below the neural tube. Moreover, the distribution of neuregulin-1 is compatible with a role in directing the neural crest either by chemotactic or haptotactic cues: initially it is present in the sclerotome of the somite, and eventually is focused in a ventral stripe adjacent to the dorsal aorta. Finally, if this receptor–ligand system is guiding the crest, then the ErbB2–ErbB3 receptor dimer should be present on early migrating neural crest cells, and it is. If neuregulins pattern neural crest migration, then neural crest cells are predicted to accumulate at sites of ectopically expressed neuregulin. Also, the ErbB2 receptor may be a marker for a sympathetic lineage that segregated early from the sensory lineage. Finally, it will be interesting to see if a similar expression pattern exists in the chick embryo, in which experimental manipulation would be possible to show directly if neuregulin-1 is a positive guidance cue that directs a subpopulation of neural crest cells ventrally.

3.3.4 Enteric lineage

The neural crest cells also migrate into the gut, and there give rise to the neurons and glial cells of the enteric nervous system. Enteric neural crest cells arise from two axial levels: somite level 1–7 (the so-called vagal crest), which populate the entire length of the gut; and somite level 28 and posterior (the sacral level), which populate the postumbilical gut alone and also form the ganglion of Remak, which is external to the gut and considered part of the parasympathetic nervous system (64–67). A fundamental question is why neural crest cells enter the gut at these two restricted sites. Is the environment at the vagal and sacral level responsible for facilitating neural crest invasion of the gut, or are the neural crest cells from these two axial levels endowed with special migratory abilities to find their way to the gut?

When is the enteric lineage specified?

At present, this is a difficult question to answer, in part because various laboratories have used chick, mouse, or rat for their experimental systems, which has undoubtedly introduced some confusion. In addition, definitive experiments have been difficult to perform without unique markers for the enteric crest. Despite these complexities, two things seem clear: molecular markers for enteric crest cells do not generally appear until after the cells get into the gut, which is at least compatible with the notion that they are not specified until they reach the gut and are exposed to that environment (142); and many of the neural crest cells isolated from the gut are multipotent and able to differentiate into both enteric-specific neurons and glial cells, and are also capable of differentiating into sympathetic and parasympathetic neurons (100, 143, 144). (It should be noted, however, that other reports suggest that some neural crest cells isolated from the gut are restricted in their lineage (145, 146), suggesting that perhaps there are different subpopulations of the neural crest that migrate into the gut. Nevertheless, all the studies purporting to show limited developmental potential of the enteric crest have used cells that have already migrated to the gut, and do not assess the developmental potential of the pre-migratory crest from the vagal and sacral levels).

Do vagal or sacral neural crest cells have unique migratory properties?

There is no direct evidence that a subpopulation of premigratory neural crest cells (that is, still contained within the neural epithelium) or even early dispersing crest cells are specified as enteric precursors. Nevertheless, it still may be possible that some neural crest cells at the vagal and sacral level have special migratory properties that allow them to reach and invade the gut mesenchyme. To test this hypothesis, we heterotopically grafted neural crest cells from the sacral axial level to the thoracic level and vice versa, and observed that the neural crest cells behave according to their new position, rather than their site of origin (143). The simplest explanation of this result is that the environment at the sacral level facilitates migration to the gut.

Our study further suggests that at least two environmental conditions at the sacral level enhance ventral migration. First, sacral neural crest cells take a ventral rather than a medial-to-lateral trajectory through the somites, and therefore the first-emigrating sacral neural crest cells arrive ventral to the dorsal aorta many hours earlier than their counterparts at the thoracic level. Because there is only a narrow window of opportunity to invade the mesentery and the gut mesenchyme (148), this early arrival assures the sacral neural crest of gaining entrance to the gut. Second, the gut is more dorsally situated at the sacral level than at the thoracic level.

We predict, based on these findings, that neural crest cells that will become enteric neurons and/or glial cells are not specified prior to migrating. It is still formally possible that at all axial levels a subpopulation of cells is specified as enteric precursors, and that they can all follow cues into the gut. However, for this to be compatible with our results, it would also have to be the case that these attractive cues are only produced at the vagal and sacral axial levels. In addition, enteric precursors should behave differently than the aggregate population of newly migrating neural crest cells. That is, one would expect all or most of the enteric crest to migrate further ventrally and accumulate in the gut, but fail to localize in more dorsal positions. But when neural crest cells are isolated from the gut and back-grafted into the early chick embryo, these cells behave like the host crest and occupy all the sites that the crest normally fill, including the sensory and sympathetic ganglia, the adrenal gland, and along the ventral root motor fibres (100). Recent studies suggest that back-grafted cells do not differentiate appropriately for these sites (145), but that their migratory behaviour is no different from early migratory crest cells. Thus neural crest cells that have already arrived in the gut appear to have the same migratory properties as the neural crest population as a whole.

A number of questions remain. Our studies focused on the sacral neural crest, but the vagal neural crest gives rise to the majority of enteric neurons. Therefore, it is possible that the vagal crest may have unusual migratory properties that we did not explore. Also once the crest cells enter the gut at its anterior end, they then migrate posteriorly (reviewed in ref. 146). What drives the crest in this directional fashion? Is there a chemotactic gradient, or do they disperse by population pressure? Conversely, the sacral crest cells migrate in a caudal-to-cranial direction once they enter the gut (66). There is at least some evidence that an electric field may drive this migration (149). The recently developed technique of culturing whole gut (142, 144, 150) should provide a convenient assay system with which to observe neural crest migratory behaviour directly and answer these questions.

4. Patterning of the cranial neural crest

This review has focused on trunk neural crest cells because they give rise to a more limited array of phenotypes than the cranial crest, and their migratory patterns are relatively simple and understood in considerable detail. The emerging story is that directed migration of different trunk neural crest lineages may be more the rule than the exception (98). What can we say about the cranial neural crest?

As in the trunk, the head crest also appears to be a multipotent population of cells, based on *in vitro* cloning studies. Some studies have shown that a single cell derived from the cranial neural folds can be capable of giving rise to all the head derivatives, including neurons, glial cells, connective tissue, and smooth muscle (106, 151). Such mulitpotency suggests that patterning is dependent upon environmental cues. Moreover, in the only study where the migratory ability of different populations of the cranial crest has been tested, when early mesencephalic crest cells and late migrating mesencephalic crest cells are exchanged, development is apparently normal (152). This latter study suggests that there is no fixed lineage restriction that affects either patterns of migration or eventual differentiation.

Nevertheless, there are several situations where some cell lineages of the cranial neural crest are known unequivocally to be restricted at the onset of migration. Neural crest cells that give rise to elements of the heart (the so-called 'cardiac crest') only arise from the postotic crest. If they are ablated, the heart is defective, and no other neural crest cells from other axial levels can substitute for the ablated crest (153). Similarly, ectomesenchyme arises only from the cranial neural folds, and trunk neural crest cells exhibit a greatly reduced ability to differentiate into ectomesenchyme when grafted to the head (154, 155). Finally, when neural crest cells from the mesencephalon are transplanted to the anterior rhombencephalon, they migrate into the second arch rather than into the first arch, but they form first-arch derivatives, suggesting they were already patterned prior to leaving the neural tube (156, 157). Thus in at least three instances, cranial neural crest cells are developmentally restricted at the time they initiate migration from the neural folds. If they are specified, might they also have cell-autonomous migratory potential to guide them to their final destination?

We are now in a position to test some of these ideas directly. Molecular tools and genetic studies have identified numerous markers for various neural crest lineages. Such markers can be used to explore the timing of lineage segregation in the head crest using the unique clonal analysis developed by Henion and Weston (2). Such studies may reveal the presence of many fate-restricted lineages, whose migratory properties could then be challenged in the embryo. The chick embryo has been a useful model for studying crest morphogenesis because it is amenable to experimental manipulation. The time is now right to test the migratory abilities of different subpopulations of the neural crest using previously developed experimental strategies (147, 152). Finally, we can now manipulate gene expression in the mouse using transgenic technology, and in the chick using viral expression and antisense approaches. When we alter the developmental end-point of certain neural crest lineages, do we also change their migratory ability or 'homing' properties (124, 141)? As we seek to understand the molecular basis of lineage specification, we should be alert to the role that early specification prior to the onset of dispersal may have in also controlling the migratory behaviour.

For some time, our view of the neural crest cells has been that their distribution and ultimate phenotype is directed by the environment through which they migrate. Perhaps the neural crest cells play a more active role in pathfinding than we have credited them. As we learn more about the timing and control of lineage segregation, we will be in a position to evaluate whether fate-restriction plays an essential role in also controlling the migratory behaviour of the neural crest.

Acknowledgements

I would like to thank my colleagues Dave McClay, Mark Reedy, and Jim Weston for their careful reading of the manuscript and their insightful comments. All unpublished research cited in this review was supported by a grant from the NIH (GM53258).

References

1. Erickson, C. A. and Reedy, M. V. (1998) Neural crest development: the interplay between morphogenesis and cell differentiation. *Curr. Top. Dev. Biol.*, **40**, 177.
2. Henion, P. D. and Weston, J. A. (1997) Timing and pattern of cell fate restrictions in the neural crest lineage. *Development*, **124**, 4351.
3. Reedy, M. V., Faraco, C. D., and Erickson, C. A. (1998) The delayed entry of thoracic neural crest cells into the dorsolateral path is a consequence of the late emigration of melanogenic neural crest cells from the neural tube. *Dev. Biol.*, **200**, 234.
4. Wakamatsu, Y., Mochii, M., Vogel, K. S., and Weston, J. A. (1998) Avian neural crest-derived neurogenic precursors undergo apoptosis on the lateral migration pathway. *Development*, **125**, 4205.
5. Richardson, M. K. and Sieber-Blum, M. (1993) Pluripotent neural crest cells in the developing skin of the quail embryo. *Dev. Biol.*, **157**, 348.
6. Erickson, C. A. and Goins, T. L. (1995) Avian neural crest cells can migrate in the dorsolateral path only if they are specified as melanocytes. *Development*, **121**, 915.
7. Hamburger, V. and Hamilton, H. L. (1951) A series of normal stages in the development of the chick embryo. *J. Morphol.*, **88**, 49.
8. Serbedzija, G. N., Bronner-Fraser, M., and Fraser, S. E. (1989) A vital dye analysis of the timing and pathways of avian trunk neural crest cell migration. *Development*, **106**, 809.
9. Erickson, C. A., Duong, T. D., and Tosney, K. W. (1992) Descriptive and experimental analysis of the dispersion of neural crest cells along the dorsolateral path and their entry into ectoderm in the chick embryo. *Dev. Biol.*, **151**, 251.
10. Oakley, R. A. and Tosney, K. W. (1991) Peanut agglutinin and chondroitin-6-sulfate are molecular markers for tissues that act as barriers to axon advance in the avian embryo. *Dev. Biol.*, **147**, 187.
11. Oakley, R. A., Lasky, C. J., Erickson, C. A., and Tosney, K. W. (1994) Glycoconjugates mark a transient barrier to neural crest migration in the chicken embryo. *Development*, **120**, 103.
12. Debby-Brafman, A., Burstyn-Cohen, T., Klar, A., and Kalcheim, C. (1999) F-Spondin, expressed in somite regions avoided by neural crest cells, mediates inhibition of distinct somite domains to neural crest migration. *Neuron*, **22**, 475.
13. Wang, H. U. and Anderson, D. J. (1997) Eph family transmembrane ligands can mediate repulsive guidance of trunk neural crest migration and motor axon outgrowth. *Neuron*, **18**, 383.
14. Kitamura, K., Takiguchi-Hayashi, K., Sezaki, M., Yamamoto, H., and Takeuchi, T. (1992) Avian neural crest cells express a melanogenic trait during early migration from the neural tube: observations with the new monoclonal antibody, 'MEBL-1'. *Development*, **114**, 367.
15. Wehrle-Haller, B., Morrison-Graham, K., and Weston, J. A. (1996) Ectopic c-kit expression affects the fate of melanocyte precursors in Patch mutant embryos. *Dev. Biol.*, **177**, 463.
16. Wehrle-Haller, B. and Weston, J. A. (1995) Soluble and cell-bound forms of steel factor activity play distinct roles in melanocyte precursor dispersal and survival on the lateral neural crest migration pathway. *Development*, **121**, 731.
17. Wehrle-Haller, B. and Weston, J. A. (1997) Receptor tyrosine kinase-dependent neural crest migration in response to differentially localized growth factors. *BioEssays*, **19**, 337.
18. Lecoin, L., Lahav, R., Martin, F. H., Teillet, M. A., and Le Douarin, N. M. (1995) Steel and c-kit in the development of avian melanocytes: a study of normally pigmented birds and of the hyperpigmented mutant silky fowl. *Dev. Dyn.*, **203**, 106.

19. Erickson, C. A., Tosney, K. W., and Weston, J. A. (1980) Analysis of migratory behavior of neural crest and fibroblastic cells in embryonic tissues. *Dev. Biol.*, **77**, 142.

20. Sears, R. and Ciment, G. (1988) Changes in the migratory properties of neural crest and early crest-derived cells *in vivo* following treatment with a phorbol ester drug. *Dev. Biol.*, **130**, 133.

21. Weiss, P. and Andres, G. (1952) Experiments on the fate of embryonic cells (chick) disseminated by the vascular route. *J. Exp. Zool.*, **121**, 449.

22. Werb, Z. (1997) ECM and cell surface proteolysis: regulating cellular ecology. *Cell*, **91**, 439.

23. Alexander, C. M. and Werb, Z. (1991). Extracellular matrix degradation. In *Cell biology of extracellular matrix* (ed. E. D. Hay), p. 255. Plenum Press, New York.

24. Hebert, C. A. and Baker, J. B. (1988) Linkage of extracellular plasminogen activator to the fibroblast cytoskeleton: colocalization of cell surface urokinase with vinculin. *J. Cell Biol.*, **106**, 1241.

25. Chen, W. T. (1989) Proteolytic activity of specialized surface protrusions formed at rosette contact sites of transformed cells. *J. Exp. Zool.*, **251**, 167.

26. Valinsky, J. E. and Le Douarin, N. M. (1985) Production of plasminogen activator by migrating cephalic neural crest cells. *EMBO J.*, **4**, 1403.

27. Erickson, C. A. and Isseroff, R. R. (1989) Plasminogen activator activity is associated with neural crest cell motility in tissue culture. *J. Exp. Zool.*, **251**, 123.

28. Agrawal, M. and Brauer, P. R. (1996) Urokinase-type plasminogen activator regulates cranial neural crest cell migration *in vitro*. *Dev. Dyn.*, **207**, 281.

29. Millichip, M. I., Dallas, D. J., Wu, E., Dale, S., and McKie, N. (1998) The metallo-disintegrin ADAM10 (MADM) from bovine kidney has type IV collagenase activity *in vitro*. *Biochem. Biophys. Res. Commun.*, **245**, 594.

30. Fambrough, D., Pan, D., Rubin, G. M., and Goodman, C. S. (1996) The cell surface metalloprotease/disintegrin Kuzbanian is required for axonal extension in *Drosophila*. *Proc. Natl Acad. Sci. USA*, **93**, 13233.

31. Beauvais, A., Erickson, C. A., Goins, T., Craig, S. E., Humphries, M. J., Thiery, J. P., and Dufour, S. (1995) Changes in the fibronectin-specific integrin expression pattern modify the migratory behavior of sarcoma S180 cells *in vitro* and in the embryonic environment. *J. Cell Biol.*, **128**, 699.

32. Takeichi, M. (1991) Cadherin cell adhesion receptors as a morphogenetic regulator. *Science*, **251**, 1451.

33. Takeichi, M. (1995) Morphogenetic roles of classic cadherins. *Curr. Opin. Cell Biol.*, **7**, 619.

34. Nakagawa, S. and Takeichi, M. (1995) Neural crest cell–cell adhesion controlled by sequential and subpopulation-specific expression of novel cadherins. *Development*, **121**, 1321.

35. Simonneau, L., Kitagawa, M., Suzuki, S., and Thiery, J. P. (1995) Cadherin 11 expression marks the mesenchymal phenotype: towards new functions for cadherins? *Cell Adhes. Commun.*, **3**, 115.

36. Nishimura, E. K., Yoshida, H., Kunisada, T., and Nishikawa, S. I. (1999) Regulation of E- and P-cadherin expression correlated with melanocyte migration and diversification. *Dev. Biol.*, **215**, 155.

37. Tessier-Lavigne, M. and Goodman, C. S. (1996) The molecular biology of axon guidance. *Science*, **274**, 1123.

38. Flenniken, A. M., Gale, N. W., Yancopoulos, G. D., and Wilkinson, D. G. (1996) Distinct and overlapping expression patterns of ligands for Eph-related receptor tyrosine kinases during mouse embryogenesis. *Dev. Biol.*, **179**, 382.

39. Krull, C. E., Lansford, R., Gale, N. W., Collazo, A., Marcelle, C., Yancopoulos, G. D., Fraser, S. E., and Bronner-Fraser, M. (1997) Interactions of Eph-related receptors and ligands confer rostrocaudal pattern to trunk neural crest migration. *Curr. Biol.*, **7**, 571.

40. Eickholt, B. J., Mackenzie, S. L., Graham, A., Walsh, F. S., and Doherty, P. (1999) Evidence for collapsin-1 functioning in the control of neural crest migration in both trunk and hindbrain regions. *Development*, **126**, 2181.

41. Van Vactor, D. V. and Lorenz, L. J. (1999) Neural development: the semantics of axon guidance. *Curr. Biol.*, **9**, R201.

42. Holmberg, J., Clarke, D. L., and Frisén, J. (2000) Regulation of repulsion versus adhesion by different slice forms of an Eph receptor. *Nature*, **408**, 203.

43. Raible, D. W. and Eisen, J. S. (1994) Restriction of neural crest cell fate in the trunk of the embryonic zebrafish. *Development*, **120**, 495.

44. Epperlein, H. H. and Lofberg, J. (1984) Xanthophores in chromatophore groups of the premigratory neural crest initiate the pigment pattern of the axolotl larva. *Roux's Arch. Dev. Biol.*, **193**, 357.

45. Schilling, T. F. and Kimmel, C. B. (1994) Segment and cell type lineage restrictions during pharyngeal arch development in the zebrafish embryo. *Development*, **120**, 483.

46. Dorsky, R. I., Moon, R. T., and Raible, D. W. (1998) Control of neural crest cell fate by the Wnt signalling pathway. *Nature*, **396**, 370.

47. Ikeya, M., Lee, S. M., Johnson, J. E., McMahon, A. P., and Takada, S. (1997) Wnt signalling required for expansion of neural crest and CNS progenitors. *Nature*, **389**, 966.

48. Jin, E.-J., Erickson, C. A., Takada, S., and Burrus, L. W. (2001) Wnt and Bmp signaling govern lineage segregation of melanocytes in the avian embryo. *Dev. Biol.*, **233**, 22.

49. Liem, K. F. J., Tremml, G., Roelink, H., and Jessell, T. M. (1995) Dorsal differentiation of neural plate cells induced by BMP-mediated signals from epidermal ectoderm. *Cell*, **82**, 969.

50. Capdevila, J., Tabin, C., and Johnson, R. L. (1998) Control of dorsoventral somite patterning by Wnt-1 and beta-catenin. *Dev. Biol.*, **193**, 182.

51. Baranski, M., Berdougo, E., Sandler, J. S., Darnell, D. K., and Burrus, L. W. (2000) The dynamic expression pattern of frzb-1 suggests multiple roles in chick development. *Dev. Biol.*, **217**, 25.

52. Duprez, D., Leyns, L., Bonnin, M.-A., Lapointe, F., Etchevers, H., De Robertis, E. M., and Le Douarin, N. (1999) Expression of Frzb-1 during chick development. *Mech. Dev.*, **89**, 179.

53. Tosney, K. W. (1978) The early migration of neural crest cells in the trunk region of the avian embryo: an electron microscopic study. *Dev. Biol.*, **62**, 317.

54. Loring, J. F. and Erickson, C. A. (1987) Neural crest cell migratory pathways in the trunk of the chick embryo. *Dev. Biol.*, **121**, 220.

55. Delannet, M. and Duband, J. L. (1992) Transforming growth factor-beta control of cell-substratum adhesion during avian neural crest cell emigration *in vitro*. *Development*, **116**, 275.

56. Ordahl, C. P. and Le Douarin, N. M. (1992) Two myogenic lineages within the developing somite. *Development*, **114**, 339.

57. Spence, S. G. and Poole, T. J. (1994) Developing blood vessels and associated extracellular matrix as substrates for neural crest migration in Japanese quail, *Coturnix coturnix japonica*. *Int. J. Dev. Biol.*, **38**, 85.

58. Rickmann, M., Fawcett, J. W., and Keynes, R. J. (1985) The migration of neural crest cells and the growth of motor axons through the rostral half of the chick somite. *J. Embryol. Exp. Morphol.*, **90**, 437.

59. Tosney, K. W., Dehnbostel, D. B., and Erickson, C. A. (1994) Neural crest cells prefer the myotome's basal lamina over the sclerotome as a substratum. *Dev. Biol.*, **163**, 389.

60. Bronner-Fraser, M. (1986) Analysis of the early stages of trunk neural crest migration in avian embryos using monoclonal antibody HNK-1. *Dev. Biol.*, **115**, 44.

61. Teillet, M. A., Kalcheim, C., and Le Douarin, N. M. (1987) Formation of the dorsal root ganglia in the avian embryo: segmental origin and migratory behavior of neural crest progenitor cells. *Dev. Biol.*, **120**, 329.

62. Vogel, K. S. and Weston, J. A. (1990) The sympathoadrenal lineage in avian embryos. I. Adrenal chromaffin cells lose neuronal traits during embryogenesis. *Dev. Biol.*, **139**, 1.

63. Vogel, K. S. and Weston, J. A. (1990) The sympathoadrenal lineage in avian embryos. II. Effects of glucocorticoids on cultured neural crest cells. *Dev. Biol.*, **139**, 13.

64. Serbedzija, G. N., Burgan, S., Fraser, S. E., and Bronner-Fraser, M. (1991) Vital dye labelling demonstrates a sacral neural crest contribution to the enteric nervous system of chick and mouse embryos. *Development*, **111**, 857.

65. Pomeranz, H. D., Rothman, T. P., and Gershon, M. D. (1991) Colonization of the post-umbilical bowel by cells derived from the sacral neural crest: direct tracing of cell migration using an intercalating probe and a replication-deficient retrovirus. *Development*, **111**, 647.

66. Burns, A. J. and Le Douarin, N. M. (1998) The sacral neural crest contributes neurons and glia to the post-umbilical gut: spatiotemporal analysis of the development of the enteric nervous system. *Development*, **125**, 4335.

67. Le Douarin, N. M. and Teillet, M. A. (1973) The migration of neural crest cells to the wall of the digestive tract in avian embryo. *J. Embryol. Exp. Morphol.*, **30**, 31.

68. Bronner-Fraser, M. and Stern, C. (1991) Effects of mesodermal tissues on avian neural crest cell migration. *Dev. Biol.*, **143**, 213.

69. Krull, C. E., Collazo, A., Fraser, S. E., and Bronner-Fraser, M. (1995) Segmental migration of trunk neural crest: time-lapse analysis reveals a role for PNA-binding molecules. *Development*, **121**, 3733.

70. Kerr, R. S. E. and Newgreen, D. F. (1997) Isolation and characterization of chondroitin sulfate proteoglycans from embryonic quail that influence neural crest cell behavior. *Dev. Biol.*, **192**, 108.

71. Tosney, K. W. and Oakley, R. A. (1990) The perinotochordal mesenchyme acts as a barrier to axon advance in the chick embryo: implications for a general mechanism of axonal guidance. *Exp. Neurol.*, **109**, 75.

72. Ring, C., Hassell, J., and Halfter, W. (1996) Expression pattern of collagen IX and potential role in the segmentation of the peripheral nervous system. *Dev. Biol.*, **180**, 41.

73. Landolt, R. M., Vaughan, L., Winterhalter, K. H., and Zimmermann, D. R. (1995) Versican is selectively expressed in embryonic tissues that act as barriers to neural crest cell migration and axon outgrowth. *Development*, **121**, 2303.

74. Ranscht, B. and Bronner-Fraser, M. (1991) T-cadherin expression alternates with migrating neural crest cells in the trunk of the avian embryo. *Development*, **111**, 15.

75. Erickson, C. A. and Perris, R. (1993) The role of cell–cell and cell–matrix interactions in the morphogenesis of the neural crest. *Dev. Biol.*, **159**, 60.

76. Fredette, B. J. and Ranscht, B. (1994) T-cadherin expression delineates specific regions of the developing motor axon-hindlimb projection pathway. *J. Neurosci.*, **14**, 7331.

77. Krull, C. E. and Kulesa, P. M. (1998) Embryonic explant and slice preparations for studies of cell migration and axon guidance. *Curr. Top. Dev. Biol.*, **36**, 145.

78. Kulesa, P. M. and Fraser, S. E. (1998) Neural crest cell dynamics revealed by time-lapse video microscopy of whole embryo chick explant cultures. *Dev. Biol.*, **204**, 327.

79. Davies, J. A., Cook, G. M., Stern, C. D., and Keynes, R. J. (1990) Isolation from chick somites of a glycoprotein fraction that causes collapse of dorsal root ganglion growth cones. *Neuron*, **4**, 11.

80. Holder, N. and Klein, R. (1999) Eph receptors and ephrins: effectors of morphogenesis. *Development*, **126**, 2033.

81. Kolodkin, A. L. and Ginty, D. D. (1997) Steering clear of semaphorins: neuropilins sound the retreat. *Neuron*, **19**, 1159.

82. Chen, H., He, Z., and Tessier-Lavigne, M. (1998) Axon guidance mechanisms: semaphorins as simultaneous repellents and anti-repellents. *Nature Neurosci.*, **1**, 436. [News; Comment.]

83. Shepherd, I. T. and Raper, J. A. (1999) Collapsin-1/semaphorin D is a repellent for chick ganglion of Remak axons. *Dev. Biol.*, **212**, 42.

84. He, Z. and Tessier-Lavigne, M. (1997) Neuropilin is a receptor for the axonal chemo-repellent Semaphorin III. *Cell*, **90**, 739.

85. Kolodkin, A. L., Levengood, D. V., Rowe, E. G., Tai, Y. T., Giger, R. J., and Ginty, D. D. (1997) Neuropilin is a semaphorin III receptor. *Cell*, **90**, 753.

86. Wang, H. U., Chen, Z. F., and Anderson, D. J. (1998) Molecular distinction and angiogenic interaction between embryonic arteries and veins revealed by ephrin-B2 and its receptor Eph-B4. *Cell*, **93**, 741. [See Comments.]

87. Behar, O., Golden, J. A., Mashimo, H., Schoen, F. J., and Fishman, M. C. (1996) Semaphorin III is needed for normal patterning and growth of nerves, bones and heart. *Nature*, **383**, 525.

88. Kitsukawa, T., Shimizu, M., Sanbo, M., Hirata, T., Taniguchi, M., Bekku, Y., Yagi, T., and Fujisawa, H. (1997) Neuropilin–semaphorin III/D-mediated chemorepulsive signals play a crucial role in peripheral nerve projection in mice. *Neuron*, **19**, 995.

89. Kil, S. H., Krull, C. E., Cann, G., Clegg, D., and Bronner-Fraser, M. (1998) The alpha4 subunit of integrin is important for neural crest cell migration. *Dev. Biol.*, **202**, 29.

90. Lallier, T., Deutzmann, R., Perris, R., and Bronner-Fraser, M. (1994) Neural crest cell interactions with laminin: structural requirements and localization of the binding site for alpha 1 beta 1 integrin. *Dev. Biol.*, **162**, 451.

91. Delannet, M., Martin, F., Bossy, B., Cheresh, D. A., Reichardt, L. F., and Duband, J. L. (1994) Specific roles of the alpha V beta 1, alpha V beta 3 and alpha V beta 5 integrins in avian neural crest cell adhesion and migration on vitronectin. *Development*, **120**, 2687.

92. Sheppard, A. M., Onken, M. D., Rosen, G. D., Noakes, P. G., and Dean, D. C. (1994) Expanding roles for alpha 4 integrin and its ligands in development. *Cell Adhes. Commun.*, **2**, 27.

93. Stepp, M. A., Urry, L. A., and Hynes, R. O. (1994) Expression of alpha 4 integrin mRNA and protein and fibronectin in the early chicken embryo. *Cell Adhes. Commun.*, **2**, 359.

94. Yang, J. T. and Hynes, R. O. (1996) Fibronectin receptor functions in embryonic cells deficient in alpha 5 beta 1 integrin can be replaced by alpha V integrins. *Mol. Biol. Cell*, **7**, 1737.

95. Tucker, R. P., Hagios, C., Chiquet-Ehrismann, R., Lawler, J., Hall, R. J., and Erickson, C. A. (1999) Thrombospondin-1 and neural crest cell migration. *Dev. Dyn.*, **214**, 312.

96. Pettway, Z., Guillory, G., and Bronner-Fraser, M. (1990) Absence of neural crest cells from the region surrounding implanted notochords in situ. *Dev. Biol.*, **142**, 335.

97. Anderson, D. J. (2000) Genes, lineages and the neural crest: a speculative review. *Phil. Trans. R. Soc. Lond. B*, **355**, 953.

98. Wehrle-Haller, B. and Weston, J. A. (1997) Receptor tyrosine kinase-dependent neural crest migration in response to differentially localized growth factors. *BioEssays*, **19**, 337.

99. Le Douarin, N. M. and Teillet, M. A. (1974) Experimental analysis of the migration and differentiation of neuroblasts of the autonomic nervous system and of neurectodermal mesenchymal derivatives, using a biological cell marking technique. *Dev. Biol.*, **41**, 162.

100. Rothman, T. P., Le Douarin, N. M., Fontaine-Perus, J. C., and Gershon, M. D. (1990) Developmental potential of neural crest-derived cells migrating from segments of developing quail bowel back-grafted into younger chick host embryos. *Development*, **109**, 411.

101. Le Lievre, C. S., Schweizer, G. G., Ziller, C. M., and Le Douarin, N. M. (1980) Restrictions of developmental capabilities in neural crest cell derivatives as tested by *in vivo* transplantation experiments. *Dev. Biol.*, **77**, 362.

102. Dupin, E. (1984) Cell division in the ciliary ganglion of quail embryos *in situ* and after back-transplantation into the neural crest migration pathways of chick embryos. *Dev. Biol.*, **105**, 288.

103. Sieber-Blum, M. and Cohen, A. M. (1980) Clonal analysis of quail neural crest cells: they are pluripotent and differentiate *in vitro* in the absence of noncrest cells. *Dev. Biol.*, **80**, 96.

104. Ito, K. and Sieber-Blum, M. (1991) *In vitro* clonal analysis of quail cardiac neural crest development. *Dev. Biol.*, **148**, 95.

105. Ziller, C., Dupin, E., Brazeau, P., Paulin, D., and Le Douarin, N. M. (1983) Early segregation of a neuronal precursor cell line in the neural crest as revealed by culture in a chemically defined medium. *Cell*, **32**, 627.

106. Baroffio, A., Dupin, E., and Le Douarin, N. M. (1988) Clone-forming ability and differentiation potential of migratory neural crest cells. *Proc. Natl Acad. Sci. USA*, **85**, 5325.

107. Bronner-Fraser, M. and Fraser, S. (1989) Developmental potential of avian trunk neural crest cells *in situ*. *Neuron*, **3**, 755.

108. Bronner-Fraser, M. and Fraser, S. E. (1988) Cell lineage analysis reveals multipotency of some avian neural crest cells. *Nature*, **335**, 161.

109. Coulombe, J. N. and Bronner-Fraser, M. (1986) Cholinergic neurones acquire adrenergic neurotransmitters when transplanted into an embryo. *Nature*, **324**, 569.

110. Stemple, D. L. and Anderson, D. J. (1993) Lineage diversification of the neural crest: *in vitro* investigations. *Dev. Biol.*, **159**, 12.

111. Le Douarin, N. M., Ziller, C., and Couly, G. F. (1993) Patterning of neural crest derivatives in the avian embryo: *in vivo* and *in vitro* studies. *Dev. Biol.*, **159**, 24.

112. Weston, J. A. (1963) A radiographic analysis of the migration and localization of trunk neural crest cells in the chick. *Dev. Biol.*, **6**, 279.

113. Gvirtzman, G., Goldstein, R. S., and Kalcheim, C. (1992) A positive correlation between permissiveness of mesoderm to neural crest migration and early DRG growth. *J. Neurobiol.*, **23**, 205.

114. Spence, M. S., Yip, J., and Erickson, C. A. (1996) The dorsal neural tube organizes the dermamyotome and induces axial myocytes in the avian embryo. *Development*, **122**, 231.

115. Kalcheim, C. and Teillet, M. A. (1989) Consequences of somite manipulation on the pattern of dorsal root ganglion development. *Development*, **106**, 85.

116. Goldstein, R. S., Teillet, M. A., and Kalcheim, C. (1990) The microenvironment created by grafting rostral half-somites is mitogenic for neural crest cells. *Proc. Natl Acad. Sci. USA*, **87**, 4476.

117. Sieber-Blum, M. (1989) Commitment of neural crest cells to the sensory neuron lineage. *Science*, **243**, 1608.

118. Fraser, S. E. and Bronner-Fraser, M. (1991) Migrating neural crest cells in the trunk of the avian embryo are multipotent. *Development*, **112**, 913.

119. Le Lievre, C. S., Schweizer, G. G., Ziller, C. M., and Le Douarin, N. M. (1980) Restrictions of developmental capabilities in neural crest cell derivatives as tested by *in vivo* transplantation experiments. *Dev. Biol.*, **77**, 362.

120. Le Douarin, N. M. (1986) Cell line segregation during peripheral nervous system ontogeny. *Science*, **231**, 1515.

121. Sieber-Blum, M. (1989) SSEA-1 is a specific marker for the spinal sensory neuron lineage in the quail embryo and in neural crest cell cultures. *Dev. Biol.*, **134**, 362.

122. Sharma, K., Korade, Z., and Frank, E. (1995) Late-migrating neuroepithelial cells from the spinal cord differentiate into sensory ganglion cells and melanocytes. *Neuron*, **14**, 143.

123. Henion, P. D., Garner, A. S., Large, T. H., and Weston, J. A. (1995) trkC-mediated NT-3 signaling is required for the early development of a subpopulation of neurogenic neural crest cells. *Dev. Biol.*, **172**, 602.

124. Perez, S. E., Rebelo, S., and Anderson, D. J. (1999) Early specification of sensory neuron fate revealed by expression and function of neurogenins in the chick embryo. *Development*, **126**, 1715.

125. Pinco, O., Carmeli, C., Rosenthal, A., and Kalcheim, C. (1993) Neurotrophin-3 affects proliferation and differentiation of distinct neural crest cells and is present in the early neural tube of avian embryos. *J. Neurobiol.*, **24**, 1626.

126. Sieber-Blum, M. (1998) Growth factor synergism and antagonism in early neural crest development. *Biochem. Cell Biol.*, **76**, 1039.

127. Kalcheim, C. and Le Douarin, N. M. (1986) Requirement of a neural tube signal for the differentiation of neural crest cells into dorsal root ganglia. *Dev. Biol.*, **116**, 451.

128. Kalcheim, C., Barde, Y. A., Thoenen, H., and Le Douarin, N. M. (1987) *In vivo* effect of brain-derived neurotrophic factor on the survival of developing dorsal root ganglion cells. *EMBO J.*, **6**, 2871.

129. Doupe, A. J., Patterson, P. H., and Landis, S. C. (1985) Small intensely fluorescent cells in culture: role of glucocorticoids and growth factors in their development and interconversions with other neural crest derivatives. *J. Neurosci.*, **5**, 2143.

130. Cohen, A. M. (1972) Factors directing the expression of sympathetic nerve traits in cells of neural crest origin. *J. Exp. Zool.*, **179**, 167.

131. Norr, S. C. (1973) *In vitro* analysis of sympathetic neuron differentiation from chick neural crest cells. *Dev. Biol.*, **34**, 16.

132. Zhang, J. M. and Sieber-Blum, M. (1992) Characterization of the norepinephrine uptake system and the role of norepinephrine in the expression of the adrenergic phenotype by quail neural crest cells in clonal culture. *Brain Res.*, **570**, 251.

133. Varley, J. E. and Maxwell, G. D. (1996) BMP-2 and BMP-4, but not BMP-6, increase the number of adrenergic cells which develop in quail trunk neural crest cultures. *Exp. Neurol.*, **140**, 84.

134. Varley, J. E., Wehby, R. G., Rueger, D. C., and Maxwell, G. D. (1995) Number of adrenergic and islet-1 immunoreactive cells is increased in avian trunk neural crest cultures in the presence of human recombinant osteogenic protein-1. *Dev. Dyn.*, **203**, 434.

135. Varley, J. E., McPherson, C. E., Zou, H., Niswander, L., and Maxwell, G. D. (1998) Expression of a constitutively active type I BMP receptor using a retroviral vector promotes the development of adrenergic cells in neural crest cultures. *Dev. Biol.*, **196**, 107.

136. Reissmann, E., Ernsberger, U., Francis-West, P. H., Rueger, D., Brickell, P. M., and Rohrer, H. (1996) Involvement of bone morphogenetic protein-4 and bone morphogenetic protein-7 in the differentiation of the adrenergic phenotype in developing sympathetic neurons. *Development*, **122**, 2079.

137. Rothman, T. P., Gershon, M. D., and Holtzer, H. (1978) The relationship of cell division to the acquisition of adrenergic characteristics by developing sympathetic ganglion cell precursors. *Dev. Biol.*, **65**, 322.

138. Lo, L., Tiveron, M. C., and Anderson, D. J. (1998) MASH1 activates expression of the paired homeodomain transcription factor Phox2a, and couples pan-neuronal and subtype-specific components of autonomic neuronal identity. *Development*, **125**, 609.

139. Stanke, M., Junghans, D., Geissen, M., Goridis, C., Ernsberger, U., and Rohrer, H. (1999) The Phox2 homeodomain proteins are sufficient to promote the development of sympathetic neurons. *Development*, **126**, 4087.

140. Sommer, L., Shah, N., Rao, M., and Anderson, D. J. (1995) The cellular function of MASH1 in autonomic neurogenesis. *Neuron*, **15**, 1245.

141. Britsch, S., Li, L., Kirchhoff, S., Theuring, F., Brinkmann, V., Birchmeier, C., and Riethmacher, D. (1998) The ErbB2 and ErbB3 receptors and their ligand, neuregulin-1, are essential for development of the sympathetic nervous system. *Genes Dev.*, **12**, 1825.

142. Young, H. M., Hearn, C. J., Ciampoli, D., Southwell, B. R., Brunet, J. F., and Newgreen, D. F. (1998) A single rostrocaudal colonization of the rodent intestine by enteric neuron precursors is revealed by the expression of Phox2b, Ret, and p75 and by explants grown under the kidney capsule or in organ culture. *Dev. Biol.*, **202**, 67.

143. Pisano, J. M. and Birren, S. J. (1999) Restriction of developmental potential during divergence of the enteric and sympathetic neuronal lineages. *Development*, **126**, 2855.

144. Natarajan, D., Grigoriou, M., Marcos-Gutierrez, C. V., Atkins, C., and Pachnis, V. (1999) Multipotential progenitors of the mammalian enteric nervous system capable of colonising aganglionic bowel in organ culture. *Development*, **126**, 157.

145. White, P. M. and Anderson, D. J. (1999) *In vivo* transplantation of mammalian neural crest cells into chick hosts reveals a new autonomic sublineage restriction. *Development*, **126**, 4351.

146. Taraviras, S. and Pachnis, V. (1999) Development of the mammalian enteric nervous system. *Curr. Opin. Genet. Dev.*, **9**, 321.

147. Erickson, C. A. and Goins, T. L. (2000) Sacral neural crest cell migration to the gut is dependent upon the migratory environment and not cell-autonomous migratory properties. *Dev. Biol.*, **217**,

148. Rothman, T. P., Le Douarin, N. M., Fontaine-Perus, J. C., and Gershon, M. D. (1993) Colonization of the bowel by neural crest-derived cells re-migrating from foregut backtransplanted to vagal or sacral regions of host embryos. *Dev. Dyn.*, **196**, 217.

149. Hotary, K. B. and Robinson, K. R. (1990) Endogenous electrical currents and the resultant voltage gradients in the chick embryo. *Dev. Biol.*, **140**, 149.

150. Durbec, P. L., Larsson-Blomberg, L. B., Schuchardt, A., Costantini, F., and Pachnis, V. (1996) Common origin and developmental dependence on c-ret of subsets of enteric and sympathetic neuroblasts. *Development*, **122**, 349.

151. Ito, K. and Sieber-Blum, M. (1993) Pluripotent and developmentally restricted neural-crest-derived cells in posterior visceral arches. *Dev. Biol.*, **156**, 191.

152. Baker, C. V., Bronner-Fraser, M., Le Douarin, N. M., and Teillet, M. A. (1997) Early- and late-migrating cranial neural crest cell populations have equivalent developmental potential *in vivo*. *Development*, **124**, 3077.

153. Kirby, M. L. (1989) Plasticity and predetermination of mesencephalic and trunk neural crest transplanted into the region of the cardiac neural crest. *Dev. Biol.*, **134**, 402.

154. Le Lievre, C. S. and Le Douarin, N. M. (1975) Mesenchymal derivatives of the neural crest: analysis of chimaeric quail and chick embryos. *J. Embryol. Exp. Morphol.*, **34**, 125.

155. Nakamura, H. and Ayer-Le Lievre, C. S. (1982) Mesectodermal capabilities of the trunk neural crest of birds. *J. Embryol. Exp. Morphol.*, **70**, 1.
156. Noden, D. M. (1983) The role of the neural crest in patterning of avian cranial skeletal, connective, and muscle tissues. *Dev. Biol.*, **96**, 144.
157. Noden, D. M. (1978) The control of avian cephalic neural crest cytodifferentiation. I. Skeletal and connective tissues. *Dev. Biol.*, **67**, 296.

9 | Insights into the molecular basis of vertebrate forelimb and hindlimb identity

YASUHIKO KAWAKAMI, TOHRU TSUKUI, JENNIFER K. NG, and JUAN CARLOS IZPISÚA-BELMONTE

1. Introduction

Whilst recently we have gained insights into the molecular and cellular mechanisms that initiate the outgrowth and patterning of the vertebrate limb (reviewed in refs 1–5), very little is known about the mechanisms that determine differences between forelimbs and hindlimbs. In all vertebrates the hindlimb is morphologically distinct from the forelimb, yet both are composed of the same components such as nerves, muscles, blood vessels, and bone. These distinctions are even more striking among species. Compare, for instance, the forelimb and hindlimb of various vertebrates, such as the human arm and leg, the chick wing and leg, and the fish pectoral and pelvic fins. The recent identification of genes specifically expressed in the developing forelimb (*Tbx5*) and hindlimb (*Pitx1* and *Tbx4*) in vertebrates, including human, mouse, chick, *Xenopus laevis*, newt, and zebrafish, has begun to aid our understanding of how limb identity is determined. Furthermore, gain- and loss-of-function studies have unveiled how these molecular cues function in the limb field to mediate the development of distinct structures. This chapter will review knowledge about the morphological and molecular basis of forelimb and hindlimb identity in vertebrates, and discuss the roles of *Tbx4*, *Tbx5*, and *Pitx1* in limb-type specification.

2. Morphological differences

The vertebrate limb arises from small buds, called 'limb buds', located at specific axial levels along the main body axis. As the limb bud grows distally, the mesenchymal cells differentiate into tissues, including cartilage and tendon, in a proximal to distal order. The cartilage elements are initially formed as a chondrogenic condensation in the core region of the limb. This initial condensation grows, bifurcates, and finally results in a stereotypic cartilage pattern (Fig. 1).

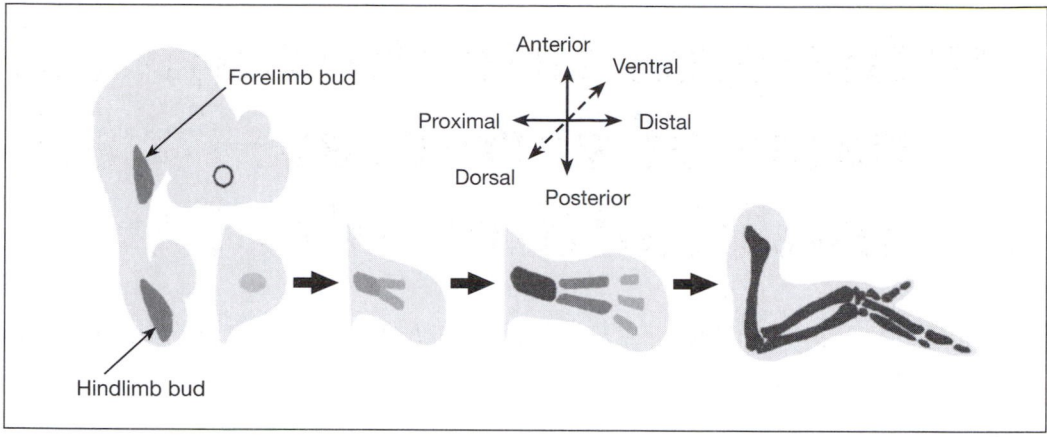

Fig. 1 Schematic representation of limb morphogenesis in the chick. Limb development and chondrogenic pattern formation in the chick forelimb is represented schematically. The darkened structures denote chondrogenesis.

The adult limb is divided into three regions based on the skeletal structure along the proximodistal axis: the stylopod that becomes the upper arm and leg, the zeugopod that becomes the lower arm and leg, and the autopod that becomes the hand and foot (Fig. 2). All the cartilage elements, except the carpals and tarsals in the autopod, grow along the proximodistal axis to form long bones with a characteristic three-layer structure: the quiescent, proliferative, and hypertrophic regions. Carpal and tarsal cartilage elements grow randomly to form short bones. Figure 2 illustrates the elements of limb cartilage and their nomenclature. Each element has a specific shape, position, and length. In general, many of the bones in the hindlimb are longer and thicker than those in the forelimb.

The chick wing has three digits, numbered II, III, IV, while the leg has four digits, numbered I, II, III, IV (Figs 3(A), (B)). In the wing, digit II is the shortest and digit III the longest. In the chick leg, digits II, III, IV are approximately equal in length, while the anterior digit I is the smallest. There are claws at the distal tip of the leg phalanges that do not exist in the wing phalanges. In the zeugopod region, the wing has two cartilage elements—the radius in the anterior and the ulna in the posterior. These are similar in size, whereas, in the hindlimb, the anterior cartilage tibia is thicker and longer than the fibula. At the knee, the leg has the patella that does not exist in the wing. In the mouse (Figs 3(C), (D)), there are no clear differences between the digits. However, mice have several forelimb- or hindlimb-specific structures like the pisiform, a characteristic element in the mammalian forelimb carpals, absent in the hindlimb. In contrast, the hindlimb has the calcareous at the heel, and the patella and the fabella at the knee. The differences between the zeugopod elements are the same as in the chick.

In the chick forelimb, the humerus articulates with both the radius and ulna, whereas in the hindlimb, the femur articulates only with the tibia (Figs 3(A), (B)). In the wrist of the wing, the autopod is posteriorly flexed relative to the zeugopod, and

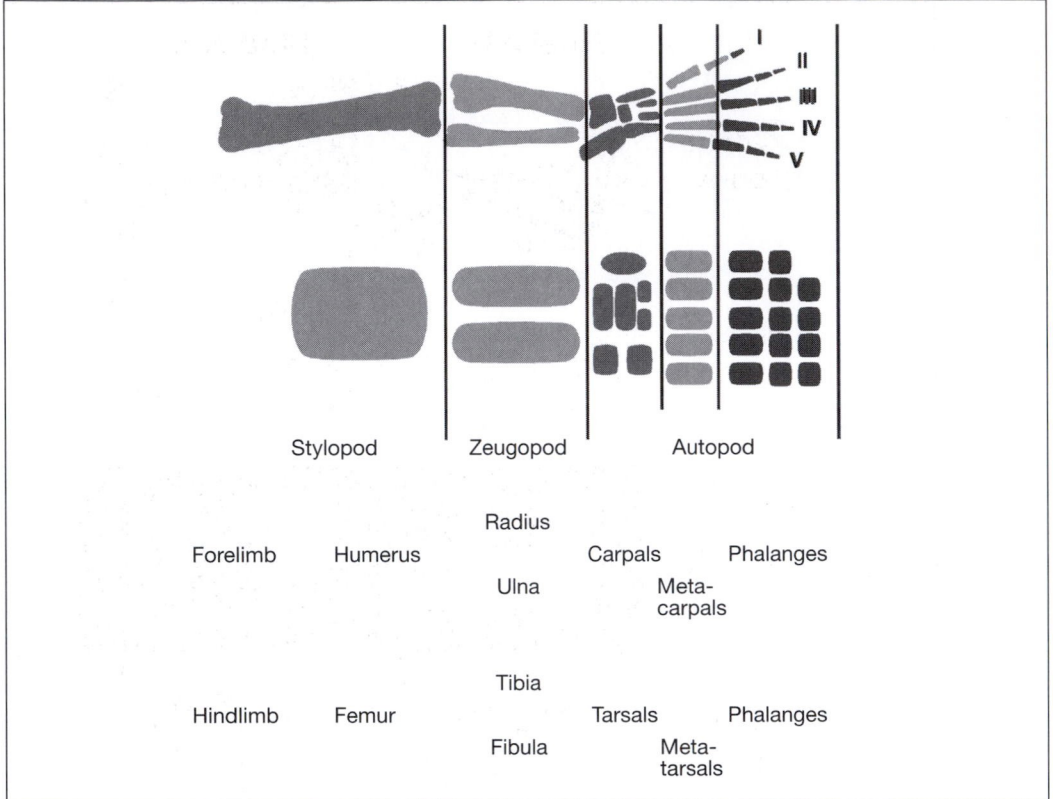

Fig. 2 Schematic drawing of limb cartilage elements and their nomenclature. Scheme and illustration of mouse forelimb cartilage pattern. The name of each cartilage element is shown.

the digits remain in the same plane in an anterior to posterior order. In contrast, in the hindlimb, there is no equivalent posterior flexure at the ankle, thus, the distal elements of the leg remain in a straight orientation. The autopod is rotated 90 degrees such that the digits are in a horizontal position, with digit I medial and the more posterior digits more lateral in position. In the mouse (Figs 3(C), (D)), the differences in the articulation between the elbow and knee are the same as in the chick. However, there are no clear differences between the wrist and the ankle.

The chick limbs exhibit a quite different ectodermal morphology (Figs 3(E), (F)). The wing is covered with feathers, whereas the distal leg displays scales. The wing forms a flap of skin, the patagium, extending between the flexed autopod and the zeugopod. This does not exist in the leg. In the chick leg, interdigital cell death results in the separation of each of the digits, whereas the wing digits III and IV remain joined surrounded by soft tissue. In contrast, the ectodermal structures and digit separation in the mouse are the same in the forelimb and hindlimb.

Limb muscle is formed after the undifferentiated myoblasts migrate into the limb bud from the somites. The myoblasts initially form dorsal and ventral muscle masses

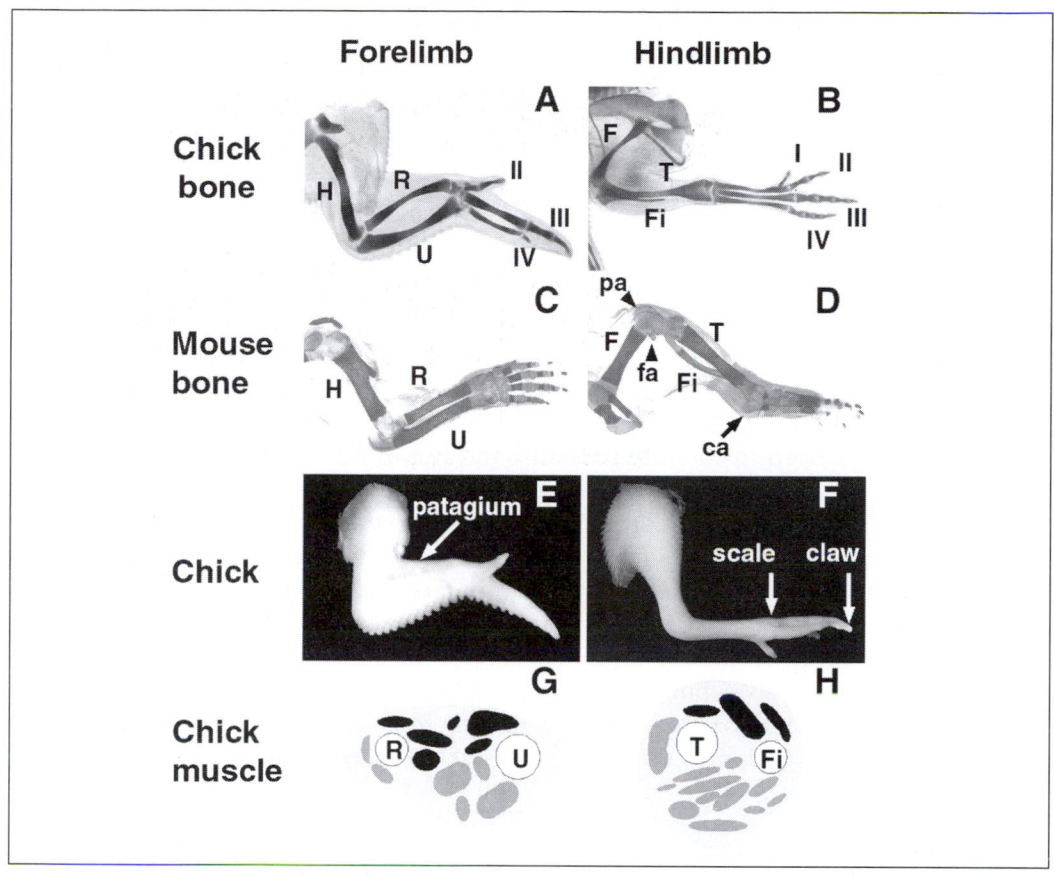

Fig. 3 Differences between the forelimb and hindlimb in the chick and mouse. Skeletal pattern of the forelimb (A) and hindlimb (B) of a 10-day-old chick embryo, and that of the forelimb (C) and hindlimb (D) of an 18.5-day postcoitus mouse embryo. Whole-mount view of the forelimb (E) and hindlimb (F) of 10-day-old chick embryo. Schematic drawing of the muscle pattern in a transverse section of the zeugopod of a 10-day chick forelimb (G) and an 8-day-old chick hindlimb (H). Muscle masses derived from the dorsal muscle mass are indicated in black, and muscle masses derived from the ventral muscle mass are shown in grey. Abbreviations for cartilage/bone elements: ca, calcareous; fa, fabella; F, femur; Fi, fibula; H, humerus; pa, patella; R, radius; T tibia; U, ulna. The numbers in panels (A) and (B) indicate each digit. (Panels (C) and (D) are modified with permission from ref. 62; panels (G) and (H) are adapted with permission from refs 63 and 64, respectively.)

that undergo sequential divisions to form the final muscle masses. The muscle pattern is thought to be determined by the connective tissue to which they migrate. Also, it is believed that the muscle pattern can be formed by the same mechanisms that determine the cartilage pattern. In addition, recent studies in the chick hindlimb have indicated the importance of reciprocal interactions between muscle and tendon for specifying and coordinating muscle and tendon patterning (6, 7). Muscles on the long bones are used for movements such as flexion, extension, abduction, adduction, pronation, and supination. The chick forelimb has flexor, extensor, pronator, and spinator muscles on the long bones, whereas the hindlimb has flexor, extensor, abductor, and adductor muscles on the long bones. Each muscle is defined by its unique shape, position, fibre orientation, origin, and insertion. As an example, the transverse sections of the chick forelimb and hindlimb in Figs 3(G) and (H) schematically show the location of muscle masses.

3. Molecular differences

If a gene is involved in the specification of forelimb or hindlimb identity leading to the formation of the differences described above, its expression pattern is likely to be restricted to either the forelimb or hindlimb buds. It is also likely that the restricted expression pattern would start before limb bud outgrowth and be maintained during limb development. Recently, three genes have been reported that fit these criteria in the chick and mouse. The two T-box transcription factors, *Tbx5* and *Tbx4*, are expressed in the forelimb bud and hindlimb bud, respectively (8–13), while the *bicoid*-related homeobox transcription factor, *Pitx1*, is expressed in the hindlimb bud (13, 14). Although other genes such as those of the *Hoxc* cluster are differentially expressed in the forelimb or hindlimb (15), none of them are expressed throughout the limb mesenchyme. In addition, their relatively late onset of expression indicates that they may not play an instructive role in the early stage of limb identity.

Tbx genes contain a highly conserved motif at the N-terminus that encodes a 180 amino-acid DNA-binding domain, called the T-domain (for review see ref. 16). The C-terminal half of the protein contains a transactivation/repression domain (17, 18). The nuclear localization signal is located between these two domains. X-ray structure analysis has revealed that the T-domain forms a dimer which binds to its target sequence (19). *Tbx* genes have been identified in a variety of vertebrates and invertebrates including human, mouse, chick, *Xenopus laevis*, zebrafish, *Caenorhabditis elegans*, and *Drosophila melanogaster*.

Pitx1, also known as *BACKFOOT* in humans (20), was first identified as a factor capable of interacting with the N-terminal activation domain of Pit-1, a pituitary-specific transcription factor (21). It also binds to a *cis*-acting element of a pituitary hormone gene promoter (22). This gene has also been identified in the human, mouse, chick, *Xenopus*, and *Drosophila* (23–25).

Tbx5, *Tbx4*, and *Pitx1* are expressed in the lateral plate mesoderm (LPM) prior to the outgrowth of the limb bud in the chick and mouse, at the time when forelimb and hindlimb territories are being specified (Fig. 4; and see ref. 26). However, their onsets of expression are not the same—*Tbx5* expression starts slightly earlier than that of *Tbx4* (8, 10). The onset of *Pitx1* expression is earlier than that of *Tbx4* (13). In the developing limb bud, the expression patterns of these genes are highly limb-type specific—*Tbx5* expression is observed throughout the mesenchyme of the forelimb bud, whereas *Tbx4* and *Pitx1* are expressed in the entire hindlimb mesenchyme. However, these genes are downregulated before adulthood in the chick and mouse.

An interesting feature of *Tbx* expression is that these genes are also expressed in regenerating newt limbs in a forelimb- or hindlimb-specific manner (27). In the *Xenopus* and newt limbs, these genes continue to be expressed at low levels in the

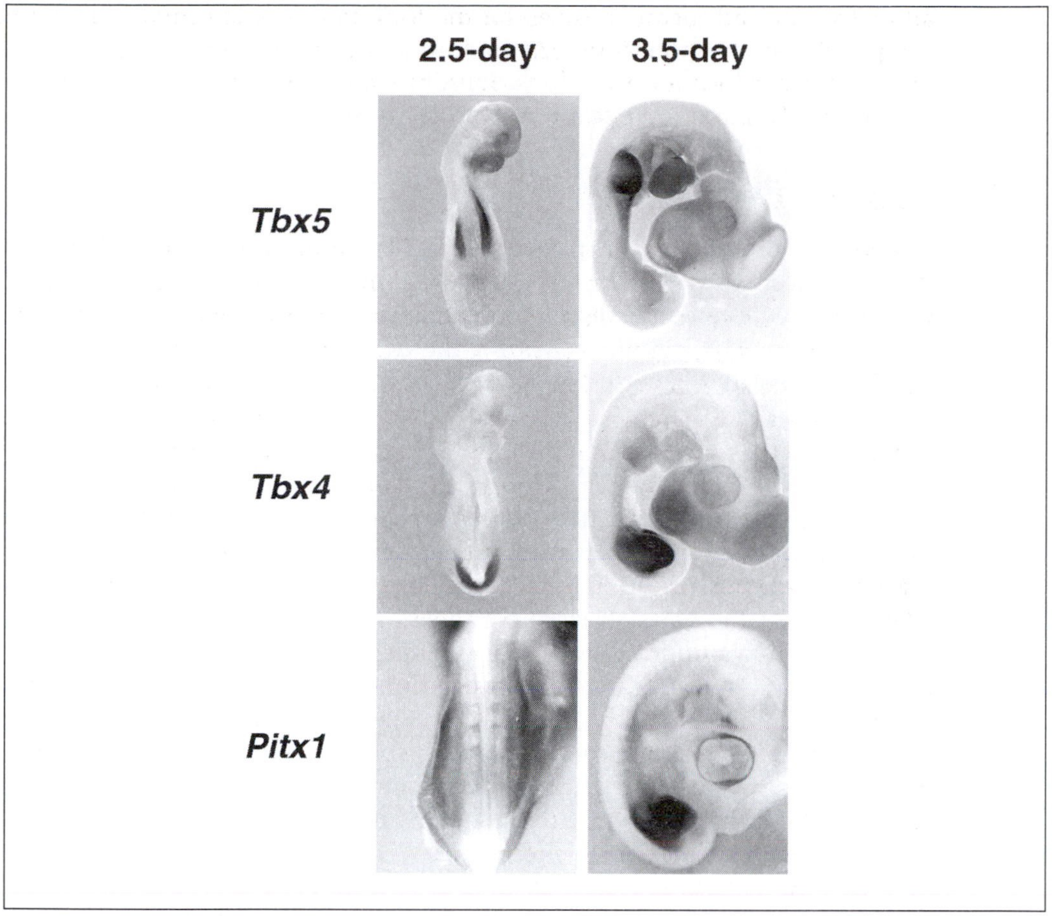

2.5-day   **3.5-day**

Tbx5

Tbx4

Pitx1

Fig. 4 Expression pattern of *Tbx5*, *Tbx4*, and *Pitx1* during chick limb development. Expression of *Tbx5*, *Tbx4*, and *Pitx1* is visualized by whole-mount *in situ* hybridization. *Tbx5* is expressed in the prospective forelimb region before limb bud outgrowth, and in the entire forelimb bud during limb development. *Tbx4* and *Pitx1* are expressed in the prospective hindlimb region prior to limb outgrowth, and in the entire hindlimb bud during limb development. All these genes are expressed in the mesenchyme, but not in the ectoderm.

adult, which is not observed in other species (27, 28). These observations suggest that *Tbx* genes might be involved not only in the specification of limb identity, but that they may also play a role during regeneration and adult limb function in these species.

Tbx genes are also present in zebrafish, which have two types of fins: the pectoral and pelvic. Restricted *Tbx5* expression is observed throughout the mesenchyme of the pectoral fin buds, but is never detected in the pelvic fins. *Tbx4* is expressed throughout the pelvic fins, but is never detected in the pectoral fin buds (29–31).

Phylogenetic analysis and chromosomal mapping of mouse and human *Tbx* genes have led to a model for the evolution of the *Tbx2* subfamily genes: *Tbx2*, *Tbx3*, *Tbx4*, and *Tbx5* (32). Agulnik and colleagues have proposed that a single primordial gene for the subfamily underwent a tandem duplication event that produced *Tbx2/3* and

Tbx4/5 progenitor genes. Whole-cluster duplication occurred to generate *Tbx2* and *Tbx4* on the same chromosome (human chromosome 11), and *Tbx3* and *Tbx5* on the same chromosome (human chromosome 5). Analyses of T-box genes in zebrafish and amphioxus suggest that the duplication of such a primordial gene might have occurred before the separation of the vertebrate and invertebrate lineages, and that the cluster duplication might have occurred after the separation of vertebrates from cephalochordates (31, 33). This is also supported by the conserved expressions of *Tbx5* in the forelimb counterpart, and *Tbx4* in the hindlimb counterpart in zebrafish, which indicate that the cluster duplication giving rise to *Tbx4* and *Tbx5* occurred before the developmental duplication of fish fins. Thus, it appears that *Tbx5* and *Tbx4* are evolutionarily selected genes for the differential specification of forelimb and hindlimb identities in vertebrates.

4. Classical tissue-graft experiments updated

Classical tissue-transplantation studies in the chick suggest the presence of forelimb- or hindlimb-specific genes. It was shown that wing tissue grafted to a hindlimb bud generates wing-like digits in the leg, and leg tissue grafted to the wing bud generates leg-like digits (34, 35). These experiments indicate that the limb-field mesoderm contains intrinsic information about the decision to develop into forelimb or hindlimb, and, once specified, the limb tissue cannot alter its identity. More recently, three groups have independently updated the experiments by using *Tbx5*, *Tbx4*, and *Pitx1* as marker genes to explain limb-identity specification (11–13). They examined the expression of these genes after grafting donor wing-bud cells to a host leg bud, or grafting donor leg cells to a host wing bud. The original gene expression was retained after the operation, indicating that the grafted cells contain the original identity and will form distal structures accordingly.

Analysis of ectopic limbs induced in the flank of chicks has also been used to understand the correlation of these genes with limb type identity (10–13). It has been shown that members of the fibroblast growth factor (FGF) family can induce an extra limb in the flank of the chick embryo when ectopically applied between the presumptive forelimb and hindlimb regions (36, 37). The limb type of the extra limb depends on the position of FGF application (36). An FGF-soaked bead placed near the wing induces a wing-like limb, whereas an FGF-soaked bead placed near the leg bud induces a leg-like limb. When the bead is placed in an intermediate region between the limb buds, it induces a mosaic limb, exhibiting both wing- and leg-like features. All the extra limb buds induced by FGF express both *Tbx5* and *Tbx4*. A wing-like extra limb bud is predominantly *Tbx5*-positive in the anterior of the bud, and *Tbx4*-positive in a small region in the posterior. An extra leg-like limb bud has a substantial *Tbx4*-positive domain in the posterior, and a *Tbx5*-positive small domain in the anterior of the bud. In the mosaic extra limb, the anterior half of the bud expresses *Tbx5*, and the posterior half of the bud expresses *Tbx4*. The expression boundaries of both genes are not fixed at a certain axial level, but vary depending on limb position. Therefore, mosaic extra limb buds maintain various ratios of *Tbx5/*

Tbx4 expression. It is likely that the ratio of *Tbx5/Tbx4* expression reflects the limb-type morphology of the ectopic limb.

The mesenchymal expression patterns of *Tbx5/Tbx4* closely correlate to the limb-specific epithelial pattern, even though these genes are not expressed in the ecto-derm. This suggests that these genes regulate secreted factors that affect epithelial patterning. This idea is supported by the fact that overlapping expression domains of both *Tbx5* and *Tbx4* sometimes exhibit characteristics of both the wing and leg (feather bud and claw).

It is likely that the initiation of limb outgrowth and the specification of the limb bud are tightly related events. The expression patterns of *Fgf8* in the intermediate mesoderm and *Fgf10* in the LPM are confined to the limb-forming level before the onset of limb outgrowth (38–40). This occurs almost at the same time as *Tbx4* and *Tbx5* expression. Further support of a relationship between *Tbx* and *Fgf* genes comes from studies in *Xenopus* embryos where *Xbra*, another T-box gene, and FGF are involved in an indirect autoregulatory loop (41, 42) (see also Chapter 3). Recently, it was demonstrated in the chick that *Wnt-2b* and *Wnt-8c*, members of the *Wnt* gene family encoding secreted proteins, are transiently expressed in the LPM at the forelimb and hindlimb levels, respectively, before limb budding. Both *Wnt-2b* and *Wnt-8c* are able to induce limb initiation through control of *Fgf10* in the LPM (43). *Wnt-2b* and *Wnt-8c* seem to be regulated by axial signals, most likely by FGFs from the somites and/or intermediate mesoderm. The fact that *Tbx4* and *Tbx5* are induced in the LPM of *Fgf10$^{-/-}$* mice (44) indicates that FGF10 alone is not sufficient to initiate *Tbx5* and *Tbx4*. However, the differentially expressed *Wnt* genes alone or in combination with FGFs may be involved in determination of limb specificity.

The experiments mentioned above suggest that limb identity seems to be controlled by the axial level at which the limbs are formed. One putative family of factors that could be involved in this process is that of the *Hox* genes. *Hox* genes are involved in patterning the vertebrate body along the anteroposterior axis (reviewed in ref. 45) (see also Chapter 4), and hence they might also contribute to the determination of limb identity. In this respect it has been shown that the expression of a certain set of *Hox9* paralogous genes in the LPM correlates with the specification of forelimb and hindlimb identity (46). According to this view, *Hox* genes could act upstream of *Tbx5*, *Tbx4*, and *Pitx1*. Several gain- and loss-of-function experiments have indeed suggested that a specific Hox code could contribute to the allocation of the limb fields (see, for instance, ref. 46 and references therein). However, most likely, regulators of *Hox* gene activity expression will also be involved (47, 48 and references therein).

At this stage, further studies are necessary to evaluate the critical factors implicated in the allocation of the limb fields around specific axial levels.

5. Roles of *Pitx1*, *Tbx4*, and *Tbx5* in limb identity

Recently, several groups have further elucidated the roles of the genes mentioned above by conducting gain- and/or loss-of-function experiments in chick and mouse models (49–53). Alteration of the normal levels of the *Pitx1*, *Tbx4*, and *Tbx5* genes during embryogenesis generated partial transformations of either the forelimb to hindlimb or vice versa, and also suggested that limb outgrowth and identity are tightly linked. What other molecules interact with these genes to confer limb identity, and how do they in turn regulate limb patterning? In this section we address these questions by discussing the molecular interactions of *Pitx1*, *Tbx4*, and *Tbx5*.

One candidate molecule for conferring hindlimb identity is *Pitx1*. As described in Section 4, *Pitx1* is expressed in the mouse and chick hindlimb. At later stages, *Pitx1* is expressed at high levels in the dorsal and distal extremes of the limb bud and at lower levels ventrally, unlike *Tbx4*, which is expressed throughout the mesenchyme (13). Misexpression of *Pitx1* in the presumptive wing field of chick embryos by using a replication-competent variant of the avian leukaemia virus (RCAS) caused several morphological changes. The perturbed outgrowth of the wing affected autopodal elements, including a reduced or absent patagium (49, 50). The *Pitx1-RCAS* virus-infected wings adopted some leg-like structures: four digits (extra digit looked like digit I of the leg), muscle pattern alterations, changes of the degree of rotation of the flexure at the wrist, and conversion of scales to feathers (49, 50). Upon molecular analyses, it was determined that *Pitx1* induced the ectopic expression of *Tbx4*, *Hoxc10*, and *Hoxc11*, which are hindlimb markers. *Tbx5* expression remained unaltered.

The complementary experiment in mice (deletion of the *Pitx1* gene by homologous recombination) resulted in the generation of forelimb-like compositions of the ulna and radius in the mouse hindlimb structures (Fig. 5) (49, 51). More distally, the mutant mice displayed a pisiform-like element, which normally is only present in the forelimb. In *Pitx1*$^{-/-}$ mice, *Tbx4* transcripts were reduced while *Tbx5* transcripts were unaffected as determined by *in situ* hybridization (49, 51). There were no transcriptional changes in several genes known to be important for limb patterning, for example, *Wnt-5a*, *Fgf8*, *Bmp-4*, *goosecoid*, *Hoxd10*, *Hoxd11*, and *Hoxd13* (49). The

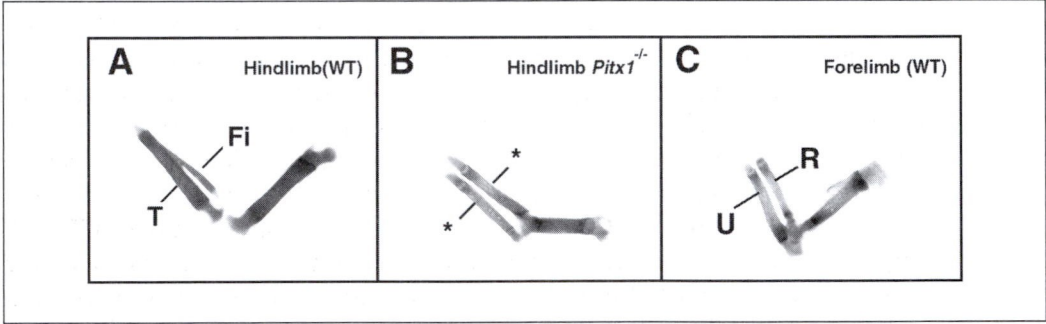

Fig. 5 Comparison of the zeugopodal elements of the hindlimb and forelimb in wild type and *Pitx1*$^{-/-}$ mice. Bones were prepared from postnatal-day 0 mice. (A) Hindlimb skeletal structures in the wild-type (WT) mouse. Tibia (T) and fibula (Fi) bones are indicated. (B) Hindlimb skeletal structures of *Pitx1*-null (*Pitx1*$^{-/-}$) mice. There are alterations in the size of the femur, and the tibia and fibula appear morphologically similar to the radius and ulna (asterisks). (C) Forelimb skeletal structures in *Pitx1*$^{-/-}$ mice. The ulna (U) and radius (R) are shown. Morphological structures of the forelimb appear normal in the null mouse. (Modified with permission from ref. 49.)

conclusions reached after gain- and loss-of-function experiments are that *Pitx1* is necessary and sufficient to induce as well as maintain a subset of hindlimb cell fates.

Another candidate gene for conferring hindlimb identity is *Tbx4*. Ectopic expression of *Tbx4* in the presumptive wing field of the chick embryo causes arrested limb outgrowth. *Tbx4-RCAS* virus-infected embryos display either truncations or disappearances of the zeugopodal and autopodal elements (52). Overexpression of *Tbx4* using electroporation technology *in ovo* occasionally generated leg-like structures, such as four distinct digits and the appearance of a claw-like structure (53). These phenotypes were similar to the ones obtained after *Pitx1* misexpression in the chick embryo (49, 50). While misexpression of *Tbx4* in the wing region cannot give rise to a complete transformation of a wing to a leg, nonetheless, these results indicate that *Tbx4* has a role in promoting leg development (Fig. 6). Indeed, when *Tbx4* was misexpressed in the presumptive wing at early stages of chick development, and, later on, when the *Tbx4-RCAS* infected wings were transplanted to the coelom of a host embryo, ectopic limbs that resembled legs were obtained.

Misexpression of both *Tbx5* and *Tbx4* also appears to affect cell proliferation, since limbs ectopically expressing *Tbx5* or *Tbx4* display reductions and truncations of the zeugopodal and autopodal elements. The details of how TBX proteins may simultaneously affect different aspects of cell proliferation and limb identity is still unknown. Several studies have led to the hypothesis that the phenotypes generated by *Tbx5* and *Tbx4* misexpression might be caused by the formation of heterodimers between exogenous and endogenous TBX proteins.

Tbx3 transcripts partially overlap with *Tbx5* and *Tbx4* transcripts in the anterior and posterior portions of the forelimb and hindlimb, respectively, in chick (11, 12) and mice (8). Analysis of other *Tbx* genes seems to support the idea of heterodimer formation. *Brachyury* and *Tbx6* have partially overlapping patterns of expression, *Tbx6* transcripts are downregulated in *Brachyury* mutant-mouse embryos (54), and ectopic *Tbx5* expression downregulates *Tbx4* transcripts (52). Furthermore, crystallographic experiments have revealed that T-box domains bind to DNA as dimers (19).

Mutations of *TBX3* in humans result in forelimb abnormalities of the posterior elements (55). Human *TBX5* mutations result in the anterior forelimb abnormalities (56, 57). In a subset of patients, besides limb abnormalities, congenital cardiac defects were also observed. The defects appear to correlate with specific mutations (56, 57). Either of the missense mutations, Arg237Gln or Arg237Trp, causes extensive forelimb malformations and less significant cardiac abnormalities. In contrast, the Gly80Arg mutation causes significant cardiac malformations (57). Structural analyses of the TBX5 protein have shown that residue 237 is located in the T-box domain that selectively binds to the minor groove of DNA. On the other hand, residue 80 is highly conserved within the T-box sequences that interact with the major groove of target DNA (57). These data imply that the TBX5 protein has different target DNA sequences and can interact with the major and minor grooves. Hence, the differences in TBX5 interactions with its targets might account for its distinct roles in forelimb patterning and heart organogenesis. Whilst definitive experimental proof is still lacking, all these results suggest that dimer formation and

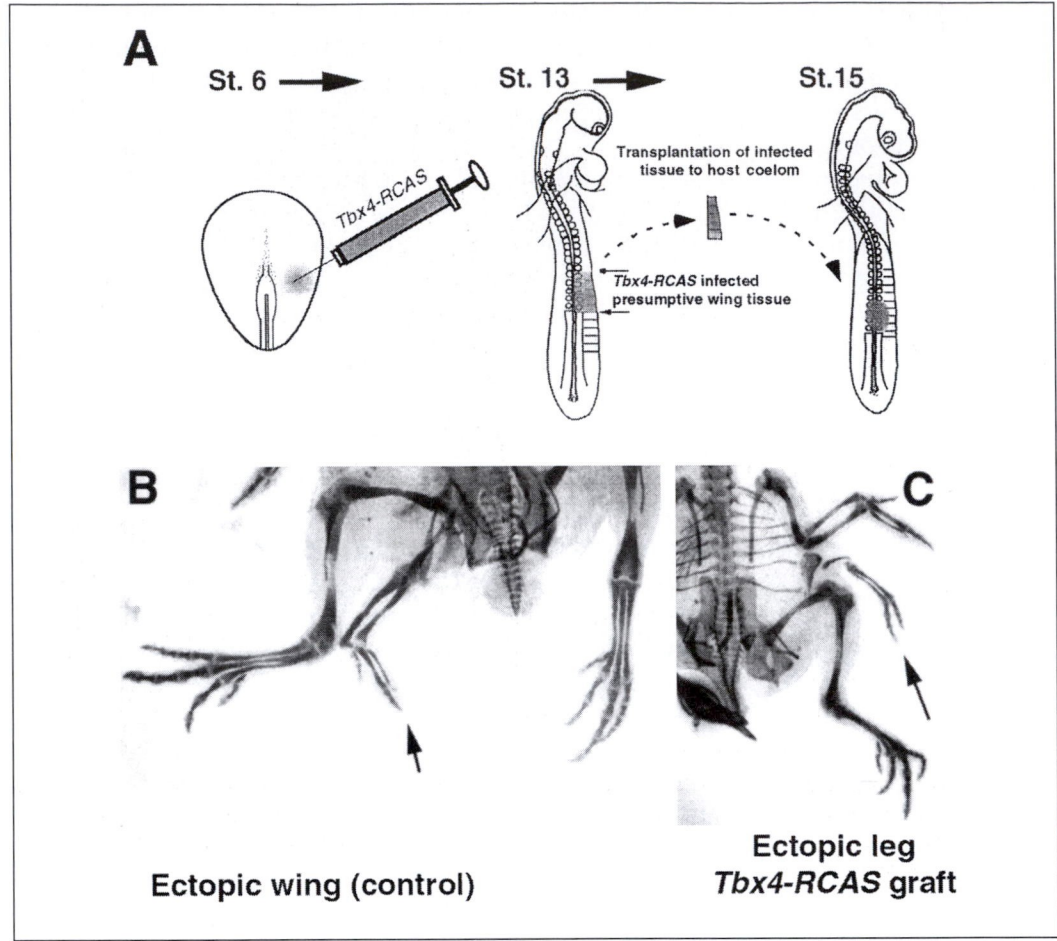

Fig. 6 *Tbx4* promotes 'legness'. (A) Scheme of experimental procedures. Grafting of early stage (Hamburger and Hamilton stage 13 (65)#) presumptive wing cells with or without *Tbx4-RCAS* infection, to the coelom of host embryos (Hamburger and Hamilton stage 15–16). (B) Most of the experimental embryos without infection generated an ectopic wing. Arrow indicates ectopic wing resulting from wing-field cell grafts. (C) Wing-field cells infected with *Tbx4-RCAS* virus prior to grafting generated an ectopic leg-like limb. Arrow indicates ectopic leg

target DNA specificity of TBX proteins might account for the appearance of specific forelimb and hindlimb structures.

6. Conclusions

Morphological studies suggest that all tetrapods evolved from a common ancestor sharing a common skeletal structure, which includes the stylopod composed of a single bone, the zeugopod composed of one or two bones, and the autopod composed of a variable number of digits (Fig. 7). How have all these divergent, but common, structures been generated? Insights into this question could be obtained by

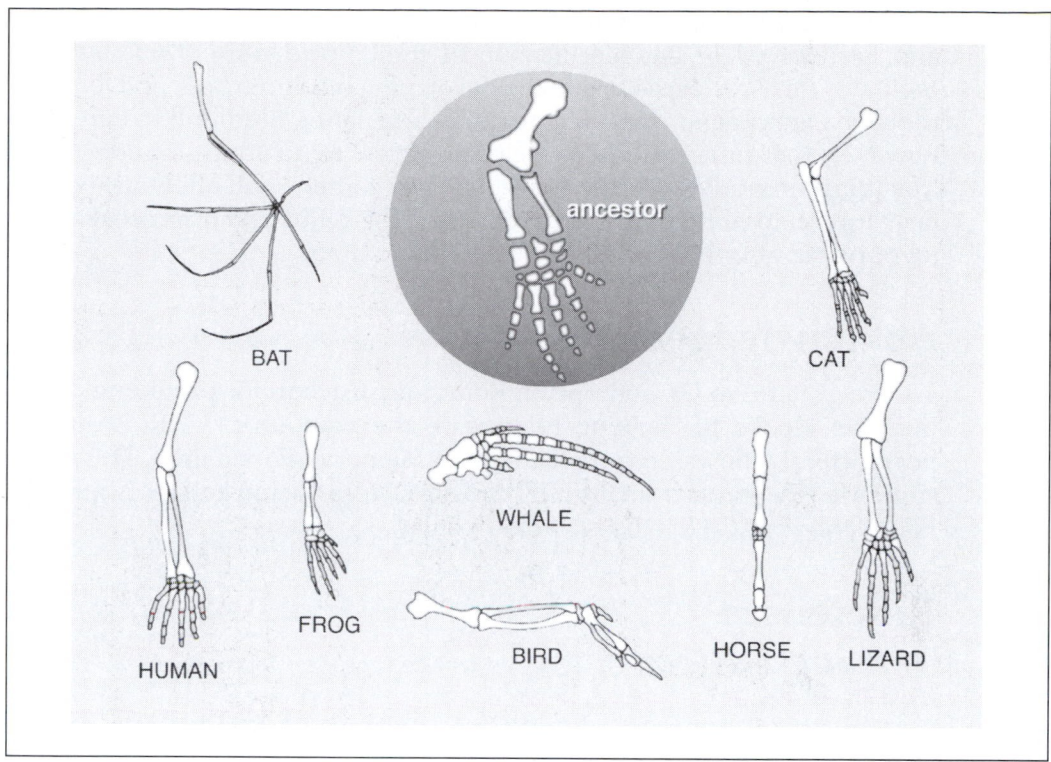

Fig. 7 Skeletal patterns of various vertebrate forelimbs and their prospective common ancestor. Illustration of the appendage bone structure of various vertebrates. They all possess similar stylopodal, zeugopodal, and autopodal elements. It has been postulated that all tetrapod vertebrates evolved from a common ancestor. (Modified with permission from ref. 66.)

assuming the existence of selector genes, acting very early in development, that might regulate the expression of target genes. These targets genes, subsequently, will generate the morphological differences observed throughout vertebrate limb evolution. Since several of the selector genes that control patterning appear to be conserved (58–60), the target genes and/or the regulatory elements might have diverged. At present, little is known about the target genes for *Tbx5*, *Tbx4*, and *Pitx1*. Yokouchi and colleagues (61) have suggested that cell adhesiveness could be involved in the determination of cartilage morphology, and have indicated that members of the *Hox* gene cluster could be involved in this process by regulating cell-surface molecules. This idea is supported, in part, by research in arthropod appendages, where differences in wing morphology between *Drosophila* and butterflies are partly due to the divergence of *Ubx*-regulated genes (60). Recent experiments in the chick have also suggested that secreted factors affecting epithelial morphology may also act downstream of the limb-identity selector genes (10, 11).

The identification of three transcription factors, *Tbx5*, *Tbx4*, and *Pitx1*, has allowed us to start elucidating some of the molecular mechanisms that determine vertebrate forelimb and hindlimb identity. *Tbx5* is expressed in the developing forelimb bud and is involved in the specification of forelimb identity. *Tbx4* and its upstream regulator *Pitx1* are expressed in the developing hindlimb bud, and loss- and gain-of-function experiments implicate processes regulating hindlimb identity. It is evident, however, that other unknown 'selector genes' regulating limb identity may exist. Identification of these genes, as well as their targets, will allow us, hopefully in the near future, to not only better understand the cellular and molecular basis of limb identity, but also the evolution of the tetrapod limb.

Acknowledgements

We are grateful to Dr Concepción Rodriguez-Esteban for providing pictures and to Lorraine Hooks for helping to prepare the manuscript. Y.K is supported by a postdoctoral fellowship from the Uehara Memorial Foundation. This work has been supported by grants from the NIH and the G. Harold and Leila Y. Mathers Charitable Foundation to J.C.I.B. who is a PEW Scholar.

References

1. Tickle, C. and Eichele, G. (1994) Vertebrate limb development. *Annu. Rev. Cell Biol.*, **10**, 121.
2. Johnson, R. L. and Tabin, C. J. (1997) Molecular models for vertebrate limb development. *Cell*, **90**, 979.
3. Schwabe, J. W., Rodriguez-Esteban, C., and Izpisua-Belmonte, J. C. (1998) Limbs are moving: where are they going? *Trends Genet.*, **14**, 229.
4. Ng, J. K., Tamura, K., Büscher, D., and Izpisua-Belmonte, J. C. (1999) Molecular and cellular basis of pattern formation during vertebrate limb development. *Curr. Top. Dev. Biol.*, **41**, 37.
5. Capdelia, J. and Izpisúa-Belmonte, J. C. (2001) Patterning mechanisms controlling vertebrate limb development. *Annu. Rev. Cell Dev. Biol.*, **17**, 87.
6. Kardon, G. (1998) Muscle and tendon morphogenesis in the avian hind limb. *Development*, **125**, 4019.
7. Ros, M. A., Rivero, F. B., Hinchliffe, J. R., and Hurle, J. M. (1995) Immunohistological and ultrastructural study of the developing tendons of the avian foot. *Anat. Embryol.*, **192**, 483.
8. Gibson-Brown, J. J., Agulnik, S. I., Chapman, D. L., Alexiou, M., Garvey, N., Silver, L. M., and Papaioannou, V. E. (1996) Evidence of a role for T-box genes in the evolution of limb morphogenesis and specification of forelimb/hindlimb identity. *Mech. Dev.*, **56**, 93.
9. Gibson-Brown, J. J., Agulnik, S. I., Silver, L. M., and Papaioannou, V. E. (1998) Expression of T-box genes *Tbx2-Tbx5* during chick organogenesis. *Mech. Dev.*, **74**, 165.
10. Ohuchi, H., Takeuchi, J., Yoshioka, H., Ishimaru, Y., Ogura, K., Takahashi, N., Ogura, T., and Noji, S. (1998) Correlation of wing-leg identity in ectopic FGF-induced chimeric limbs with the differential expression of chick *Tbx5* and *Tbx4*. *Development*, **125**, 51.
11. Gibson-Brown, J. J., Agulnik, S. I., Silver, L. M., Niswander, L., and Papaioannou, V. E. (1998) Involvement of T-box genes *Tbx2-Tbx5* in vertebrate limb specification and development. *Development*, **125**, 2499.

12. Isaac, A., Rodriguez-Esteban, C., Ryan, A., Altabef, M., Tsukui, T., Patel, K., Tickle, C., and Izpisua-Belmonte, J. C. (1998) Tbx genes and limb identity in chick embryo development. *Development*, **125**, 1867.

13. Logan, M., Simon, H. G., and Tabin, C. (1998) Differential regulation of T-box and homeobox transcription factors suggests roles in controlling chick limb-type identity. *Development*, **125**, 2825.

14. Lanctot, C., Lamolet, B., and Drouin, J. (1997) The *bicoid*-related homeoprotein *Ptx1* defines the most anterior domain of the embryo and differentiates posterior from anterior lateral mesoderm. *Development*, **124**, 2807.

15. Nelson, C. E., Morgan, B. A., Burke, A. C., Laufer, E., DiMambro, E., Murtaugh, L. C., Gonzales, E., Tessarollo, L., Parada, L. F., and Tabin, C. (1996) Analysis of *Hox* gene expression in the chick limb bud. *Development*, **122**, 1449.

16. Papaioannou, V. E., and Silver, L. M. (1998) The T-box gene family. *BioEssays*, **20**, 9.

17. Kispert, A. (1995) The Brachyury protein: a T-domain transcription factor. *Semin. Dev. Biol.*, **6**, 395.

18. He, M. L., Wen, L., Campbell, C. E., Wu, J. Y., and Rao, Y. (1999) Transcription repression by *Xenopus* ET and its human ortholog TBX3, a gene involved in ulnar-mammary syndrome. *Proc. Natl Acad. Sci. USA*, **96**, 10212.

19. Muller, C. W. and Herrmann, B. G. (1997) Crystallographic structure of the T domain-DNA complex of the *Brachyury* transcription factor. *Nature*, **389**, 884.

20. Shang, J., Li, X., Ring, H. Z., Clayton, D. A., and Francke, U. (1997) Backfoot, a novel homeobox gene, maps to human chromosome 5 (*BFT*) and mouse chromosome 13 (*Bft*). *Genomics*, **40**, 108.

21. Szeto, D. P., Ryan, A. K., O'Connell, S. M., and Rosenfeld, M. G. (1996) P-OTX: a PIT-1-interacting homeodomain factor expressed during anterior pituitary gland development. *Proc. Natl Acad. Sci. USA*, **93**, 7706.

22. Lamonerie, T., Tremblay, J. J., Lanctot, C., Therrien, M., Gauthier, Y., and Drouin, J. (1996) Ptx1, a *bicoid*-related homeo box transcription factor involved in transcription of the pro-opiomelanocortin gene. *Genes Dev.*, **10**, 1284.

23. Gage, P. J., Suh, H., Camper, S. A. (1999) The *bicoid*-related Pitx gene family in development. *Mamm. Genome*, **10**, 197.

24. Hollemann, T. and Pieler, T. (1999) Xpitx-1: a homeobox gene expressed during pituitary and cement gland formation of Xenopus embryos. *Mech. Dev.* **88**, 249.

25. Vorbruggen, G., Constien, R., Zilian, O., Wimmer, E. A., Dowe, G., Taubert, H., Noll, M., and Jackle, H. (1997) Embryonic expression and characterization of a *Ptx1* homolog in *Drosophila*. *Mech. Dev.* **68**, 139.

26. Stephens, T. D., Beier, R. L., Bringhurst, D. C., Hiatt, S. R., Prestridge, M., Pugmire, D. E., and Willis, H. J. (1989) Limbness in the early chick embryo lateral plate. *Dev. Biol.*, **133**, 1.

27. Simon, H. G., Kittappa, R., Khan, P. A., Tsilfidis, C., Liversage, R. A., and Oppenheimer, S. (1997) A novel family of T-box genes in urodele amphibian limb development and regeneration: candidate genes involved in vertebrate forelimb/hindlimb patterning. *Development*, **124**, 1355.

28. Takabatake, Y., Takabatake, T., and Takeshima, K. (2000) Conserved and divergent expression of T-box genes Tbx2-Tbx5 in *Xenopus*. *Mech. Dev.*, **91**, 433.

29. Tamura, K., Yonei-Tamura, S., and Izpisua-Belmonte, J. C. (1999) Differential expression of *Tbx4* and *Tbx5* in zebrafish fin buds. *Mech. Dev.*, **87**, 181.

30. Begemann, G. and Ingham, P. W. (2000) Developmental regulation of Tbx5 in zebrafish embryogenesis. *Mech. Dev.*, **90**, 299.

31. Ruvinsky, I., Oates, A. C., Silver, L. M., and Ho, R. K. (2000) The evolution of paired appendages in vertebrates: T-box genes in the zebrafish. *Dev. Genes Evol.*, **210**, 82.

32. Agulnik, S. I., Garvey, N., Hancock, S., Ruvinsky, I., Chapman, D. L., Agulnik, I., Bollag, R., Papaioannou, V., and Silver, L. M. (1996) Evolution of mouse *T-box* genes by tandem duplication and cluster dispersion. *Genetics*, **144**, 249.

33. Ruvinsky, I., Silver, L. M., and Gibson-Brown, J. J. (2000) Phylogenetic analysis of T-box genes demonstrates the importance of amphioxus for understanding evolution of the vertebrate genome. *Genetics*, **156**, 1249.

34. Saunders, J. W. and Gasseling, M. T. (1959) The differentiation of prospective thigh mesoderm grafted beneath the apical ectodermal ridge of the wing bud in the chick embryo. *Dev. Biol.*, **1**, 281.

35. Summerbell, D. and Tickle, C. (1977) Pattern formation along the anterior-posterior axis of the chick limb bud. In *Vertebrate limb and somite morphogenesis* (ed. D. A. Ede, J. R. Hinckliffe, and M. Balls), p. 41. Cambridge University Press, Cambridge.

36. Cohn, M. J., Izpisua-Belmonte, J. C., Abud, H., Heath, J. K., and Tickle, C. (1995) Fibroblast growth factors induce additional limb development from the flank of chick embryos. *Cell*, **80**, 739.

37. Ohuchi, H., Nakagawa, T., Yamauchi, M., Ohata, T., Yoshioka, H., Kuwana, T., Mima, T., Mikawa, T., Nohno, T., and Noji, S. (1995) An additional limb can be induced from the flank of the chick embryo by FGF4. *Biochem. Biophys. Res. Commun.*, **209**, 809.

38. Crossley, P. H., Minowada, G., MacArthur, C. A., and Martin, G. R. (1996) Roles for FGF8 in the induction, initiation, and maintenance of chick limb development. *Cell*, **84**, 127.

39. Vogel, A., Rodriguez, C., and Izpisua-Belmonte, J. C. (1996) Involvement of FGF-8 in initiation, outgrowth and patterning of the vertebrate limb. *Development*, **122**, 1737.

40. Ohuchi, H., Nakagawa, T., Yamamoto, A., Araga, A., Ohata, T., Ishimaru, Y., Yoshioka, H., Kuwana, T, Nohno, T., Yamasaki, M., Itoh, N., and Noji, S. (1997) The mesenchymal factor, FGF10, initiates and maintains the outgrowth of the chick limb bud through interaction with FGF8, an apical ectodermal factor. *Development*, **124**, 2235.

41. Isaacs, H. V., Pownall, M. E., and Slack, J. M. (1994) eFGF regulates Xbra expression during Xenopus gastrulation. *EMBO J.*, **13**, 4469.

42. Schulte-Merker, S. and Smith, J. C. (1995) Mesoderm formation in response to Brachyury requires FGF signalling. *Curr. Biol.*, **5**, 62.

43. Kawakami, Y., Capdevila, J., Buscher, D., Itoh, T., Rodriguez Esteban, C., and Izpisua Belmonte, J. C. (2001) WNT signals control FGF-dependent limb initiation and AER induction in the chick embryo. *Cell*, **104**, 891.

44. Sekine, K., Ohuchi, H., Fujiwara, M., Yamasaki, M., Yoshizawa, T., Sato, T., Yagishita, N., Matsui, D., Koga, Y., Itoh, N., and Kato, S. (1999) Fgf10 is essential for limb and lung formation. *Nat. Genet.*, **21**, 138.

45. Krumlauf, R. (1994) Hox genes in vertebrate development. *Cell*, **78**, 191.

46. Cohn, M. J., Patel, K., Krumlauf, R., and Wilkinson, D. G., Clarke, J. D., and Tickle, C. (1997) *Hox9* genes and vertebrate limb specification. *Nature*, **387**, 97.

47. Cohn, M. J. and Tickle, C. (1999) Developmental basis of limblessness and axial patterning in snakes. *Nature*, **399**, 474.

48. Carroll, R. L. (1988) *Vertebrate paleontology*. Freeman, San Francisco, CA.

49. Szeto, D. P., Rodriguez-Esteban, C., Ryan, A. K., O'Connell, S. M., Liu, F., Kioussi, C., Gleiberman, A. S., Izpisua-Belmonte, J. C., and Rosenfeld, M. G. (1999) Role of the Bicoid-related homeodomain factor Pitx1 in specifying hindlimb morphogenesis and pituitary development. *Genes Dev.*, **13**, 484.

50. Logan, M. and Tabin, C. J. (1999) Role of Pitx1 upstream of Tbx4 in specification of hindlimb identity. *Science*, **283**, 1736.

51. Lanctot, C., Moreau, A., Chamberland, M., Tremblay, M. L., and Drouin, J. (1999) Hindlimb patterning and mandible development require the *Ptx1* gene. *Development*, **126**, 1805.

52. Rodriguez-Esteban, C., Tsukui, T., Yonei, S., Magallon, J., Tamura, K., and Izpisua-Belmonte, J. C. (1999) The T-box genes *Tbx4* and *Tbx5* regulate limb outgrowth and identity. *Nature*, **398**, 814.

53. Takeuchi, J. K., Koshiba-Takeuchi, K., Matsumoto, K., Vogel-Hopker, A., Naitoh-Matsuo, M., Ogura, K., Takahashi, N., Yasuda, K., and Ogura, T. (1999) *Tbx5* and *Tbx4* genes determine the wing/leg identity of limb buds. *Nature*, **398**, 810.

54. Chapman, D. L., Agulnik, I., Hancock, S., Silver, L. M., and Papaioannou, V. E. (1996) Tbx6, a mouse T-box gene implicated in paraxial mesoderm formation at gastrulation. *Dev. Biol.*, **180**, 534.

55. Bamshad, M., Lin, R. C., Law, D. J., Watkins, W. C., Krakowiak, P. A., Moore, M. E., Franceschini, P., Lala, R., Holmes, L. B., Gebuhr, T. C., Bruneau, B. G., Schinzel, A., Seidman, J. G., Seidman, C. E., and Jorde, L. B. (1997) Mutations in human *TBX3* alter limb, apocrine and genital development in ulnar-mammary syndrome. *Nat. Genet.*, **16**, 311.

56. Basson, C. T., Bachinsky, D. R., Lin, R. C., Levi, T., Elkins, J. A., Soults, J., Grayzel, D., Kroumpouzou, E., Traill, T. A., Leblanc-Straceski, J., Renault, B., Kucherlapati, R., Seidman, J. G., and Seidman, C. E. (1997) Mutations in human *TBX5* cause limb and cardiac malformation in Holt-Oram syndrome. *Nat. Genet.*, **15**, 30.

57. Basson, C. T., Huang, T., Lin, R. C., Bachinsky, D. R., Weremowicz, S., Vaglio, A., Bruzzone, R., Quadrelli, R., Lerone, M., Romeo, G., Silengo, M., Pereira, A., Krieger, J., Mesquita, S. F., Kamisago, M., Morton, C. C., Pierpont, M. E., Muller, C. W., Seidman, J. G., and Seidman, C. E. (1999) Different *TBX5* interactions in heart and limb defined by Holt-Oram syndrome mutations. *Proc. Natl Acad. Sci. USA*, **96**, 2919.

58. Weatherbee, S. D., Halder, G., Kim, J., Hudson, A., and Carroll, S. (1998) *Ultrabithorax* regulates genes at several levels of the wing-patterning hierarchy to shape the development of the *Drosophila* haltere. *Genes Dev.*, **12**, 1474.

59. Weatherbee, S. D. and Carroll, S. B. (1999) Selector genes and limb identity in arthropods and vertebrates. *Cell*, **97**, 283.

60. Weatherbee, S. D., Nijhout, H. F., Grunert, L. W., Halder, G., Galant, R., Selegue, J., and Carroll, S. (1999) *Ultrabithorax* function in butterfly wings and the evolution of insect wing patterns. *Curr. Biol.*, **9**, 109.

61. Yokouchi, Y., Nakazato, S., Yamamoto, M., Goto, Y., Kameda, T., Iba, H., and Kuroiwa, A. (1995) Misexpression of Hoxa-13 induces cartilage homeotic transformation and changes cell adhesiveness in chick limb buds. *Genes Dev.*, **9**, 2509.

62. Kaufman, M. K. (1992) Differentiation of the skeletal system. In *The atlas of mouse development* (ed. M. K. Kaufman), p. 504. Academic Press, London.

63. Shellswell, G. B. and Wolpert, L. (1977) The pattern of muscle and tendon development in the chick wing. In *Vertebrate limb and somite morphogenesis* (ed. D. A. Ede, J. R. Hinckliffe, and M. Balls), p. 71. Cambridge University Press, Cambridge.

64. Wortham, R. A. (1948) The development of the muscles and tendons in the lower leg and foot of chick embryos. *J. Morph.*, **83**, 105.

65. Hamburger, V. and Hamilton, H. L. (1951) A series of normal stages in the development of the chick embryo. *J. Exp. Morphol.*, **88**, 49.

66. Wallace, R. A., King, J. L., and Sanders, G. P. (1988) In *The realm of life*. Scott, Foresman and Company, Illinois.

10 | Evolutionary aspects of vertebrate patterning

SEBASTIAN M. SHIMELD

1. Introduction

1.1 The relationship between evolution and development

In the later half of the nineteenth century and early decades of the twentieth century, embryology and evolutionary biology where intricately linked, as the work of scientists such as Haeckel, von Baer, and Garstang testifies. However, during the early twentieth century evolutionary embryology essentially reached a technical boundary and, since the New Synthesis, when Mendelian genetics were unified with Darwinian natural selection, evolutionary biology has been largely quantitative. Principal themes have been the study of variation in natural populations, the effect of selection, and the establishment of phylogenies of living and extinct animals using molecular and morphological systematics. These studies essentially focus on diversity. Conversely, the model system approach of modern developmental biology has emphasized conservation, i.e. the similarity of developmental mechanisms between taxa. Modern evolutionary developmental biology attempts to reunite these two areas, using modern molecular and developmental techniques to investigate questions that were often first posed over 100 years ago.

The morphology of an individual has two histories, one developmental and the other evolutionary. To fully understand what has shaped individual morphology, both histories have to be considered. Exploring the limits of molecular developmental conservation can give insight into the underlying causes of morphological diversity by determining how organisms differ at the molecular developmental level. Likewise, evolutionary analysis is essential to provide a rigorous phylogenetic framework, and thereby both interpret differences and appropriately extrapolate between model organisms. A recent example of this is the placing of both *Drosophila* and *Caenorhabditis elegans* in the Superphylum Ecdysozoa, and therefore being more closely related than previously recognized (1). In vertebrates, palaeontology, morphological systematics, and molecular phylogenetics can provide the historical framework needed to determine the order of evolutionary change.

The study of the evolution of vertebrate patterning is as broad as vertebrate patterning itself. This chapter will discuss some general concepts, then focus on a number of examples where recent molecular developmental analyses have shed new light on evolutionary process. These will include the origin of vertebrates, the origin of some of the key characters of higher vertebrates, and the role of *Hox* genes in the evolution of vertebrate patterning. By addressing these examples, this chapter aims to show how comparative studies of vertebrate development can give insight into how vertebrate morphology has evolved.

1.2 Concepts and definitions

There are two often-debated concepts central to evolutionary developmental biology. The first is homology. In its strict historical sense, structures in different species are homologous if they are descended from a common ancestral structure in the common ancestor. Genes and gene pathways in different species can also be described as homologous, in the sense that they are descended from a common ancestor. It is important to recognize that genes are frequently duplicated or recruited and reused in evolution, and that therefore morphological and molecular homology are often (and probably usually) uncoupled.

The second is the debate concerning evolution by small or large morphological changes. The vast majority of variation in natural populations is of minor phenotypic effect, and we can therefore predict that much evolutionary change, including the large differences between many higher level lineages, has occurred by the accumulation of many small phenotypic changes over time. Conversely, the hierarchical nature of developmental control does create the theoretical possibility that small genetic changes could lead to large phenotypic ones. This may be especially true of small changes early in development, which can have large consequences in the adult (discussed in detail in ref. 2). We should therefore treat with caution the extreme gradualist view that *all* evolution happens by the accumulation of very small phenotypic changes. Ultimately, the key lies in the relative importance of large and small changes, and we have currently far too few data to assess this.

2. The origin of vertebrates

2.1 The chordate bodyplan and conserved features of chordate patterning

All chordates possess a number of unique characters that unite them in one phylum, including a dorsal, hollow neural tube and a notochord. Living vertebrates (here used as equivalent to craniates, i.e. the clade (gnathostomes plus lampreys plus hagfish)) are distinguished from other members of the phylum Chordata by a number of other characters, including the neural crest and many of the tissues it forms and a cartilaginous or ossified endoskeleton. Placed in a phylogenetic context established by independent, non-morphological methods, this can be viewed as the acquisition of characters along particular evolutionary lineages (Fig. 1).

Analysis of basal chordate development has added molecular and developmental characters to this picture. Notochord development is at least partly controlled by members of the same gene families, including *Brachyury* and *HNF-3*, in both vertebrates and basal chordates (3–8). Similarly, patterning of the neural plate along both anteroposterior (AP) and dorsoventral (DV) axes shares many similarities with patterning of the vertebrate neural plate (Fig. 2). This includes the involvement of *Hox*, *Pax-2/5/8*, and *Otx* genes in AP patterning (reviewed in ref. 9) (see also Chapter 6) and of *HNF-3*, *Hh*, *Msx*, *Lim*, *Snail*, *Dll*, and *Bmp* genes in DV patterning (reviewed in ref. 10). This presumably reflects homology (i.e. descent from a common ancestor possessing a neural plate patterned by these mechanisms), and suggests considerable complexity of neural patterning in the chordate ancestor. Detailed analysis of nervous system development in *Drosophila* has shown that some, but not all, of these

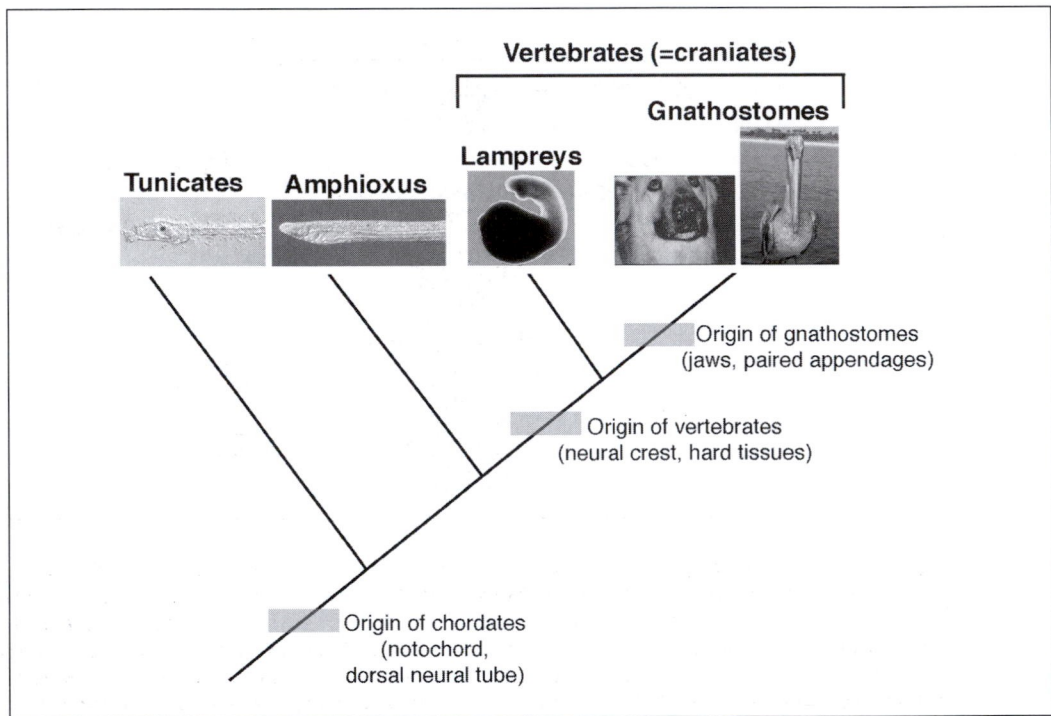

Fig. 1 Relationships of the major chordate groups as established by molecular phylogenetics (see refs 73, 74). Hagfish have been omitted due to the paucity of developmental data; their precise position is debatable, but falls within the craniates. Tunicates, of which the ascidians are the best studied, are most basal, then the cephalochordates (amphioxus). These are united with the vertebrates by the notochord and dorsal neural tube. Amphioxus also possesses overtly segmented mesoderm, which forms myotomes. The vertebrates are separated from these basal chordates by a number of characters, including neural crest-derived structures, a segmented hindbrain, and a cartilaginous or bony endoskeleton.

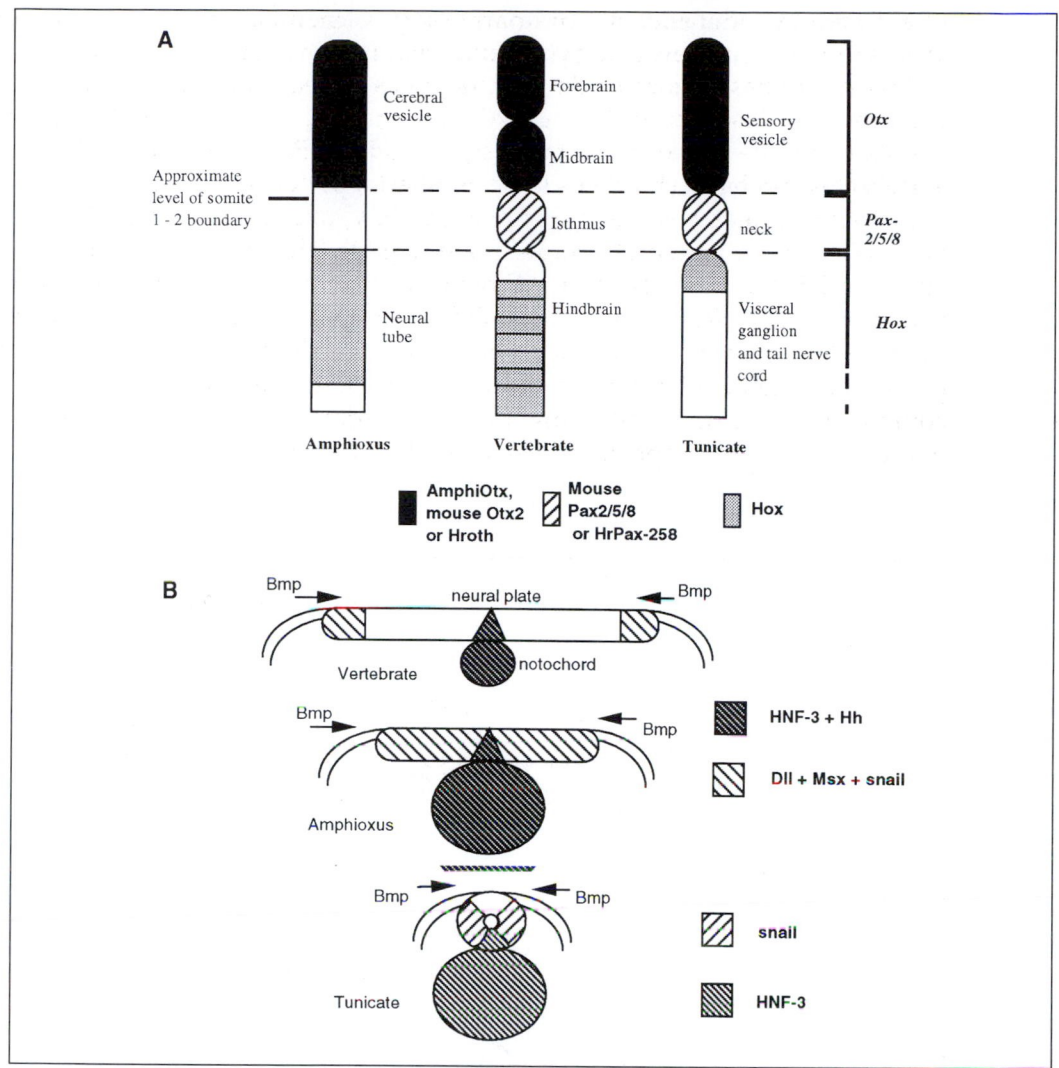

Fig. 2 (A) Basic molecular ground plan of the chordate neural tube (modified from refs 9 and 30). An anterior *Otx*-expressing domain (possible equal to the forebrain/midbrain) is separated from a posterior *Hox*-expressing domain by a *Pax-2/5/8*-expressing domain, equivalent to the isthmus of vertebrates, a well-established organizing centre (see also Chapter 6). Recent analysis of *Pax-2/5/8* expression in amphioxus, however, failed to identify an equivalent expression domain, suggesting the isthmus region of vertebrates may be a novelty and not homologous to the *Pax-2/5/8* domain of the tunicate tadpole (see ref. 75). Segmentation in the vertebrate hindbrain is also a novel character (see ref. 63) (see Chapter 6). (B) Comparison of the expression patterns of genes responsible for dorsoventral patterning of the vertebrate neural tube, with the expression patterns of homologous genes in amphioxus and tunicates (here represented by an ascidian, which has just four cells spanning the dorsoventral axis of the neural tube, as shown). All three taxa have a specific *HNF-3*-expressing cell population at the ventral midline, and in amphioxus and vertebrates these cells also express *hedgehog*. In all three taxa the neural plate is bounded by *Bmp-2/4*-expressing ectodermal cells. The vertebrate dorsally expressed genes of the *Distaless*, *Msx*, and *snail/slug* families have amphioxus homologues expressed throughout the neural plate. (See ref. 10 for references to individual gene-expression patterns.)

genes perform what may also be homologous roles in arthropods, suggesting a much deeper evolutionary origin (reviewed in ref. 11).

2.2 Molecular evolution of developmental genes in early vertebrate evolution

2.2.1 Gene duplication and vertebrate origins

A recent estimate of gene number in the tunicate *Ciona intestinalis* gave a figure of 15 500 ($\pm$3700) (12). In contrast, gene number estimates in jawed vertebrates suggest at least twice as many genes. Placed in a phylogenetic context and compared to gene number estimates in other invertebrates, these data suggest a large increase in gene number during early vertebrate evolution. There are currently insufficient data to precisely determine how much of this increase occurred before and how much after the divergence of amphioxus, lampreys, and hagfish. Table 1 shows the number of vertebrate gene families involved in embryonic patterning which have been conclusively shown, by analysis of amphioxus genes, to contain extra genes in vertebrates. This shows that much of the increase occurred after the divergence of amphioxus, i.e. specifically in the vertebrate lineage. Molecular phylogenetic analysis of individual gene families shows these extra vertebrate genes have arisen by duplication, and some evidence suggests that this may have occurred *en masse* as the result of tetraploidy (discussed in ref. 13).

Table 1 Number of genes found in developmental gene familes in amphioxus and vertebrates[a]

Gene family	Gene number in amphioxus	Gene number in vertebrates
Hox cluster	1	4–6
Otx	1	2–3
Emx	1	2–3
Msx	1	3–5
Engrailed	1	2
Pax-3/7	1	2
Pax-2/5/8	1	3
Pax-1/9	1	2
Hedgehog	1	3–5
Bmp-2/4	1	2
HNF-3	1	3
Gli	1	3
Zic	1	4
Netrin	1	3
Cdx	1	3
Gsx	1	2
Xlox	1	1
Snail	1	2–3
Krox	1	4

[a]For sources of data see references 5, 6, 9, 23, 45, 46, 63, 76–78 plus unpublished data.

Theoretically, duplicated genes need to diverge, at least partly, for both to be maintained (reviewed in ref. 14). Genes can diverge both by changes in the protein-coding sequence that affect function (biochemical changes) and by changes in the elements controlling tissue-specific expression (regulatory changes). While both demonstrably occur, their relative importance is unknown, though it has been speculated that developmental genes may be particularly susceptible to divergence through regulatory changes (discussed in refs 15, 16). Studies in *Drosophila*, where analysis of regulatory elements can be performed with relative ease, have shown that the regulatory elements of some developmental genes evolve at a surprisingly rapid rate, in part due to a degree of redundancy between multiple transcription-factor binding sites in individual enhancer modules (17–19). Studies in vertebrates are not sufficiently advanced to determine if vertebrate developmental genes evolve in a similar manner, although the modular nature of the elements controlling the expression of many vertebrate developmental genes suggests this is likely.

2.2.2 The use and reuse of patterning genes in vertebrate evolution

The extra genetic material produced by gene duplications early in vertebrate evolution has been proposed to have been instrumental in the evolution of vertebrate-specific features, an attractive hypothesis linking increased molecular and morphological complexity via the control of patterning (20, 21). The assumption behind this hypothesis is that extra genes form a sort of 'raw material' which can be used in the evolution of new morphology, with one copy of a duplicated gene retaining ancestral functions. Other copies are therefore freed from the constraint of having to perform this function and are able to take on new functions. An apparent example of this can be seen in the evolution of the *hedgehog* gene family in chordates, where a single ancestral gene in amphioxus performs a conserved role in patterning the dorsoventral axis of the nerve cord (6). In vertebrates, this role has been retained by *sonic hedgehog*, while other vertebrate *hedgehog* genes have taken on different roles in development, including different roles for orthologous genes in different vertebrate taxa (6; reviewed in ref. 22). Similar patterns of evolution have been noted for the *Otx* and *Msx* gene families, with ancestral roles in head development and dorsal neural patterning, respectively (9, 23, 24). However, as discussed more generally above, these new roles probably evolved primarily by changes in where the genes are expressed, rather than in the biochemical properties of the proteins they encode. Such reuse of developmental patterning genes in development is extremely common in vertebrates, for instance with a rather small number of signalling molecules of the transforming growth factor-β (TGF-β), hedgehog, Wnt, and FGF families used in multiple sites in the developing embryo (see earlier chapters for examples). These multiple functions necessarily evolved by modifying the regulatory hierarchy controlling tissue-specific expression, such that regulation of some vertebrate genes has become extremely complicated. This tells us that diversification of gene function and accompanying morphological evolution can precede gene duplication and divergence, with multiple functions of an ancestral gene divided up between duplicated copies by differential mutation of regulatory elements. This hypothesis has the advantage of a firm theoretical rooting in population genetics (25), something all 'evolutionary developmental' theories must eventually achieve if they are to become widely accepted by evolutionary biologists. Discriminating between these two hypotheses may be possible in some instances, for example if a single ancestral gene is identified with the relevant multiple functions. Conversely, in the absence of the now-extinct intermediate populations in which the gene duplications occurred, absolute proof of duplication-preceded-divergence hypotheses will be impossible. In summary, while extensive gene duplication has occurred in early vertebrate evolution, it is hard to test if this was instrumental in the origin of vertebrates. Experimental studies in *Drosophila* have shown, however, that regulatory regions can evolve very quickly—explaining how duplicated genes could rapidly evolve new functions and therefore be retained, and how the complexity of regulation of many vertebrate developmental genes has evolved.

2.3 Evolution of vertebrate characters: neural crest, placodes, and endoskeleton

2.3.1 The neural crest and placodes

The importance of placodes and neural crest in the evolution of vertebrates was emphasized by Gans and Northcutt in 1983 (26). These tissues contribute to many of the structures considered to be vertebrate novelties, including the complex head with prominent sensory organs and cranial skeleton. Placodes themselves may not be specific to vertebrates, as protochordates may possess homologues of some of the sensory placodes. The olfactory placode seems to have a direct homologue in basal chordates, most notably as the Organs de Quatrefages of amphioxus (24, 27–29). Ascidians also have a possible homologue of the otic placode (30). There is no evidence for neurogenic placodes in amphioxus or tunicates. Neural crest cells also have no obvious homologues in tunicates or amphioxus. Reports of migratory neural precursor cells in ascidians are intriguing, although their origin, in the adult, is quite different from that of vertebrate neural crest cells (29, 31). These form at the boundary between the neural plate and ectoderm, under the control of *Bmp*s and in a territory marked by the expression of *Msx*, *Snail*, and *Dll* family members. The conserved expression of these patterning genes in basal chordates (see Fig. 2(B)) makes it unlikely that they have directly contributed to the evolution of neural crest

Neural crest cells possess a number of key characters, including migration and the ability to form non-neural cell types (see Chapter 8). Currently, we have little idea of the molecular changes that underlay the origin of these new cellular characters. However, analysis of some vertebrate mutants in which neural crest development is disrupted has provided recent support for one scenario of neural crest evolution, discussed by Fritsch and Northcutt (32). Neural crest cells in vertebrates migrate in successive waves, with early emerging cells taking the medial path and contributing to the spinal ganglia. In the zebrafish mutant *narrowminded*, development of these early cells is disrupted, while the development of later emerging cells, which take the dorsolateral pathway, is not (33). This mutation also affects a dorsal cell population known as Rohon-Beard neurons, a mechanosensory cell population that is also present in amphioxus (32). This developmental link, between an ancient dorsal neural population of Rohon-Beard cells and the early emerging neural crest that contribute to spinal ganglia, suggests that one of the first steps in crest evolution may have been the emergence of a neuronal, or neuronal supporting, cell population from the central nervous system to contribute to the peripheral nervous system. Other neural crest cell fates, particularly the non-neural connective tissue fates of cephalic crest, would therefore be of later origin.

2.3.2 The evolution of vertebrate hard tissues and the origin of the axial endoskeleton

The evolution of vertebrate hard and endoskeletal tissues (enamel, dentine, bone, and cartilage) has had a major impact on vertebrate evolution. There are several theories concerning the origins of these tissues, including an initial origin in the head as supporting structures for the branchial region, followed by co-option for loco-motory and/or predatory usage (discussed in more detail in ref. 34). There is little molecular or developmental evidence that sheds light on this and other evolutionary scenarios, nevertheless the separate developmental origins of the cranial skeleton (largely neural crest-derived with some mesodermal components) and the axial skel-eton (mesoderm-derived) supports their separate origins. Palaeontological evidence suggests that at least mineralization of skeletal tissues first evolved in the head, as seen in the extensive predatory apparatus of the conodonts (35). These and other basal vertebrates show no evidence of a mineralized axial skeleton, though it is poss-ible they had a cartilaginous axial skeleton. Such structures are often poorly preserved in the fossil record, so it is still possible that a cartilaginous axial skeleton preceded the head skeleton (36, 37). In summary, it is not yet possible to determine definitely if neural crest-derived or mesoderm-derived skeletal tissues evolved first, though it is likely that both skeletal elements and mineralization did first evolve in the head.

Only the paraxial mesoderm of vertebrates is overtly segmented, except in lampreys where the lateral plate mesoderm is also segmented in register with the somites (38). Embryonically, the axial skeleton of vertebrates develops from the sclerotome, a ventromedial compartment of the somite that forms under the control of *hedgehog* signalling from the notochord and signalling by *Bmp*s from the dorsal ectoderm and lateral plate (39–41) (see Chapter 4). Although secondarily reduced in some species, the ventromedial sclerotome compartment has been identified in the embryos of all vertebrates, with the possible exception of the jawless hagfish (42). In amphioxus, all the mesoderm segments into somites (Fig. 3). Dorsoventral sub-division of somites also occurs during amphioxus development, with amphioxus somites subdivided into at least three zones that can be visualized early in develop-ment by the expression of muscle-related genes (43) and of *Gli* and *Zic* transcription factors (author's unpublished observations; see Fig. 3 for a discussion of the possible relationships of these zones with vertebrate mesoderm). In common with verte-brates, amphioxus ventral somite cells are adjacent to *hedgehog*-expressing cells of the

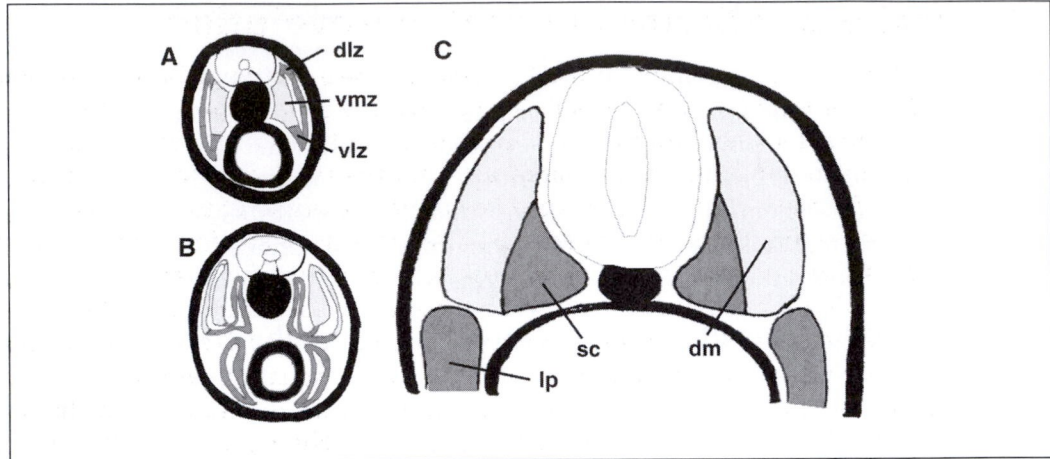

Fig. 3 Comparison of mesoderm subdivision in amphioxus (adapted from ref. 44) and a generalized vertebrate. Ectoderm, endoderm, and notochord are black, the neural tube is white, and the mesoderm is shaded. All the amphioxus mesoderm is segmented and this may be the ancestral state for chordate mesoderm. (A) In a transverse section of a late amphioxus neurula, three mesoderm zones are visible in each somite. The ventro-medial zone (vmz) forms myotome. The dorsolateral and ventrolateral zones (dlz and vlz) also eventually differen-tiate into muscle, but also move down around the gut and notochord to form the perivisceral and perinotochordal coeloms, shown in (B). No equivalent of sclerotome (sc) is seen. (C) This panel shows a section of a generalized vertebrate in the same orientation (see also Chapter 4). It is possible that the ventrolateral zone of amphioxus mesoderm, which forms perivisceral coelom, is equivalent to the lateral plate (lp) and/or splanchnic mesoderm of vertebrates because of its position and fate. A similar argument could be made for homology of the amphioxus dlz to vertebrate dermatome. However, as yet, no molecular data confirm this. (dm, dermomyotome.)

notochord and dorsal endoderm (6). These cells do not form sclerotome and do not express the sclerotome markers *Pax-1* and *Pax9*; instead they form myotome (44, 45). An amphioxus homologue of *Bmp-2* and *Bmp-4* is expressed throughout the somite, and is not laterally restricted as in vertebrates (46). In summary, these data suggest a number of steps in the evolution of vertebrate mesoderm in general, and the axial skeleton in particular, including (and not necessarily in order):

- the progressive restriction of segmentation from the whole mesoderm (amphioxus and lampreys) to paraxial mesoderm (gnathostomes);
- the restriction of BMP-2 and -4 signalling to the lateral plate;
- the origin of sclerotome, possible by co-option of a developmental pathway originally used to make skeletal tissues in the head, and formation of an axial cartilaginous endoskeleton; and
- mineralization.

Following its origin, the axial skeleton has continued to evolve, and displays great diversity both between different vertebrate species and between different antero-posterior positions in the same species. One aspect of this, regionalization along the anteroposterior axis, is discussed in more detail in Section 4.

3. Major transitions in vertebrate evolution

The term 'major transition' is used to describe the evolution of a structure which has had a profound effect on the subsequent evolution of a taxon, and the definition is therefore somewhat arbitrary. In vertebrate evolution, examples of major transitions might include the origin of paired appendages (fins) and their modification into tetrapod limbs, the origin of hinged jaws, and the origin of the amniote egg. It must be stressed that these do not necessarily reflect single, large phenotypic changes (as discussed in Section 1.2; and, as the fossil record often demonstrates, more probably an extended series of small changes) but are denoted as major transitions because of their legacy in the form of surviving lineages, emphasized by the extinction of intermediates. To the evolutionary biologist, such transitions are an attractive field of research because of their subsequent significance. Most have been the subject of considerable speculation over the last 150 years. The following will focus on some examples where recent advances in our understanding of the cellular and molecular mechanisms of development have shed new light on their evolutionary origin.

3.1 The origin of paired fins

From our anthropocentric viewpoint, the origin of limbs was a key event in vertebrate evolution. This can be broken down into two stages, the origin of paired appendages (fins) in primitive fish and the origin of tetrapod limbs. Both these transitions have the advantage of a considerable body of palaeontological data that can be used to test the predictions of developmental hypotheses of origin.

Teleost fish have two sets of paired fins, the pectoral and pelvic. These are homologous to tetrapod fore- and hindlimbs, respectively (see Chapter 9). Two main theories of paired fin origins are found in vertebrate anatomy textbooks (see, for example, ref. 47). The first is by modification of posterior branchial arches; this is extremely unlikely considering the very different morphology and embryology of branchial arches and fins. The second is derivation from an ancestrally continuous lateral fin fold. Many other scenarios could be imagined; here discussion will be restricted to instances where recent molecular and developmental analysis has been informative.

The mechanisms that pattern the pelvic fins of teleost fish are similar to those that pattern the pectoral fins, including the expression of *sonic hedgehog* and posterior *Hox* genes. However, fossil evidence suggests that pelvic and pectoral fins primitively differ and that a single pair of fins evolved first, although it is not possible to say whether these are homologous to either the pelvic fins or pectoral fins of modern fish (48). Laufer and Tabin (49) followed the suggestion of a paired-fin origin from an ancestrally continuous lateral fin fold, with mesoderm of the pelvic fin expressing posterior *Hox* genes (i.e. *Hox9* to *Hox13*) and mesoderm of the pectoral fin expressing more anterior *Hox* genes. They then proposed that a homeotic transformation of the pectoral fin took place, with pelvic patterning mechanisms (particularly posterior *Hox* genes) transferred to the pectoral fin bud. However, a detailed analysis of *Hox* gene expression and regulation in lateral plate mesoderm (50) led Coates and Cohn (51, 52) to propose an alternative hypothesis. They pointed out that early in development the mesoderm in the region where the forelimb of the chick would develop expressed *Hoxd9*, and therefore that posterior *Hox* gene expression was a normal feature of early forelimb development and did not require *de novo* activation by homeotic transformation. Increased axial regionalization of *Hox* gene expression in the mesoderm could therefore have provided both the cues to localize outgrowth promoting signals and the *Hox* gene complement necessary for the separate evolutionary origin but similar development of pelvic and pectoral fins. Furthermore, this correlates with the evolution of increased complexity of gut regionalization, providing a developmental basis for why lateral plate mesodermal *Hox* expression should have become regionalized prior to fin evolution.

3.2 The origin of the tetrapod limb

The invasion of the land by vertebrates required several key adaptations, including a respiratory system that could function in air, a new locomotory apparatus, and ultimately reproductive changes negating the need to return to an aqueous environment to breed. The tetrapod limb is one of these adaptations. Living tetrapods are primitively pentadactyl, with taxa with smaller numbers of digits (for example, birds with their four digits) demonstrably derived (see Chapter 9). No natural, living tetrapods populations have more than five digits (for instance the '6th digit' of the giant panda is a modified carpel), although some mutations can produce polydactylous phenotypes. Fossil evidence, however, shows that this was not always so, with stem tetrapods such as *Acanthostega* having as many as eight digits (53).

Comparison of *Hox* gene expression between teleost and tetrapod limb buds has provided evidence that the digits are newly evolved structures, and are not directly homologous to any structures in the teleost fin (54). These authors noted that *Hox* gene expression in a tetrapod limb is biphasic, with an initial expression phase with anteroposterior orientation during which the proximal limb is formed, followed by a second phase with proximodistal orientation during which the distal digits are formed. Teleost fish show only the first phase, suggesting that the second phase of *Hox* expression, which correlates with continued proliferation of the limb-bud mesoderm, was essential for the evolution of the digits.

Analysis of *Hox* gene mutant phenotypes in mice has also given further insight into the origin and patterning of tetrapod digits. Progressive reduction of posterior *Hox* gene dosage results in the progressive loss of digit length, such that in animals mutant for *Hoxd11* to *Hoxd13* and hemizygous for a mutation in *Hoxa13*, little is left of the digits (55). Interestingly, these studies also identified an intermediate stage of *Hox* gene dosage, which manifested in a polydactylous short-digit phenotype, reminiscent of that seen in *Acanthostega*. These data link changes in the regulation of posterior *Hox* gene expression with control of proliferation, and thereby responsibility for both the evolution of the digits and the subsequent stabilization of a polydactylous form into a pentadactylous form. They also render it likely that a previous model for constrained pentadactyly based on the five nested expression domains of *Hoxd* genes in the early limb bud is too simplistic, since this can not account for the shortened digit and polydactylous phenotypes found in *Hox* gene mutants (56, 57).

3.3 The evolution of jaws and modification of the branchial arches

Most living vertebrates have hinged jaws; the so-called 'gnathostomes' (see Fig. 1). Lampreys and hagfish do not (Fig. 4(A)), and the fossil record shows that these two taxa split from the lineage leading to jawed vertebrates before jaws evolved (58) (see Fig. 1). Hinged jaws develop primarily from the first branchial arch, with second-arch components modified into supporting structures. This suggests that the origin of jaws involved the modification of the simpler branchial skeleton of a single arch, with proximal skeletal elements evolving into the dorsal cartilages and distal elements into ventral, mandibular cartilages (Figs 4(C) and (D)). If this scenario is correct, it implies homology between skeletal components of the gnathostome jaw

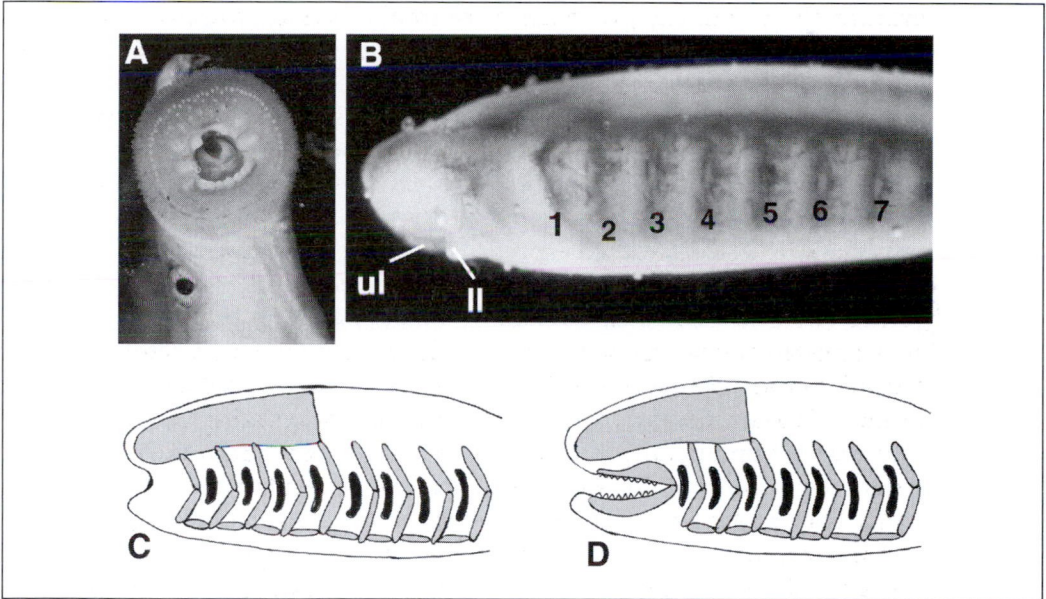

Fig. 4 (A) An adult lamprey showing the circular mouth with rasping teeth but no hinged jaws. (B) The head of a larval lamprey (ammocete) showing the upper lip (ul), lower lip (ll), and the seven pharyngeal gill slits (numbered). (C) and (D) Schematic diagrams of one theory of jaw evolution (adapted from ref. 47). (C) This panel shows the branchial skeleton of a hypothetical jawless vertebrate ancestor, superficially similar to the branchial skeleton of a lamprey. In (D), one arch has been modified such that the distal element becomes the mandible, and proximal element the maxilla. This scenario predicts homology between vertebrate-jaw skeletal elements and lamprey first-arch skeletal elements. In most modern vertebrates the second-arch skeleton is also modified to form supporting structures for the jaws.

apparatus and proximal and distal elements of the lamprey first-arch branchial skeleton, and predicts that the mandibular and maxillary processes are not homologous to the lower and upper lips of the lamprey. Scanning electron micrography of lamprey embryonic development supports this view, showing that the lamprey upper lip is not of first-arch origin (59). However, patterning of arch-derived structures in lampreys has yet to be tested at the cell-lineage or molecular levels. Analysis in lamprey embryos of genes confined to specific gnathostome arches might be informative in this respect.

The entire facial and branchial region of vertebrates is an extremely complicated structure, and its development involves orchestrated patterning of ectoderm, endoderm, mesoderm, and neural crest to develop a patterned arrangement of bone, tendon, muscle, nerve, and gut. The organization of branchial connective and skeletal tissue has been greatly modified in vertebrate evolution, both before and after the evolution of jaws. This is apparent from the great variety of head and facial form exhibited by both living and fossil vertebrates. A major question is how this has occurred while retaining the functional interconnectivity essential to produce a working structure? The molecular mechanisms controlling this patterning are still too poorly understood to answer this at the gene expression level. At the cellular level, however, studies by Kontges and Lumsden (60) have provided a partial answer. They performed rhombomere transplants from quail to chick hindbrains, and monitored the fate of rhombomere-specific neural crest cells into late development. Here, they found a surprising degree of coherence in the origin of individual muscle connective tissues and their attachment sites to the skeleton, with both being formed of crest from the same axial level. They also found sharp, cryptic boundaries between crest populations in individual crest-derived skeletal elements. These results suggest that a certain degree of connectivity between skeletal and muscular tissues in the vertebrate head is preprogrammed by the neural crest. This provides an explanation for how the shapes of vertebrate head tissues could be radically altered during evolution while still retaining functional connections between muscle and bone, since these are ensured by specific crest populations. It will be interesting to see how widely distributed this is in vertebrates (specifically, is it ancestral?) and to test if and how alterations in patterning genes have altered head structures within this connectivity. Again, these results emphasize the importance of neural crest in vertebrate evolution, and suggest a link between rhombencephalic segmentation, neural crest specification, and the evolvability of branchial-arch structures.

4. *Hox* genes and the modification of axial skeletal regionalization

The ancestral site of *Hox* gene expression in chordates is probably the nervous system, as this is where both amphioxus and ascidian *Hox* genes are expressed (61, 62). Expression in vertebrate mesoderm should therefore be considered a derived feature, superimposed on an already segmented tissue (in contrast to the nervous system, where rhombomeric segmentation has been imposed on an ancestral site of *Hox* expression: see ref. 63).

A common feature of the phenotypes of mice mutant for *Hox* genes is transformation in the identity of vertebrae (64) (see Chapter 4). This typically manifests in changes in the position of the boundaries between different types of vertebrae, and may be phenocopied by mutations in genes that regulate *Hox* gene expression, such as members of the *polycomb* group of genes (64). These results implicate murine *Hox* genes in specifying the axial identity of vertebrae.

Comparative anatomy suggests progressive regionalization of the axial skeleton on the lineage leading to amniotes—with basal fish having less overt regionalization than amniotes, and basal vertebrates such as the lamprey very little. Are changes in the axial expression of *Hox* genes responsible for the modification of vertebral identity in vertebrate evolution? Comparative studies by Burke and co-workers and Gaunt and co-workers (65, 66) have suggested that boundaries of *Hox* gene expression in different taxa indeed respect morphological boundaries, and not numbers of segmental units. For instance, *Hoxc6* marks the cervical to thoracic transition in both mice and chicks, though the former has seven cervical vertebrae and the latter fourteen. An extensive analysis of *Hox* gene expression in the zebrafish found that the anterior mesodermal boundaries of expression were compressed compared to those of amniotes, correlating with the lower axial complexity of zebrafish (67). These data link regionalized *Hox* expression with the control of vertebral identity, and suggest that molecular changes that affect the boundaries of *Hox* gene expression in the paraxial mesoderm can cause evolutionary change in the patterning of the axial skeleton.

Confirming this, simplification of mesodermal *Hox* gene expression is also associated with secondary loss of axial complexity. During the evolution of snakes, thoracic vertebral identity has expanded into the region that was ancestrally cervical. This has been accompanied by an anterior expansion in the distribution of HOXC6 and HOXC8 protein, although HOXC8 maintains a posterior boundary at the level of the most posterior thoracic vertebra (68). Such axial homogenization, accompanied by forelimb loss, is quite common in vertebrate evolution, having convergently evolved in several taxa including snakes, legless lizards, and the amphibian *Caecillians*. Determining if these convergently evolved morphologies stem from similar changes to underlying patterning mechanisms, particularly *Hox* genes, will be an exciting if challenging task. In summary, there is now good comparative evidence supporting a role for changes in *Hox* gene regulation in the evolution of patterning of the vertebrate axial skeleton.

5. Summary and conclusions

Evolutionary developmental biology is not a new area of research, but a re-emergence of an old one. This re-emergence is largely technology based, as techniques such as degenerate polymerase chain reaction (PCR) and *in situ* hybridization have made it possible to address previously intractable questions, first asked over 100 years ago.

Recent progress in answering some of these questions is described above, though this is by no means exhaustive; for more general coverage of many issues not described here, the reader is referred to the second edition of Hall's *Evolutionary developmental biology*, to Raff's *The shape of life*, and to *From DNA to diversity* by Carroll and colleagues (69–71).

The aim of this field of research is not to compare every vertebrate with every other vertebrate to determine how and why they are different. More realistically, two focused themes are central to investigating the relationship between vertebrate development and evolution. The first is to understand, as far as is possible, those developmental changes underlying evolutionary transitions that have had a major subsequent effect on vertebrate origins and morphological evolution. Recent progress towards understanding the molecular and developmental basis for some of these is described above. The second is to try to draw general principles governing the detailed evolution of developmental genes and the role of this in generating diversity. Some aspects of the evolution of vertebrate developmental genes, particularly with respect to gene duplication, have been well characterized. However, we have no idea how (or even if) population-level variation in developmental genes gives rise to the phenotypic variation seen in natural vertebrate populations (see also Chapter 9). Experiments investigating this are underway in *Drosophila*, made possible by advanced molecular techniques (17–19, 72). Similar experiments have yet to be carried out in vertebrates; but this will be made possible with the development of new techniques (particularly rapid, cheap transgenesis) plus a more detailed understanding of how the morphology of model organisms develops. In summary, by combining the detailed investigation of transitions with the elucidation of more general principles, it should be possible to understand both the details of how vertebrate populations evolve and the basis of the evolution of key structures.

References

1. Aguinaldo, A., Turbeville, J., Linford, L., Rivera, M., Garey, J., Raff, R., and Lake, J. (1997) Evidence for a clade of nematodes, arthropods and other moulting animals. *Nature*, **387**, 489.
2. Richardson, M. K. (1999) Vertebrate evolution: the developmental origins of adult variation. *BioEssays*, **21**, 604.
3. Corbo, J. C., Erives, A., Di Gregorio, A., Chang, A., and Levine, M. (1997) Dorsoventral patterning of the vertebrate neural tube is conserved in a protochordate. *Development*, **124**, 2335.
4. Holland, P. W. H., Koschorz, B., Holland, L. Z., and Herrmann, B. G. (1995) Conservation of Brachyury (T) genes in amphioxus and vertebrates: developmental and evolutionary implications. *Development*, **121**, 4283.
5. Shimeld, S. M. (1997) Characterisation of amphioxus HNF-3 genes: conserved expression in the notochord and floor plate. *Dev. Biol.*, **183**, 74.
6. Shimeld, S. M. (1999) The evolution of the hedgehog gene family in chordates: insights from amphioxus hedgehog. *Dev. Genes Evol.*, **209**, 40.

7. Terazawa, K. and Satoh, N. (1997) Formation of the chordamesoderm in the amphioxus embryo: analysis of *Brachyury* and *fork head/HNF-3* genes. *Dev. Genes. Evol.*, **207**, 1.

8. Yasuo, H. and Satoh, N. (1993) Function of the vertebrate T gene. *Nature*, **364**, 582.

9. Williams, N. A. and Holland, P. W. A. (1998) Molecular evolution of the brain of chordates. *Brain Behav. Evol.*, **52**, 177.

10. Shimeld, S. M. (1999) The evolution of dorsoventral patterning in the vertebrate neural tube. *Am. Zool.*, **39**, 641.

11. Arendt, D. and Nübler-Jung, K. (1999) Comparison of early nerve cord development in insects and vertebrates. *Development*, **126**, 2309.

12. Simmen, M. W., Leitgeb, S., Clark, V. H., Jones, S. J. M., and Bird, A. (1998) Gene number in an invertebrate chordate, *Ciona intestinalis*. *Proc. Natl Acad. Sci. USA*, **95**, 4437.

13. Sharman, A. C. and Holland, P. W. H. (1996) Conservation, duplication and divergence of developmental genes during chordate evolution. *Neth. J. Zool.*, **46**, 47.

14. Krakauer, D. C. and Nowak, M. (1999) Evolutionary preservation of redundant duplicated genes. *Semin. Dev. Biol.*, **10**, 555.

15. Cooke, J., Nowak, M. A., Boerlijst, M., and Maynard-Smith, J. (1997) Evolutionary origins and maintenance of redundant gene expression during metazoan development. *Trends Genet.*, **13**, 360.

16. Shimeld, S. M. (1999) Gene function, gene networks and the fate of duplicated genes. *Semin. Cell Dev. Biol.*, **10**, 549.

17. Hancock, J. M., Shaw, P. J., Bonneton, F., and Dover, G. A. (1999) High sequence turnover in the regulatory regions of the developmental gene *hunchback*. *Mol. Biol. Evol.*, **16**, 253.

18. Ludwig, M. Z., Patel, N. H., and Kreitmen, M. (1998) Functional analysis of eve stripe 2 enhancer evolution in *Drosophila*: rules governing conservation and change. *Development*, **125**, 949.

19. Piano, F., Parisi, M. J., Karess, R., and Kambysellis, M. P. (1999) Evidence for redundancy but not trans factor-cis element coevolution in the regulation of *Drosophila* Yp genes. *Genetics*, **152**, 605.

20. Holland, P. W. H., Garcia-Fernàndez, J., Williams, N. A., and Sidow, A. (1994) Gene duplications and the origins of vertebrate development. *Development*, **Suppl.**, 125.

21. Holland, P. W. H. (1992) Homeobox genes in vertebrate evolution. *BioEssays*, **14**, 267.

22. Hammerschmidt, M., Brook, A., and McMahon, A. P. (1997) The world according to *hedgehog*. *Trends Genet.*, **13**, 14.

23. Sharman, A. C., Shimeld, S. M., and Holland, P. W. H. (1999) An amphioxus Msx gene expressed predominantly in the dorsal neural tube. *Dev. Genes Evol.*, **209**, 260.

24. Williams, N. A. and Holland, P. W. H. (1996) Old head on young shoulders. *Nature*, **383**, 490.

25. Lynch, M. and Force, A. (2000) The probability of duplicate gene preservation by subfunctionalisation. *Genetics*, **154**, 459.

26. Gans, C. and Northcutt, R. G. (1983) Neural crest and the origin of vertebrates: a new head. *Science*, **220**, 268.

27. Powell, J. F. F., Reska-Skinner, S. M., Om Prakash, M., Fischer, W. H., Park, M., Rivier, J. E., Craig, A. G., Mackie, G. O., and Sherwood, N. (1996) Two new forms of gonadotropin-releasing hormone in a protochordate and the evolutionary implications. *Proc. Natl Acad. Sci. USA*, **93**, 10461.

28. Mackie, G. O. (1995) On the visceral nervous system of Ciona. *J. Mar. Biol. Assoc.*, **75**, 141.

29. Bollner, T., Beesley, P. W., and Thorndyke, M. C. (1997) Investigation of the contribution from peripheral GnRH-immunoreactive neuroblasts to the regenerating nervous system in the protochordate Ciona intestinalis. *Proc. R. Soc. Lond.*, **264**, 1117.

30. Wada, H., Saiga, H., Satoh, N., and Holland, P. W. H. (1998) Tripartite organization of the ancestral chordate brain and the antiquity of placodes: insights from ascidan *Pax-2/5/8*, *Hox* and *Otx* genes. *Development*, **125**, 1113.

31. Manni, L., Lane, L. J., Sorrentino, M., Zanziolo, G., and Burighel, P. (1999) Mechanism of neurogenesis during the embryonic development of a tunicate. *J. Comp. Neurol.*, **412**, 527.

32. Fritsch, B. and Northcutt, R. G. (1993) Cranial and spinal nerve organization in amphioxus and lampreys: evidence for an ancestral craniate pattern. *Acta Anat.*, **148**, 96.

33. Artinger, K. B., Chitnis, A. B., Mercola, M., and Driever, W. (1999) Zebrafish *narrowminded* suggests a genetic link between formation of neural crest and primary sensory neurons. *Development*, **126**, 3969.

34. Gans, C. (1993). Evolutionary origin of the vertebrate skull. In *The skull* (ed. J. Hanken and B. K. Hall), Volume 2 p. 1–35. University of Chicago Press, Chicago.

35. Donoghue, P. C. J. (1998) Growth and patterning in the conodont skeleton. *Phil. Trans. R. Soc. Lond. B.*, **353**, 633.

36. Bardack, D. and Zangerl, R. (1968) First fossil lamprey: a record from the Pennsylvanian of Illinois. *Science*, **162**, 1265.

37. Bardack, D. (1991) First fossil hagfish (Myxinoidea): a record from the Pennsylvanian of Illinois. *Science*, **254**, 701.

38. Damas, H. (1944) Recherches sur le dévelopment de *Lampetra Fluviatilis* L. Contribution à l'étude de la céphalogenèse des Vertébrés. *Arch. Biol.*, **55**, 1.

39. Pourquié, O., Fan, C.-M., Coltey, M., Hirsinger, E., Watanabe, Y., Bréant, C., Francis-West, P., Brickell, P., Tessier-Lavigne, M., and Le Douarin, N. M. (1996) Lateral and axial signals involved in avian somite patterning: a role for BMP4. *Cell*, **84**, 461.

40. Fan, C. and Tessier-Lavigne, M. (1994) Patterning of mammalian somites by surface ectoderm and notochord: evidence for sclerotome induction by a hedgehog homolog. *Cell*, **79**, 1175.

41. Fan, C., Porter, J., Chiang, C., Chang, D., Beachy, P., and Tessier-Lavigne, M. (1995) Long-range sclerotome induction by sonic hedgehog: direct role of the amino-terminal cleavage product and modulation by the cyclic AMP signaling pathway. *Cell*, **81**, 457.

42. Gorbman, A. (1997) Hagfish development. *Zool. Sci.*, **14**, 375.

43. Holland, L., Pace, D., Blink, M., Kene, M., and Holland, N. (1995) Sequence and expression of amphioxus alkali myosin light chain (AmphiMLC-alk) throughout development: implications for vertebrate myogenesis. *Dev. Biol.*, **171**, 665.

44. Holland, L. Z., Pace, D. A., Blink, M. L., Kene, M., and Holland, N. D. (1995) Sequence and expression of amphioxus alkali myosin light chain (AmphiMLC-alk) throughout development: implications for vertebrate myogenesis. *Dev. Biol.*, **171**, 665.

45. Holland, N., Holland, L., and Kozmik, Z. (1995) An amphioxus Pax gene, AmphiPax-1, expressed in embryonic endoderm, but not in mesoderm: implications for the evolution of class I paired box genes. *Mol. Mar. Biol. Biotechnol.*, **4**, 206.

46. Panopoulou, G. D., Clark, M. D., Holland, L. Z., Lehrach, H., and Holland, N. D. (1998) AmphiBMP2/4, an amphioxus bone morphogenetic protein closely related to *Drosophila* decapentaplegic and vertebrate BMP2 and BMP4: insights into evolution of dorsoventral axis specification. *Dev. Dyn.*, **213**, 130.

47. Romer, A. S. (1970) *The vertebrate body*. W. B. Saunders, Philadelphia, PA.

48. Wilson, M. V. H. and Caldwell, M. W. (1993) New Silurian and Devonian fork-tailed 'thelodonts' are jawless vertebrates with stomachs and deep bodies. *Nature*, **361**, 442.

49. Laufer, E. and Tabin, C. (1993) Hox genes and serial homology. *Nature*, **361**, 692.

50. Cohn, M. J., Patel, K., Krumlauf, R., Wilkinson, D. G., Clarke, J. D. W., and Tickle, C. (1997) *Hox9* genes and vertebrate limb specification. *Nature*, **387**, 97.

51. Coates, M. I. and Cohn, M. J. (1998) Fins, limbs and tails: outgrowths and axial patterning in vertebrate evolution. *BioEssays*, **20**, 371.

52. Coates, M. I. and Cohn, M. J. (1999). Vertebrate axial and appendicular patterning: the early development of paired appendages. *Am. Zool.* **39**, 676.

53. Coates, M. I. and Clack J. A. (1990) Polydactyly in the earliest known tetrapod limbs. *Nature*, **347**, 66.

54. Sordino, P., van der Hoeven, F., and Duboule, D. (1995) Hox gene expression in teleost fins and the origin of vertebrate digits. *Nature*, **375**, 678.

55. Zákány, J., Fromental-Ramain, C., Warot, X., and Duboule, D. (1997) Regulation of number and size of digits by posterior *Hox* genes: a dose dependent mechanism with potential evolutionary implications. *Proc. Natl Acad. Sci. USA*, **94**, 13695.

56. Izpisúa-Belmonte, J. C., Tickle, C., Dollé, P., Wolpert, L. and Duboule, D. (1991) Expression of the homeobox Hox-4 genes and the specification of position in chick wing development. *Nature*, **350**, 585.

57. Tabin, C. J. (1992). Why we have (only) five fingers per hand: Hox genes and the evolution of paired limbs. *Development*, **116**, 289.

58. Forey, P. and Janvier, P. (1993) Agnathans and the origin of jawed vertebrates. *Nature*, **361**, 129.

59. Kuratani, S., Horigome, N., and Hirano, S. (1999) Developmental morphology of the head mesoderm and reevaluation of segmental theories of the vertebrate head: evidence from embryos of an agnathan vertebrate, *Lampetra japonicum*. *Dev. Biol.*, **210**, 381.

60. Kontges, G. and Lumsden, A. (1996) Rhombencephalin neural crest segmentation is preserved throughout craniofacial ontogeny. *Development*, **122**, 3229.

61. Katsuyama, Y., Wada, S., Yasugi, S., and Saiga, H. (1995) Expression of the *labial* group Hox gene *HrHox-1* and its alteration by retinoic acid in development of the ascidian *Halocynthia roretzi*. *Development*, **121**, 3197.

62. Wada, H., Garcia-Fernàndez, J., and Holland, P. W. H. (1999) Colinear and segmental expression of amphioxus Hox genes. *Dev.. Biol.*, **213**, 131.

63. Knight, R. D., Panopoulou, G. D., Holland, P. W. H., and Shimeld, S. M. (2000). An amphioxus Krox gene: insights into vertebrate hindbrain evolution. *Dev. Genes Evol.*, **210**, 518.

64. van Lohuizen, M. (1998) Functional analysis of mouse polycomb group genes. *Cell. Mol. Life Sci.*, **54**, 71.

65. Burke, A. C., Nelson, C. E., Morgan, B. A., and Tabin, C. (1995). Hox genes and the evolution of vertebrate axial morphology. *Development*, **121**, 333.

66. Gaunt, S. J. (1994) Conservation in the Hox code during morphological evolution. *Int. J. Dev. Biol.*, **38**, 549.

67. Prince, V. E., Joly, L., Ekker, M., and Ho, R. K. (1998) Zebrafish *hox* genes: genomic organization and modified colinear expression patterns n the trunk. *Development*, **125**, 407.

68. Cohn, M. J. and Tickle, C. (1999) Developmental basis of limblessness and axial patterning in snakes. *Nature*, **399**, 474.

69. Hall, B. K. (1998). *Evolutionary developmental biology*. Chapman and Hall, London

70. Raff, R. A. (1996). *The shape of life*. University of Chicago Press, Chicago.

71. Carroll, S., Grenier, J., and Wetherbee, S. (2000) From DNA to Diversity. Blackwell, Oxford.

72. Stern, D. L. (1998). A role of *Ultrabithorax* in morphological differences between *Drosophila*

species. *Nature*, **396**, 463.

73. Turbeville, J. M., Schulz, J. R., and Raff, R. A. (1994) Deuterostome phylogeny and the sister group of the chordates: evidence from molecules and morphology. *Mol. Biol. Evol.*, **11**, 648.

74. Wada, H. and Satoh, N. (1994) Details of the evolutionary history from invertebrates to vertebrates, as deduced from the sequences of 18S rDNA. *Proc. Natl Acad. Sci. USA*, **91**, 1801.

75. Kozmik, Z., Holland, N. D., Kalousova, A., Paces, J., Schubert, M., and Holland, L. Z. (1999) Characterization of an amphioxus paired box gene, AmphiPax2/5/8: developmental expression patterns in optic support cells, nephridium, thyroid-like structures and pharyngeal gill slits, but not in the midbrain-hindbrain boundary region. *Development*, **126**, 1295.

76. Garcia-Fernàndez, J. and Holland, P. W. H. (1994) Archetypal organization of the amphioxus Hox gene cluster. *Nature*, **370**, 563.

77. Langeland, J. A., Tomsa, J. M., Jackman, W. R., and Kimmel, C. B. (1998) An amphioxus snail gene: expression in paraxial mesoderm and neural plate suggests a conserved role in patterning the vertebrate embryo. *Dev. Genes Evol.*, **208**, 569.

78. Brooke, N. M., Garcia-Fernàndez, J., and Holland, P. W. H. (1998) The ParaHox gene cluster is an evolutionary sister of the Hox gene cluster. *Nature*, **392**, 920.

Index